Joachim Diederich, Cengiz Günay, and James M. Hogan

Recruitment Learning

Studies in Computational Intelligence, Volume 303

Further volumes of this series can be found on our homepage: springer.com

Vol. 281. Robert Babuska and Frans C.A. Groen (Eds.)
Interactive Collaborative Information Systems, 2010
ISBN 978-3-642-11687-2

Vol. 282. Husrev Taha Sencar, Sergio Velastin, Nikolaos Nikolaidis, and Shiguo Lian (Eds.)
Intelligent Multimedia Analysis for Security Applications, 2010
ISBN 978-3-642-11754-1

Vol. 283. Ngoc Thanh Nguyen, Radoslaw Katarzyniak, and Shyi-Ming Chen (Eds.)
Advances in Intelligent Information and Database Systems, 2010
ISBN 978-3-642-12089-3

Vol. 284. Juan R. González, David Alejandro Pelta, Carlos Cruz, Germán Terrazas, and Natalio Krasnogor (Eds.)
Nature Inspired Cooperative Strategies for Optimization (NICSO 2010), 2010
ISBN 978-3-642-12537-9

Vol. 285. Roberto Cipolla, Sebastiano Battiato, and Giovanni Maria Farinella (Eds.)
Computer Vision, 2010
ISBN 978-3-642-12847-9

Vol. 286. Zeev Volkovich, Alexander Bolshoy, Valery Kirzhner, and Zeev Barzily
Genome Clustering, 2010
ISBN 978-3-642-12951-3

Vol. 287. Dan Schonfeld, Caifeng Shan, Dacheng Tao, and Liang Wang (Eds.)
Video Search and Mining, 2010
ISBN 978-3-642-12899-8

Vol. 288. I-Hsien Ting, Hui-Ju Wu, Tien-Hwa Ho (Eds.)
Mining and Analyzing Social Networks, 2010
ISBN 978-3-642-13421-0

Vol. 289. Anne Håkansson, Ronald Hartung, and Ngoc Thanh Nguyen (Eds.)
Agent and Multi-agent Technology for Internet and Enterprise Systems, 2010
ISBN 978-3-642-13525-5

Vol. 290. Weiliang Xu and John Bronlund
Mastication Robots, 2010
ISBN 978-3-540-93902-3

Vol. 291. Shimon Whiteson
Adaptive Representations for Reinforcement Learning, 2010
ISBN 978-3-642-13931-4

Vol. 292. Fabrice Guillet, Gilbert Ritschard, Henri Briand, Djamel A. Zighed (Eds.)
Advances in Knowledge Discovery and Management, 2010
ISBN 978-3-642-00579-4

Vol. 293. Anthony Brabazon, Michael O'Neill, and Dietmar Maringer (Eds.)
Natural Computing in Computational Finance, 2010
ISBN 978-3-642-13949-9

Vol. 294. Manuel F.M. Barros, Jorge M.C. Guilherme, and Nuno C.G. Horta
Analog Circuits and Systems Optimization based on Evolutionary Computation Techniques, 2010
ISBN 978-3-642-12345-0

Vol. 295. Roger Lee (Ed.)
Software Engineering, Artificial Intelligence, Networking and Parallel/Distributed Computing, 2010
ISBN 978-3-642-13264-3

Vol. 296. Roger Lee (Ed.)
Software Engineering Research, Management and Applications, 2010
ISBN 978-3-642-13272-8

Vol. 297. Tania Tronco (Ed.)
New Network Architectures, 2010
ISBN 978-3-642-13246-9

Vol. 298. Adam Wierzbicki
Trust and Fairness in Open, Distributed Systems, 2010
ISBN 978-3-642-13450-0

Vol. 299. Vassil Sgurev, Mincho Hadjiski, and Janusz Kacprzyk (Eds.)
Intelligent Systems: From Theory to Practice, 2010
ISBN 978-3-642-13427-2

Vol. 300. Baoding Liu (Ed.)
Uncertainty Theory, 2010
ISBN 978-3-642-13958-1

Vol. 301. Giuliano Armano, Marco de Gemmis, Giovanni Semeraro, and Eloisa Vargiu (Eds.)
Intelligent Information Access, 2010
ISBN 978-3-642-13999-4

Vol. 302. Bijaya Ketan Panigrahi, Ajith Abraham, and Swagatam Das (Eds.)
Computational Intelligence in Power Engineering, 2010
ISBN 978-3-642-14012-9

Vol. 303. Joachim Diederich, Cengiz Günay, and James M. Hogan
Recruitment Learning, 2010
ISBN 978-3-642-14027-3

Joachim Diederich, Cengiz Günay, and James M. Hogan

Recruitment Learning

Springer

Dr. Joachim Diederich
School of Information Technology and
Electrical Engineering
School of Medicine -
Central Clinical Division
The University of Queensland
Brisbane Q 4072
Australia
E-mail: j.diederich@uq.edu.au

Dr. James M. Hogan
School of Software Engineering and
Data Communications
Queensland University of Technology
Brisbane Q 4000
Australia
E-mail : j.hogan@qut.edu.au

Dr. Cengiz Günay
Department of Biology
Emory University
Atlanta, GA 30322
U.S.A.
E-mail: cgunay@emory.edu

ISBN 978-3-642-14027-3 e-ISBN 978-3-642-14028-0

DOI 10.1007/978-3-642-14028-0

Studies in Computational Intelligence ISSN 1860-949X

Library of Congress Control Number: 2010931383

Typeset & Cover Design: Scientific Publishing Services Pvt. Ltd., Chennai, India.

Printed on acid-free paper

9 8 7 6 5 4 3 2 1

springer.com

Preface

This monograph provides an overview of recruitment learning approaches from a computational perspective. Recruitment learning is a unique machine learning technique that: (1) explains the physical or functional acquisition of new neurons in sparsely connected networks as a biologically plausible neural network method; (2) facilitates the acquisition of new knowledge to build and extend knowledge bases and ontologies as an artificial intelligence technique; (3) allows learning by use of background knowledge and a limited number of observations, consistent with psychological theory.

Connectionist learning is almost synonymous with "learning by examples", i.e. induction on the basis of the repeated presentation of a large number of training examples. In contrast, symbolic artificial intelligence developed "learning by instruction" or explanation-based learning methods. In this context, recruitment offers an important advantage: "one-shot learning" or learning with a limited number of examples. Recruitment learning also offers other advantages in areas that are not sufficiently covered by neural network techniques and that are essential for human learning (Levelt, 1990, p.501): (1) learning new knowledge *without impairing existing knowledge*, e.g. children learn to multiply after they have learned to add and the ability to multiply does not negatively affect the skill to add numbers. (2) learning by instruction - as explained by Levelt: If a person is being told that a dialing code is extended by the number 2, *then this rule can be applied immediately* without re-learning all existing or known telephone numbers.

The idea of learning as the result of a single presentation of a stimulus is by no means new and neither artificial intelligence nor cognitive psychology introduced the concept. Guthrie (1959) assumed already that a single co-occurrence of stimulus and response is sufficient for forming an association. Gestalt theorists such as Koehler (1917) and psychologists such as Yerkes (1927) examined "learning by problem solving" or "learning by insight" at the beginning of the last century. This uniquely human learning capability is not adequately represented by statistical neural network learning systems.

As a machine learning system, recruitment learning typically operates in sparsely connected neural networks, utilizes Hebbian and competitive learning mechanisms such as "release of inhibition", and hence claims a high degree of biological plausibility. The current research on adult neurogenesis is particularly exciting, supporting the core concepts of recruitment learning, which is, after all, *the* neural network learning technique based on the physical and functional acquisition of new neurons.

This volume provides an overview of the field and introduces a number of different approaches for concept learning. The idea of recruitment learning is traced from early cognitive psychology to current connectionist systems. In addition, successful algorithms, techniques and applications are presented and future research opportunities are discussed.

The book is a reference for researchers and graduate students, and since the first chapter provides an introduction to the topic and others outline the foundations, the book should be an asset in the classroom as well. Due to the interdisciplinary nature of recruitment learning, this volume aims to be a resource for scholars in computer science and electrical engineering, biology, psychology and the social sciences.

Professor Jerome Feldman, Dr. Thomas Wennekers and Dr. Christian Huyck provided feedback on early versions of the book and the authors are grateful for their comments.

The authors would like to thank Professor Janusz Kacprzyk for including this volume in Springer Verlag's "Studies on Computational Intelligence" series. Dr. Thomas Dillinger and Heather King provided advice and support during the writing and editing process of this book and the authors appreciate their contributions. Ms. Belinda Lee assisted in the preparation of the final manuscript.

Joachim Diederich would like to thank Dr. Dale Anderson, Professor Tim Roberts, Dr. Denise Dillon and the friends and colleagues at James Cook University Australia, Singapore Campus, for their commitment to a research culture that encourages the writing of books and acknowledges their value. The first author is grateful to Professor Paul Bailes from the School of Information Technology and Electrical Engineering at the University of Queensland, Australia, for his ongoing support and the use of the excellent facilities of the University. Joachim Diederich's interest in recruitment learning developed during several visits to the International Computer Science Institute in Berkeley, California and his collaboration with Professor Jerome Feldman. Finally, Joachim Diederich would like to thank the most successful academic book author he knows, Professor Susan K. Wright, for her encouragement and tolerance.

Cengiz Günay would like to thank his parents and his partner Anca Doloc-Mihu for their support during the preparation of this monograph. He is also appreciative of his postdoctoral supervisor Astrid A. Prinz for letting him proceed with this volume. He acknowledges that his Ph.D. supervisor Anthony S. Maida contributed to most of the work included in this volume and Dr. Günay expresses his gratitude to him for giving permission to publish them within this volume.

James M. Hogan would like to thank his family – Tertia and Adrian - for their support during the preparation of this volume, particularly for their tolerance of interruptions during a number of holiday periods. A number of researchers have contributed greatly to the development of his understanding of recruitment learning and its relationship to the computational and cognitive neurosciences. While

Dr Hogan would like to express his gratitude to all of them, he would particularly like to acknowledge Professor Diederich, his doctoral supervisor, who introduced him to the area many years ago, and guided his early explorations of the topic.

March 2010

Joachim Diederich
Singapore

Cengiz Günay
Atlanta

James M. Hogan
Brisbane

Contents

Part I
Recruitment in Discrete-time Neural Networks

Chapter 1
Recruitment Learning – An Introduction

Joachim Diederich

1.1 Introduction

Computer science is the science of abstraction. Aho & Ullman (1992, p.3) call computer science the "mechanization of abstraction", i.e. an appropriate model is sought for a given problem and is realized by use of a computational device. Abstraction often means simplification: a complex and real problem is replaced by a model that can solve a given problem (Aho & Ullman, 1992, p.3).

In this chapter, a model for neural processing is introduced that is still an abstraction but one with substantial biological plausibility, as research in the neurosciences increasingly demonstrates. Models of recruitment learning have roots in "structured connectionist systems"; however, a multitude of research streams have contributed to this field. Structured connectionist models are a synthesis of artificial intelligence and neural network research. As a subsection of artificial intelligence, and thus computer science, connectionist systems focus on algorithms and data structures. The algorithms model processes in real neural networks, most fundamentally *excitation* and *inhibition*. Special attention is dedicated to learning: the change of the neural network structure, and in particular the weighted connections, triggered by the presentation of one or several training examples.

Recruitment learning models have two objectives: (1) to develop learning models at a level of abstraction that is increasingly biologically plausible, and (2) to realize rapid learning of complex concepts – a problem notoriously difficult in perceptron-type neural networks.

The progress towards a tighter link with the neurosciences has been dramatic in recent years. "Neurogenesis" is one of the most active and vibrant areas in brain research. While it was believed for decades that no new neurons are formed in the adult mammalian brain, it has now been confirmed that the generation of new neurons and synapses does occur in both the hippocampus and the olfactory bulb. Even more important, new neurons are recruited into functional circuits and the process appears to be highly relevant for theories of learning and memory. As Prickaerts et al. (2004) write:

> … learning may increase cell proliferation, yet those newly produced cells are not ready for use until they have differentiated and integrated into a neural network. Thus it might be possible that the presence of new undifferentiated cells increase the ability for future learning as more cells are available for recruitment when necessary (cf. Kempermann, 2002; in Prickaerts et al. (2004, p.7).

J. Diederich, C. Günay, and J.M. Hogan: Recruitment Learning, SCI 303, pp. 3–36.
springerlink.com

Impaired neurogenesis on the one side, and the increased generation of neurons as a result of certain drugs on the other side have been associated with mental health problems such as depression and schizophrenia. Hence, the physical and functional recruitment of new neurons is of highest relevance for neural network models that claim biological plausibility.

The functional acquisition of individual or groups of neurons lies at the heart of the class of neural network models associated with recruitment learning. This approach has been developed for more than three decades under the assumption that no new neurons are physically generated in the adult brain, and traditionally addresses the problem of the *functional recruitment of neurons into circuits.* Even as a biologically plausible learning method, recruitment learning *does not* depend on neurogenesis; it requires only the existence of *functionally available neurons (units).* Nevertheless, the recent breakthrough in the neurosciences with regard to neurogenesis provides a significant boost for recruitment learning models.

This chapter is an introduction to recruitment learning in neural networks, in particular to recruitment within structured connectionist systems, with a special emphasis on the biological foundations and the *historical development* of the method. Recruitment learning systems typically operate in sparsely connected neural networks and hence allow a claim of some biological plausibility. In this chapter, the idea of recruitment learning is traced from early cognitive psychology to current connectionist systems. Several approaches are highlighted and potential future research directions are outlined.

1.1.2 The Motivation for Recruitment Learning

Essentially all work in neural network learning has been devoted to induction from examples. The majority of neural network models lack neurobiological plausibility due to a (1) high degree of connectivity, (2) simplified processing units and (3) unreasonably powerful learning mechanisms. Yet, brains are equipped with a rich and flexible structure that may well be the foundation of intelligence:

> Humans are intelligent because evolution has equipped them with a richly structured brain. This structure, while serving a variety of functions, in particular enables them to learn. A certain critical degree of structural complexity is required of a network before it can become self-modifying -- no matter how sophisticated its reinforcement rules -- in a way that we could consider intelligent (Arbib, M.A. 1987, p.70).

Neural network learning techniques that physically or functionally change the network structure (where this change is the core learning mechanism) are comparatively rare, since it was previously assumed that the full neural structure is available soon after birth and that topological changes are degenerative in nature. However, topological changes based on altering the strength of association are biologically plausible, as is neurogenesis, as confirmed recently in the most convincing fashion. It is now well established that structural modifications play an important role in learning during animal life (Leuner et al., 2006), in addition to the well known changes at the synaptic level.

Learning through structural change first appeared in artificial neural systems more than two decades ago. This class of methods has some overlap with the weight change only models that dominate the neural network field: the recruitment of previously uncommitted units is part of well established models such as ART (Carpenter & Grossberg, 1990) and it has been used in the context of backpropagation learning (Fahlman & Lebiere, 1990) for the allocation of new hidden units.

In recruitment learning, a neural network is composed of two classes of units:

- "Committed units" are network nodes that already represent some sort of information or function (i.e., conceptual information). Committed units are connected to other committed units and their simultaneous activation represents a meaningful state of the network. Committed units are also connected to "free units", though this link may not initially encode any functional relationship.
- Free units form a kind of "primordial network" (Shastri, 1988) or *tabula rasa* (Valiant, 1994) available for recruitment, and are connected to each other as well as to the previously mentioned committed units.

Recruitment learning is based on the principle of strengthening the connections between a group of committed units and one or more free units. This weight change results in the transformation of free units into committed units that are part of the concept representation. The term "chunking" is used if "a free unit becomes committed and functions as a representation for the cluster of committed units (i.e., the activation of units in the cluster results in the activation of the chunking node, and conversely the activation of the chunking nodes activates all the units in the cluster" (Shastri 1988)). It is possible to view the sub-network of committed units embedded in the network of free units.

1.1.3 The Organization of the Chapter

Having now distinguished recruitment learning from more commonly understood notions of neural network learning, it is useful to provide some more extensive coverage of traditional connectionist systems. This provides a *historical context* for recruitment learning models and simultaneously acts as a tutorial for readers unfamiliar with the broader connectionist movement.

Section 1.1.2 thus provides an introduction to connectionist systems, leading into section 1.1.3, which traces the history of recruitment learning from early psychological models to learning in sparsely connected networks as introduced by Feldman (1982). A number of different neural network models are introduced which are at least partially based on Feldman (1982). The overview continues with Valiant's "Circuits of the Mind" (1994) to more recent approaches based on oscillatory neural networks. The focus here is on the computational aspects while later sections examine the biological foundations of recruitment learning (section 1.4) and focus on very recent results on neuro- and synaptogenesis (section 1.5).

The chapter concludes with a discussion of the role of adult neurogenesis in mental health (section 1.6). Neurogenesis appears to be impaired in persons with mental health problems and these disorders therefore shed light on the mechanisms underlying neurogenesis and its effect onlearning and memory.

1.2 Connectionist Systems – The Basics

Connectionist systems try to solve problems by the simultaneous interaction of a multiplicity of simple components. The processing model is similar to the human brain, however, connectionist systems represent an abstraction level in their own right. Connectionist systems *converge* autonomously through the local interaction of individual elements into a stable state which represents the output of the system.

Structured connectionist models are constrained by the so-called "100 step" rule (Feldman, 1986). This rule is based on a comparison of the speed of human thinking and perception (millisecond range) with the speed of neural processing. The relatively slow transmission rate of nerve fibers (up to 100 m/sec.) and the refractory periods in which a neuron cannot release action potentials (10ms) means that as the brain responds to a stimulus, each nerve cell can "fire" about 100 times only. This is a powerful argument for massively parallel approaches in the cognitive sciences.

Connectionist systems are thus characterized by massive parallelism, processing efficiency and the need for convergence, but their distributed structure also delivers error tolerance. Connectionist systems permit the dynamic reconfiguration of parts of the system even if there is a complete loss of components. This error tolerance is also characteristic of the operation of the brain, and provides a strong justification for connectionist models of cognition. Although components fail constantly (individual neurons or neural groups), the brain continues to function except in extreme pathological cases. Even the impairment of whole brain centers (for instance in aphasia) can be compensated to a considerable degree with appropriate, fast and early treatment (e.g. essentially re-learning).

1.2.1 Components of Connectionist Systems

Connectionist systems consist of one or more networks with units that are connected by weighted links. A node (or "unit") includes one or more functions as well as a finite number of continuous and discrete variables.

One of the first formal specifications of artificial neurons appeared in Feldman (1982). This specification is to some degree biologically realistic and defines processing units with the following structure:

- {q}, a set of discrete states, < 10
- p, the unit “potential”; a continuous variable in the range [-10, 10], though with limited precision
- v, an integer output value in the range $0 \leq v \leq 9$
- i, an input vector $i_i, \ldots, i_n$.

There are three functions over these variables:

$$p <- f(\mathbf{i}, p, q)$$
$$q <- g(\mathbf{i}, p, q)$$
$$v <- h(\mathbf{i}, p, q)$$

reflecting assignment of values based on state and input. The “<–” symbol is chosen to mimic assignment in conventional programming languages; *f, g* and *h* are usually very simple functions.

Two features are emphasized: A restriction over the output variable permits only a small set of integer values due to the limited precision of biological neurons. Furthermore, there is a small set of discrete state variables which can be used for the modeling of biological phenomena such as refractory periods.

However, in many models, the output of a unit can take arbitrary values within a given interval and there may be no restrictions on the precision of the potential. The following description is based on Feldman (1982).

Each unit consists of the following components:

1. The Site: Each unit has one or more sites (potential connection points for inputs). Each site has a “site function” which performs computations over input links locally and independently. The result is stored by use of a variable (the site value).
2. Input link: the physical entrance to the unit.
3. Input: a continuous variable which is associated with an input link.
4. Weight: a continuous variable associated with a link.

The unit specification is realized via three types of functions:

1. Unit function: a function of the site values; it computes the output, potential and state of the unit.
2. Side function: as described above, a function of the local inputs and weights; it computes the site value.
3. Link function: a function of the local input and of the potential/output; it computes the new weight during learning.

Each unit has a single unit function as well as a finite number of sites. The introduction of the construct “site” permits the modeling of dendrite trees in biologically realistic simulations. In particular, it permits the introduction of enabling inputs, i.e. special input links which must be active (input $\neq 0$), in order to permit a non-zero output (see also Shastri, 1988, p.118).

Structured connectionist models with the above components use *pre-coded structures in initial networks* in order to represent prior knowledge. With these initial structures, the search space is significantly limited in both problem solving and learning. This allows fast learning consistent with the "100 step" rule and the set of the functions or concepts to be learned is significantly constrained.

1.2.2 Learning in Connectionist Systems

Learning is the most important feature of artificial neural networks which are *adaptive*, i.e. networks change internal representations as a response to training data, sometimes combined with a teaching input. Since all knowledge is encoded in weights; that is, the numeric values associated with links connecting units, learning is performed by weight change. A weight represents the strength of association, the co-occurrence of the connected features, concepts, propositions, or events during a learning period. On the network level, a weight represents how frequently the receiving unit has been active simultaneously with the sending unit. Hence, weight change between two units depends on the frequency with which both units exhibit positive output simultaneously.

This form of weight change is known as Hebbian learning (Hebb, 1949), which provides a simple mathematical model for synaptic modification in biological networks. Its most general form is expressed by the equation:

$$w_{ij} = ky_i y_j$$

i.e. the weight change is proportional to the product of the output of unit i and unit j. Several important modifications of this simple weight change rule have been proposed. The basic principle - that local weight change depends on the outputs, states or potentials of the connected units - is very widely accepted.

There are additional rules for weight change in ANNs: learning should be local; that is, the only available information is the weight itself, the current input and the potential, state or output of the receiving unit. Next, no programming is allowed; no external program or other network shall modify weights or network structures directly. Weight change is slow in most cases and the connected units have to be active simultaneously for some period to get a strong association.

The simple Hebbian learning rule can be generalized to more complex networks where some units are used to represent input and output, and some internal (hidden) units are used to encode important features of the environment as the result of learning. Core features of training instances must not be known in advance; it is possible to "extract" features from training data.

The learning techniques outlined above represent weight-change-only methods: the network topology remains unmodified. It is also possible, and sometimes necessary, to use structure-changing learning schemes. Two methods are available: (1) Learning procedures that physically change the network topology (e.g., by simulating neurogenesis) and (2), learning techniques that use weight-change-only and allocate previously unused connections and units. Recruitment learning has traditionally been focused on the second alternative and the following section provides a brief overview of neural network learning.

1.2.2.1 Neural Network Learning Techniques

There are obviously several ways to classify neural network learning techniques. The distinction between supervised, reinforcement and unsupervised learning is often considered the most important organizing principle. In supervised learning, precise teaching data are available and are used to calculate weight change, e.g., when the training data consists of input-output pairs and the task is to predict the output given the input. Error correction learning of this type requires a "desired state," which is used as a reference value in the calculation of the global performance of a network. In other words, the relation between input cases and their classification must be known in advance, roughly corresponding to discriminant analysis in statistics.

The central idea in reinforcement learning is to use a global signal, which represents only *reward* or *punishment*, and to measure how local decisions correlate with this reinforcement input. Local decisions are restricted to changes in variables such as weights and states. The reinforcement signal is not a complex teaching input; it only represents "good" or "bad" performance (i.e., "positive reinforcement" and "punishment"). The network can effectively perform gradient ascent as the probability distribution associated with each local variable is changed in order to increase the expected reinforcement. In general, reinforcement methods have advantages if there are a few local variables only, but they might get stuck in local minima.

Unsupervised learning is roughly equivalent to cluster analysis and does not require a teaching or reinforcement input. In competitive learning, a technique which is most representative for unsupervised learning, the procedure classifies a set of input vectors into disjoint clusters based on their mutual similarity. Competition appears within a layer, where a set of units compete with each other to become active and to perform the weight change. The winning unit increases its weights on input-lines with strong input and decreases weights on links with weak input. Since there is a constraint on each weight vector that the sum of weights has to be constant, the winning hidden unit will be selective to the input vector where it was active, and not to other input patterns.

1.2.3 "Winner Take All" Networks

Hardly any connectionist approach is as well studied as "winner take all" (WTA) networks, competitive learning structures in which stronger units inhibit the activation of adjacent units, ultimately ensuring that only the strongest unit has non-zero activation. Theoretical computer science offered convergence and stability proofs, both for simple WTA networks, which have only one active unit after convergence, and for k-WTA-networks, which permit several active units in a stable activation pattern (Majani, Erlanson & Abu Mostafa, 1989; Wolf et al., 1991). In addition WTAs have been used in a many applications, including VLSI implementations (Wyatt & Standley, 1989; Lazzaro et al., 1989).

Competitive systems were originally studied in completely different contexts such as ecology and economics, resulting in formal models which cannot readily be applied to connectionist systems. Ecology is concerned with animal populations which compete for resources, move freely and form new groups. Economics studies complex organizations which compete in the market. These systems have complex dynamics different from those of connectionist networks. Connectionist networks are defined physically by the link structure, with units computing simple functions in the absence of "purposeful" or directed behavior (see Hirsch, 1989, p.344).

1.2.3.1 The Biological Background of WTA Networks

Winner take all networks represent the connectionist abstraction of *lateral inhibition,* the mechanism known to control what would otherwise become an avalanche-like propagation of excitation towards higher levels of the central nervous system (Schmidt et al., 1977, p.54-55). Without the concentrating effect of lateral inhibition, localized stimuli would become more and more diffuse with each synaptic connection. Inhibition works against this diffuse propagation of excitation.

Lateral inhibition can be observed on all levels of a sensory system, including the thalamus and the cortex. It is thus a feature of all sensory systems, particularly the specific systems (Schmidt et al., 1977, p.55). Starting with the first synapses of afferent connections, each excitatory neuron inhibits its environment via interneurons. The neuron with the strongest afferent input has the strongest inhibitory impact on its neighborhood. This leads to a reduction of excitation outside of the centre (Schmidt et al., 1977, p.54-55). Lateral inhibition is regarded as the functional compensation of excitation divergence which is the result of neural (or neuro-anatomical) organization. A well-known effect of lateral inhibition is the increase of sharpness in visual processing. An analysis of how lateral inhibition can be useful in recruitment learning is addressed later in Chapter 8.

An important corollary to this mechanism results when the activation of a previously dominant neuron is disrupted in some way, thereby changing the result of the competition. This effect, known as 'release of inhibition' is an important mechanism in the remapping of neural representations – for example in the visual cortex (Gilbert and Wiesel, 1992) – and in a more abstract form, as a mechanism to facilitate rapid or even one-shot recruitment learning. The connectionist version of these ideas is considered in the following sections.

1.2.3.2 Competitive Learning

WTA networks have a critical role in the realization of competitive learning systems that utilize the components introduced in section 1.2.1. In "competitive learning" a network with at least two layers is used, with activating links projecting from the lower layer towards the immediately higher layer which includes groups of units with lateral connections supporting mutual inhibition. Usually all units of a layer receive inputs from all (or a defined subset) of the units of the lower layer and project an output to all (or a defined subset) of the units of the higher layer. Mutual inhibition within a layer realizes a WTA network: only the unit with the strongest input remains active and can adjust the weights of its input links. Learning is a re-distribution of weights towards active links. The learning unit receives most activation after completion of the learning phase, i.e. during decision making, and thus has the best chance to win the competition. Since the winning unit represents the solution to the problem to be solved, the active unit after competition represents the output of the system.

1.2.3.3 Convergence and Stability

There are convergence and stability proofs for certain classes of inhibitory networks. Hirsch (1989, p.345) emphasized that only special classes of inhibitory networks have convergent activation dynamics. Simultaneously, Hirsch (1989) underlines the positive convergence characteristics of excitatory networks.

Hirsch (1989) is also concerned with convergent cascades, a generalization of a layered network. A cascade consists of several separate networks N_1, N_2... N_n, each of which can be a fully recurrent, inhibitory network. Some units of the network N_m have links to units of the network N_{m+1}, $m < n$. Hirsch investigates the question of whether the entire cascade is globally asymptotically stable, if all networks of the cascade are globally asymptotically stable. While this is the case, it cannot be assumed, however, that a cascade of convergent systems is indeed always convergent. This applies only if a set of corollary assumptions are satisfied (Hirsch, 1989, p.342).

Wolf et al. (1991) investigate the network dynamics of "interactive activation models" (McClelland & Rumelhard, 1981), in particular, the influence of initial activations and the effect of external input on the "k-more winner take all" behavior. In the absence of external input, the system converges and maintains the order of the initial activation values, i.e. the order of activation values remains stable over several time steps. Majani et al. (1989) prove this for sigmoidal units as well. If there is external input, the convergence proof applies with limitations only: the activation values can be initialized in such a way that no genuine "k-more winner take all" behavior results. In particular, "wrong" units can win if there is strong external input. Wolf et al. (1991) conclude from simulation results that the correct group of units wins given appropriate initial activation values.

In structured connectionist systems, *designed* WTA networks are used which converge under all initial conditions. An interesting but extreme case is the "convergence in two steps" network (Feldman & Ballard (1982) even describe a "convergence" in one step). In the first step, each unit of the WTA network examines whether it is the unit with the largest potential and/or output. This takes place by comparing the own output with the inputs, i.e. the outputs of other units in the layer – a comparison possible since every unit has inhibitory connections to every other unit in the WTA network. Note that this requires the unit architecture outlined above. In a second step, the unit with the largest output sets its own potential/output to the maximum value, while all other units set their potential/output to zero. The competition is thereby finalized. It is obvious that in this extreme case the dynamics of the network are determined by the initial conditions, i.e. the initial potentials and output values, the input values as well as the weights. The complexity of the units makes this fast convergence possible.

Following this brief overview of the principles of connectionist systems, an historical overview of recruitment learning is provided in the next section. The core concepts of recruitment learning emerged in psychology in the second half of last century, even though it would be easy to find earlier roots. After discussing the ideas of Wickelgren (1979) on learning in the hippocampus, several early models of recruitment learning are introduced, with a clear focus on the cornerstones provided by Feldman (1982), Valiant (1988, 1994), Hogan (2002) and others.

1.3 Recruitment Learning – An Historical Overview

Learning through structural change has a long history in psychology, in particular in cognitive science. The fundamental concept can be traced back to early association theories, as for instance in Thorndike's "Connectionism" (Thorndike, 1932, 1949, in Bower & Hilgard, 1981). According to Thorndike, the basis of learning is the association between sensory impressions and action impulses. Associations can be strengthened or weakened by experience. It is assumed that complex habits consist of simple associations and that learning recruits simple, already existing associations to form more complex habits. Learning capacity is then defined by the number of existing associations and their availability.

Wickelgren (1979, in Bower & Hilgard, 1981, pp.503) then develops this principle, which re-emerges in Feldman (1982) as recruitment learning. Wickelgren (1979) notes that patients with damage to the hippocampus show a dichotomy in their learning abilities: these patients can learn simple stimulus – response (S-R) habits, yet they are incapable of higher forms of learning. Wickelgren (1979) assumes that for these forms of learning, i.e. to form complex memories, a *vague* activation from the hippocampus is necessary.

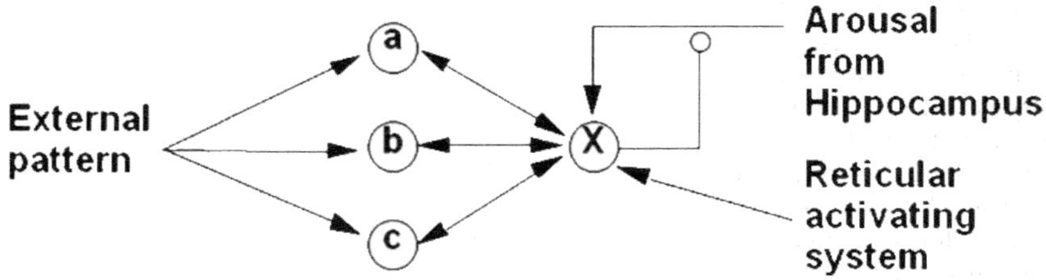

Fig. 1.1. Recruitment learning after Wickelgren (1979; Illustration 14,7 in Bower & Hilgard, 1981).

According to Wickelgren (1979), a number of neurons exist in the nervous system representing stimulus characteristics, events, concepts etc. These neurons are excited whenever the appropriate stimulus is present, but there is also a quantity of "free" neurons at higher levels, which have numerous, but weak, connections to cells with simpler receptive fields.

If a stimulus is presented, representative cells are activated and excite some of the free neurons. Wickelgren (1979) assumes the hippocampus is *vaguely excitatory:* it is a system providing diffuse activation, a motivational system that would activate free neurons. If, for example, the stimulus characteristics a, b, c activate the neuron X at a particular time, in which X receives excitation from the hippocampus, then by application of the Hebb learning rule (Hebb, 1949) an associative "cluster" is formed in which X represents the characteristics a, b and c. In other words, the synaptic efficiency of the links of a, b and c to X increases, and thus X "binds" the input features a, b and c. X becomes a "chunk node" (Wickelgren, 1979) or unit. It is important to note that even the partial presentation of input features (for instance a, b) can excite other input cells via X (for instance c).

An important element of Wickelgren (1979) is the inhibitory feedback between the "recruited" node X and the hippocampus. Without this inhibitory feedback, any diffuse activation from the hippocampus would activate X, even if the stimulus characteristics a, b and c are not present. If the Hebb rule is applied, the strongly activating connections in the "cluster" would weaken. This is prevented by inhibitory feedback; X is firmly recruited into the associative "cluster" and the renewed activation of the "cluster" must take place by means of a second excitatory system, for instance the "reticular activating system".

While the acquisition of free units and the process of chunking are key elements of recruitment learning, it is important to highlight that these processes are assumed to operate in *random networks*. The fundamental ideas of learning in random networks are outlined next.

1.3.1 Learning in Random Networks

Recruitment learning (as envisioned by Feldman, 1982) is realized in a multi-layer neural network. The links are either forward or bi-directional and there may be lateral, inhibitory connections as well. Occasionally, it is permitted that a link "jumps" over several layers to form *short-cut* connections. Furthermore, sparse random connectivity is assumed: each unit of a layer k with n units has log (n) or $\sqrt{n}$ links to the layer k+1 (see Feldman, 1982). Only rarely is full connectivity between layers permitted, as this lacks biological plausibility and is against the "spirit" of the original recruitment learning idea.

1.3.1.1 Sparse Random Networks

As outlined previously, recruitment learning operates with two classes of units: "free units" and "committed units". Figure 1.2 demonstrates Feldman's (1982) original idea of recruitment learning in networks with limited connectivity[1]. The network in Figure 1.2 does not have a genuine layered structure. There are N=16 nodes and each is linked with about $\sqrt{N}$ other units. All units have the same activation function and all weights have the same value (see Feldman 1982, 1990, Reprint p.56). Again, these are simplifying assumptions from the model described above.

Nodes F and I in Figure 1.2 are committed units and in this illustration both are input units. Coincidentally, both F and I have a link to unit O. External activation of F and I thus results in the activation of this unit. If at time zero, F and I have an activation potential of 10, and this potential is held for six time units, then after six "updates" of the network, O has an activation potential of 10 as well (Feldman, 1982, Illustration 8). Since F, I and O are active at the same time; local weight change can be performed by use of Hebb's learning rule over the connecting links. O develops strong positive links to F and I, and thereby to the group of committed units in this network. At the conclusion of the weight change process, O will change an internal state variable from "free" to "committed".

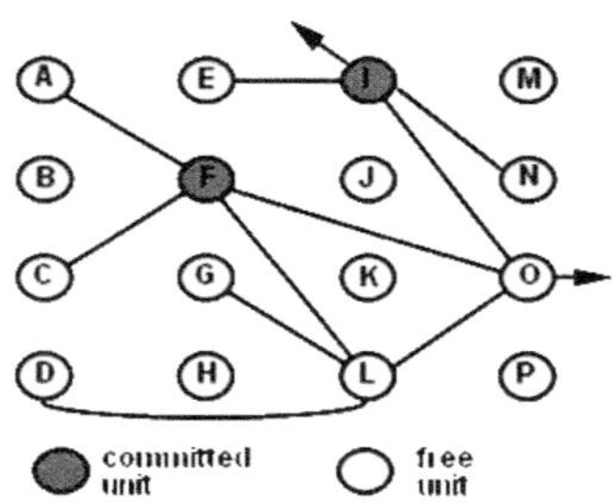

Fig. 1.2. Recruitment learning (after Feldman, 1982, Fig. 8).

Feldman (1982) examines in detail the size of a random network (number of units) as well as the degree of connectivity necessary to guarantee that a group of committed units can recruit one or more free units reliably. In particular, the probability that recruitment fails due to missing links is determined. The case of "pair recruitment" is studied, that is the question of whether in a random network with a large number of units in several layers, including "intermediate layers", a path between two end units can be found.

Figure 1.3 shows that under certain conditions, the probability that *no* path exists between the two end units x and y is very small. Recruitment of the path by means of a local learning rule and the external activation of the end units x and y is, however, far from guaranteed. For the avoidance of "crosstalk" (i.e. an intermediate unit is part of more than one path), Feldman (1982) proposes the use of (1) lateral inhibition, (2) specialized activation functions similar to those outlined in section 1.2.1 and (3) appropriate thresholds.

[1] The illustration is somewhat unusual since there are units without links and the degree of connectivity is too low.

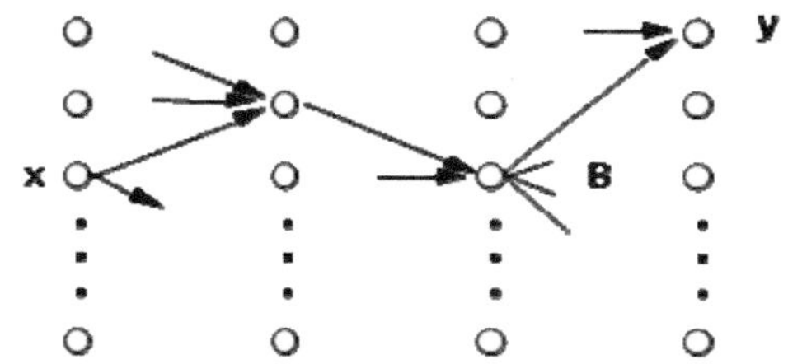

$$\bar{P} = (1 - F)^{B^K}$$

$\bar{P}$ = probability of a missing link between x and y

N = number of units per layer

B = number of (outgoing) links per unit

F = B/N (branching factor)

K = number of intermediate layers

$\bar{P}$ for B = 1000

K= \ N=	10^6	10^7	10^8
0	.999	.9999	.99999
1	.367	.905	.989
2	10^{-440}	10^{-44}	10^{-5}

Fig. 1.3. "Pair recruitment learning" after Feldman (1982, Fig. 7, Reprint p.61)

Figure 1.4 demonstrates the principle of recruitment learning in a structured random network (Shastri & Feldman, 1984). The committed units "color" and "size" as well as "red" and "large" are embedded in a randomly connected network with links to free "binding units" (here shown as triangles). These binding units in turn can be either committed or free and serve as intermediate units that realize a connection to a third class of units that represent conceptual knowledge. These concept units are also embedded in the random network, are subject to potential recruitment, and are activated by binding units. Recruitment is realized by strictly local weight change which leads to the allocation of binding units. The external activation of "color" and "size" as well as "red" and "large", as part of a training example, will result in the activation of some of the binding units, and Hebbian learning leads to the recruitment of units with the appropriate connections.

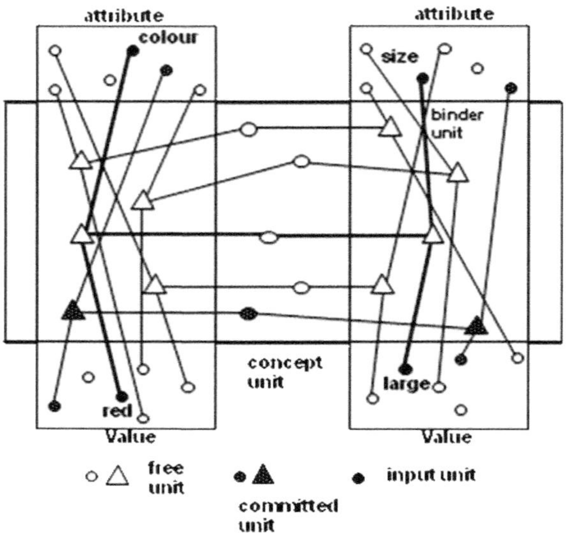

Fig. 1.4. Recruitment learning after Shastri & Feldman (1984)

After weight change concludes, there are one or more recruited concept units, which are connected via binding units with the attributes "color" and "size" as well as "red" and "large" in a sparse (not fully connected) random network. Shastri & Feldman (1984) extended this recruitment technique and introduced a "chunking" mechanism. A super ordinate unit can replace several concept units and their characteristics. In this way, hierarchy formation in a sparsely connected, random network is possible.

The method of Shastri & Feldman (1984) invites several comments: Several different sub-networks are used for attributes and their values, essentially the same *type of information*. In the example below, if "color" and "size" as well as "red" and "large" exist in the same sub-network, then these units can recruit the same binder node. This would result in crosstalk, unless the sub-network is so large that this is unlikely to happen. The use of several sub-networks for the same type of information (attribute/ attribute values) in order to avoid crosstalk can lead to the extreme situation of "one unit per sub-network".

Any neural network learning system, including recruitment learning, must answer two questions: *When* will learning and *how* will a unit or group of nodes be recruited? Wickelgren (1979) stressed that learning is triggered by non-specific activation from the hippocampus which results in the recruitment of a free unit with the "correct" connectivity. Feldman (1982) emphases the degree of connectivity of the underlying random network. Shastri & Feldman (1984) proceed from a suitable arrangement of input units which allows learning of the right binder and concept units. Valiant (1988) developed the concept of learning in random networks further and developed models for concept as well as episodic learning.

1.3.1.2 Valiant's Neuroidal Architecture for Cognitive Computation

Valiant (1988a,b) studied the functional characteristics of neural networks with sparse connectivity and classical threshold value elements. Valiant principally examines the following question: to what extent do limited connectivity, and the resulting restriction on communication, impact on learning in neural networks? In particular, the supervised and unsupervised learning of "episodes" and "concepts" is analyzed, including conjunctions and disjunctions of features. Each new episode and each new concept can be expressed in terms of existing features and concepts, so that hierarchical, incremental learning becomes possible, and unwanted interference between new and old knowledge can be avoided. Learning results in the acquisition of groups of nodes ('circuits') for the representation and computation of Boolean expressions. The space complexity of the acquisition of each new episode and concept is minimal, i.e. linear in the number of relevant parameters (associated with each node), and the algorithm requires only a few steps with asynchronous activations ("updates").

Episodic learning refers to the memorization of an individual instance (an input). Concept learning is based on the presentation of several instances with the objective of generating a rule which can classify new instances reliably (generalization). Learning is either supervised, if the classification of an input is given, or unsupervised, if this classification is missing. A special case of unsupervised learning is correlational learning which aims at the identification of correlations between groups of features in the input.

Formally the system can be described as follows (based on Valiant, 1988a, pp.630). The NTR ("neural tabula rasa") is a random graph of N identical nodes and can be specified as 6-tuple (G, W, $\sum$, I, λ, ∂). Here G is a directed graph with nodes and edges. Each edge has a weight w_{ij}, a real number, where w_{ij} and w_{ji}, are independent; O(log N) bits are required for the representation of each weight. The initial state of a node is determined by the weights of links leading to it and a state triple $s \in \sum$. $\sum$ is a subset of Q x T x {F, ¬F} with Q as the set of all states and T as the set of all threshold values (real numbers). F is a firing, active node; ¬F an inactive node. The function ∂ determines the new state triple, and the learning function λ determines the new weights. State triple and weights are computed as follows:

$$s_j = \partial\,(s_j, \textstyle\sum\{w_{ij} \mid i \text{ fires}\})$$

$$w_{ij} = \lambda(s_j, w_{ij}, f_i)$$

Nodes thus interact by firing, and an active node can change the weights of the edges leading to it. If T_j is the threshold of node j, then the function ∂ guarantees that j fires whenever $\sum\{w_{ij} \mid i \text{ fires}\} \geq T_j$.

The NTR does not have a global clock; the fundamental time unit is a cycle, and each node fires at most once per cycle. The algorithm consists of steps with asynchronous updates, in which the "peripheral system" stimulates a group of nodes which can force other nodes in the NTR to fire. The activation of nodes in the NTR will lead to a stable pattern of activation within a cycle. A node in state F will continuously fire until a stable state is reached; a node in a state "permanent firing" (F*) will still fire in the next cycle. Within a step the network is run until a stable pattern of activation is reached.

The random graph has N nodes and each node has √N edges to other units. A simple condition is that nodes i and j are linked to a third node, which can be recruited or is a functional node. Error tolerance is guaranteed by replication, i.e. each concept is represented by approximately 10 nodes. This relationship is to remain even if new nodes are recruited.

A node is either a relay unit (R) or a function node. There are episodic function nodes (E), conjunctive concepts (C), disjunctive concepts (D) as well as the correlation category (L). Each non-correlative node is initially available (A), and represents meaning after recruitment ("busy", B). A busy node cannot change back into state A. An available function node is initially recruited in the mode "unsupervised learning (U)" and keeps its function in the mode "supervised learning" (S). F stands for firing nodes, ¬F for inactive nodes, and F* for permanently active nodes. This completes the description; SEF, for instance, designates the set of episodic nodes which are active in supervised learning.

The following describes the learning rules for the supervised learning of episodic conjunctions. Learning is indeed episodic, since it requires one instance only and is nothing but the memorization of a conjunction of attributes which are true for the given example. The learning procedure is specified by the set of input nodes ("triggering set"), the objective and the algorithm:

Input node: J and $\{i\}$, $i \in UE$
Goal state: In future interactions $i \in SE$ if $J \subseteq F$, $i \in F$

Algorithm:

Step 1: Input nodes: J and $\{i\}$
$\{k \in ARF\} \Rightarrow k \in AR_1F^*$.
$\{l \in UEF, k \in F, j \in \neg F\} \Rightarrow w_{kl} := 1, w_{jl} := 0, l \in UE_1F^*$.

Step 2: Input node: $\{i\}$
$\{k \in AR_1F, \sum \{w_{jk} \mid j \in F\} \geq 1, j \in F\} \Rightarrow k \in AR_2F^*, w_{jk} := 0$.
$\{l \in UE_1F\} \Rightarrow l \in UE_2F^*$.

Step 3: Input node: J and $\{i\}$
$\{k \in AR_2F, s \in \neg F\} \Rightarrow w_{sk} := 0, k \in BR, T_k := 1$.
$\{k \in AR_1\neg F\} \Rightarrow k \in AR$.
$\{l \in UE_2F\} \Rightarrow T_l := \sum \{w_{kl} \mid k \in F\}, l \in SE$.

Valiant (1988) confirms that state triples of $O(|J|)$ units only are concerned. The weights of nodes which are not neighbors are not changed.

The idea behind this algorithm is as follows (Valiant, 1988): Since all initially available relay nodes have a threshold of 2, and arriving links from function nodes a weight of 1, all relay nodes which fire in step 1 must have edges from at least two nodes from J and {i}. For nodes in R which receive edges from i, the weight w_{ki} is changed from 0 to 1. By permanent firing it is guaranteed that node i in step 3 is in state UE_2F, and as a result the threshold is set to the sum of the incoming weights from firing nodes. Meanwhile in step 2, the set of firing nodes in R is limited to the subset of relay nodes which are neighbors of i. Nodes in R which are linked to units in J only will stop firing in step 2 and will return to state AR (step 3). In steps 2 and 3 nodes will change in such a way that the objective is reached.

Valiant developed and extended this model over two decades, most notably in Valiant (1994, 2005, and 2006). These recent models will be discussed in more detail in the following chapters; hence, a brief introduction is sufficient here.

Valiant (2005) revisits the tasks of memorization and association by considering four quantitative parameters of cortex: the neuron and synapse numbers, synapse strengths and switching times. Valiant's (2005) model allows for both disjoint and shared representations and in one version, both memorization and association require one step only. Interference, robustness to noise, and the stability of a hierarchical memorization process are addressed.

The first of two basic functions in Valiant (2005), JOIN, implements memory formation of a new item in terms of two established concepts: If A and B are already known, JOIN modifies a circuit so that in the future there is a representation of a new item C that will fire if and only if A and B are both firing. C is an "equal citizen" with A and B. JOIN is the basic form of memorization of an item. LINK, in turn, realizes association: If two items A and B are already represented in the neural system, the task of LINK is to modify the circuit so that at subsequent times whenever the representation of item A fires, the modified circuit will cause B to fire also.

Similar to Feldman (1982) and others, Valiant's (2005) model is based on random connectivity. Valiant (2005) goes beyond the question of existence of association in a random network: For any pair of items there must be a potential chain of communication to form associations:

> These chains have to exist from anywhere to anywhere in the network, rather than just somewhere in the network. A second constraint is that we want explicit computational mechanisms for enabling the chain to be invoked to perform the association, and the passive existence of the chain in the network is not enough. A third requirement is that for memory formation we need connectivity of an item to two others (Valiant, 2005, Preprint p.6).

An interesting property of Valiant's (2005) model is memory formation for shared representations. In a shared representation each neuron can represent more than one item. *There is no longer a distinction between nodes that have been allocated and those that are free.* The items each node will be assigned by memorization are determined by network connections *without any training process.*

While Valiant (2005) considers one-step memory formation, future research will address inductive learning of items with shred representation as well as multiple concept learning.

1.3.1.3 The Recruitment of Complete Networks

An approach of particular interest here have been developed by James Hogan since 1994 (leading up to Hogan, 2002, and the following chapters in this book). This work is important for a number of reasons: First, Hogan's studies are directly based on Wickelgren (1979), Feldman (1982) and Valiant (1988), extending this earlier work to investigate the limits of representation and recruitment in sparsely connected networks and offering both theoretical existence results and extensive learning simulations. This work is secondly an interesting complement to the neuroidal assemblies at the heart of Valiant's neuroidal systems, exploring the recruitment not merely of single units in support of a reinforcing concept but that of *entire networks*, capable of supporting richer functionality. Highly specialized networks are avoided, and for all these reasons, the level of biological and psychological plausibility is high.

Development of these networks and their connectivity models is explored in chapter 3, part I of the present volume, with a more extensive discussion of representation and learning taking place in chapter 4. Chapter 4 includes detailed coverage of both Hebbian and more robust approaches to learning, along with actual recruitment results for Boolean concepts and benchmark problems from the UCI Machine Learning Repository. Finally, in chapter 5, recruitment learning is applied to a restricted miniature language acquisition task, demonstrating the effectiveness of the approach for a more substantial problem.

1.3.1.4 One-Shot Learning or Learning by Instruction

Historically, the concept of sparse, random neural networks is at the core of the recruitment learning idea. Equally important is the understanding that successful learning in humans, animals and machines should be based on the presentation of a limited number of examples. That is, learning should be possible when the number of examples is insufficient from a purely statistical perspective. While repeated presentation of samples is possible, the number of repetitions is also constrained. In an extreme case, learning succeeds *with a single example presented once only*. Clearly, background knowledge is required to allow learning in this case.

Similarity-based learning, or "learning by example," is the original domain for learning in connectionist networks. The disadvantage of these procedures is that they make rather strong assumptions about the environment, e.g. the availability of large and structured data sets. One-shot learning, in contrast, is sometimes necessary and has advantages, for example when only a single new fact is available. Diederich (1989) describes a connectionist learning system that transforms an instruction expressed in a simple description language into an input for a connectionist knowledge representation system, which in turn changes the network in order to integrate this new knowledge. The newly acquired and integrated knowledge can be used in conjunction with prior knowledge.

The objective of Diederich (1989) was to demonstrate how high-level learning techniques such as "learning by being told" can be realized in connectionist systems. One-shot learning methods, which require a single training instance only, have been explored in symbolic artificial intelligence (Mitchell et al., 1986), but there are limited efforts to realize instruction in connectionist networks. The learning process in Diederich (1989) is vaguely modeled on explanation-based generalization (EBG), however, due to the absence of fast-changing variable binding and processes such as resolution in neural networks; a full implementation of EBG was not possible.

1.3.1.5 Temporal Reasoning and Recruitment Learning

In a series of papers, Shastri (2002) investigates the biological grounding of recruitment learning. The focus is on long term potentiation (LTP; a collection of biological and chemical processes that form the foundation of Hebbian learning) and a computational model that is an abstraction of LTP. As in previous work, the emphasis is on the recruitment of binder units, however, this time with a strong emphasis on the temporal dimension and the fast acquisition of episodic memory.

Similar to Diederich (1989), Shastri (2002) develops a model for one-shot learning, observing that several properties of LTP make it suitable to serve as the foundation of recruitment learning. Shastri (2002) notes that LTP is induced rapidly, within a few seconds. In addition, it is long lasting and the cooperative property of LTP makes it a good candidate for forming temporal expressions of a relationship between one or more items. A computational model based on LTP can model short-term memory as the activity over an ensemble of units, while it is easy to model long-term memory by use of weight change. Finally, LTP is synapse specific and hence can be used for the recruitment of very specific binder cells. Obviously, there is an analog process to LTP and that is long-term depression.

Shastri (2002) aims at modeling the memorization and retrieval of everyday events, i.e. episodic memory. Since this form of memory is about facts and events that may occur only once, one-shot learning is important. It is generally acknowledged that the hippocampus plays an important role in episodic memory formation. Shastri simulates the encoding of an event as a "role-entity binding". Entities are agents or objects participating in an event, and in order to avoid confusion between the roles they play in an episode, temporal binding between entities and their roles is necessary. The necessary structure is realized by "binding-detector cells".

Shastri's (2002) model includes three regions: "role", "entity" and "bind". The role and entity regions are assumed to have 750,000 cells each while the bind region is assumed to have 15 million cells. The role and entity regions have "dense projections" to the bind region; each role and entity cell makes about 17,000 connections with cells in the bind region. In Shastri's model, it is assumed that bindings realizing an event are expressed as temporal patterns of rhythmic activity over highly distributed groups of units in the three relevant regions. Memory formation is modeled via LTP and long-term depression.

Gunay & Maida (2001, 2006b) also focus on temporal encoding and recruitment learning for the binding of stimulus features. These papers address the problem that synchronized spikes travel over cortical circuits of different sizes and length. The problem is how initial stimulus-dependent synchrony may be preserved at a binding site. Gunay & Maida's (2001, 2006b) simulations are based on Valiant's (1994) neuroidal architecture and implement a continuous-time learning procedure allowing computation with spiking neurons. The results indicate that if tolerance and segregation parameters are within certain constraints, the assemblies of correct bindings are dominant over groups of spurious recruitments.

Another mechanism that can be borrowed from biological neuronal networks is, as mentioned above, lateral inhibition. Integration of this mechanism into a continuous-time recruitment learning algorithm was explored by Gunay & Maida (2006a). This is significant, because it proposes lateral inhibition as a natural, biologically-realistic mechanism to stabilize propagation of recruited concept sets across simulated cortical hierarchies. These ideas and the extension of recruitment learning to the domain of temporal binding are developed further in part II of this book.

1.4 The Biological Foundation of Recruitment Learning

The following is an overview of recent studies in neurobiology on structural modifications of neural representations. These results provide the foundation for the claim of biological plausibility for recruitment learning, and they may be summarized as follows: Neurons in the neocortex are not densely connected. There is an experience-related change of receptive fields even at an adult age during normal processing (e.g. perception) and as a result of sensory deprivation. Neurons change due to sensory input and there are processes generally referred to as heterosynaptic regulation, i.e. interaction between synapses. Neural representations are *relatively* compact. The function of non-specific activation in the nervous system of the adult has been confirmed, including its role in cortical plasticity. Neurogenesis in adult mammalian species has been confirmed and is likely to play a role in mental health problems such as schizophrenia and bi-polar disorder.

In the following sections, the core components of recruitment learning are investigated from a biological perspective. This analysis starts with the question of how well the connectivity structure of the brain is optimized. If the brain is a model for learning in artificial intelligence and connectionist systems, then this is a valid question. This is followed by a discussion of the density of the connectivity of the neocortex, neural population coding and experience-based changes in receptive fields. Additional core concepts such as the existence of diffuse activation in the nervous system and its contribution to learning, the generation of new synapses and neurons as well as fast learning are discussed next. Finally, the focus moves to adult neurogenesis and how results from clinical neuroscience allow the investigation of concepts in recruitment learning.

Neurogenesis and mental disorders are discussed in the final part of this chapter *because this topic clarifies and reinforces the role that new neurons play in both learning and memory*. Neurogenesis is relevant for the models of recruitment learning because an unlimited supply of new neurons would naturally overcome the "cross-talk" problem, i.e. the problem of catastrophic interference in neural activation. The chapter ends with a brief outline of future research directions.

1.4.1 How Optimized Is the Brain?

Before the biological foundations of recruitment learning are reviewed, it makes sense to ask to what extent the brain itself represents an optimized structure. A valid question from the viewpoint of artificial intelligence: If the human brain is the model, how optimal is it? The focus is on the density of connectivity, its structure (in particular long-distance projections) and optimality of the modularity of the brain.

Kaiser & Hilgetag (2006) investigate the optimality of brain modularity as well as the role of long-distance projection in neural systems. In particular, they explore (p.0805) the assumption that neural systems across several scales show optimal component placement, so that any spatial rearrangement of the components would lead to an increase of total wiring. Like the earlier work of Murre & Sturdy (1995), this study explores neural packing arrangements based on anatomical connectivity data. However, Kaiser & Hilgetag (2006) differ from previous authors by casting the analysis formally as an optimization problem, marrying connectivity data with spatial coordinates for network nodes to produce well-defined network layouts. They then applied an optimization algorithm to these network layouts, in order to search for wire-saving component rearrangements. The optimized component rearrangements could substantially reduce total wiring length in all tested neural networks.

Kaiser & Hilgetag (2006) summarize their results as follows:

> Wiring length reductions were possible due to the existence of long distance projections in neural networks. We explored the role of these projections by comparing the original networks with minimally rewired networks of the same size, which possessed only the shortest possible connections. In the minimally rewired networks, the number of processing steps along the shortest paths between components was significantly increased compared to the original networks. Additional benchmark comparisons also indicated that neural networks are more similar to network layouts that minimize the length of processing paths, rather than wiring length. These findings suggest that neural systems are not exclusively optimized for minimal global wiring, but for a variety of factors including the minimization of processing steps." Kaiser & Hilgetag (2006, p.0805).

Kaiser & Hilgetag's (2006) study is based on the total wiring of 95 primate (Macaque) cortical areas - which could be decreased by 32% as a result of optimization, the nematode *Caenorhabditis elegans* - in which case the wiring could be reduced by 48% on the global level and by 49% for neurons within frontal ganglia. While this study is based on computer simulation, it suggests the possibility that the brain is optimized for minimal processing steps. Historically, this is one of the assumptions of recruitment learning; Feldman's (1986) "100 step rule" was discussed earlier in this chapter.

1.4.2 The Density of Connectivity in the Neocortex

An important assumption of recruitment learning is sparse, random connectivity, which in turn is based on neurobiological investigations on the density of connections between neurons in the neocortex. These neurons are not densely connected, despite the relatively high number of synapses per neuron in the neocortex (approx. 4 x 10^3, approx. 1,7 x 10^3 for primates, AREA 17; Stevens, 1989, p.476), and receive synaptic input from a small subset of neighboring neurons only. Stevens (1989) reports that each neuron in the neocortex receives direct synaptic inputs from fewer than 3% of the neurons in the surrounding square millimeter. This level of connectivity is constant for higher species and does not increase with the size of the brain. Stevens (1989) concludes that circuits in the brain are highly modular, and that the complexity of the cortex is achieved by the inclusion of additional modules.

In a more recent review, Kaiser (2007) observes that cortical networks are not completely connected, and that only approximately 30% of all possible connections between areas actually exist. Highly connected clusters are found, which correspond to the functional differentiation of areas. In addition, approximately 20% of the connections are unidirectional. While there is a preference for short connections to spatially neighboring areas, approximately 10% of the connections cover a long distance, sometimes close to the maximum possible distance (69 mm) between two areas of one hemisphere (Kaiser & Hilgetag 2004b, in Kaiser, 2007, p.3035).

Kaiser (2007) then reviews the degree of neighborhood connectivity in biological networks. In this study, local connectivity is expressed in terms of the clustering coefficient, a well-known quantity from graph theory. Central to this idea is the concept of the neighborhood of a node, those cells connected to a particular neuron. The clustering coefficient is effectively the ratio of those connections *actually realized* amongst the members of such a neighborhood to the number of possible connections. While the clustering coefficient for the macaque is 49% (16% in comparable random networks), the average shortest path is comparably short with 2.2 (2.0 in random networks), i.e. on average, only one or two intermediate areas are on the shortest path between two areas (Kaiser, 2007, p.3036). Inter-cluster connections are critical for neural networks, removal may lead to a functional loss. There is a higher level of redundancy within clusters and removal of a local link is less critical. Kaiser (2007) summarizes his findings as follows:

> Cortical networks show an innate ability to compensate for and recover from damage to the network. Whereas removing the few highly connected nodes has a large effect on the network structure, a random removal of nodes or edges has a small effect in most of the cases. In addition, the spatial layout of cortical and neuronal networks exhibiting several long-distance connections ensures few processing steps and thus a faster response time. (Kaiser, 2007, p.3043).

These observations support several of the corner-stones of recruitment learning: sparse connectivity and fast processing times realized through a limited number of processing steps, while providing connectivity sufficient to support non-trivial information processing. This issue is examined in detail in chapter 3.

1.4.3 Functionally Equivalent Neural Units

The search for functionally equivalent neural units, i.e. groups of neurons, which implement similar functions, is a core topic in the neurosciences. A study of functionally equivalent neural units may help to clarify how memory contents are represented. In particular, the size (i.e. the number of neurons) of these neural units is of interest and there are clearly two extreme positions:

On the one side is the assumption that each individual neuron represents memory contents such as a "microfeature" or a stimulus. The extreme variant of this hypothesis is called "punctuate representation" (Feldman, 1986); the assumption of a relatively small group of neurons forming a functional unit is called local or "compact representation". The extreme opposite position is that the entire brain or a very large number of neurons is involved in each representation ("holographic memory", holographic representation) The term "distributed representation" is used if neural groups of a significant size participate in a representation (Palm, 1990).

In the neurosciences, the search for functionally equivalent neural units relies on elementary results from neuroanatomy. Lorente de No (1938; quoted after Squire, 1987, p.70) noted for the first time the vertical organization of the cortex. There are columns which constitute physiologically functional units: Cells in these columns often have similar response features (Fujita et al., 1992). If columns extend over many cortical layers, they are called macrocolumns. Squire (1987) states there are about 10^6 in the human brain; each column has about 10^3 to 10^4 cells.

Even if columns or macrocolumns represent an idealization, they are a starting point for the search for functionally equivalent neural units. Furthermore, it must be noted that cortical columns are functionally inhomogeneous (Squire, 1987, p.72). In the visual cortex there are macrocolumns with cells that have a different orientation (e.g. adjustment to one eye only). Furthermore, each column includes color sensitive cells (Livingstone & Hubel, 1984; after Squire, 1987, p.72). Functionally equivalent neural units must be significantly smaller than the 10^3 to 10^4 cells generally accepted for columns. Bullock (1980; after Squire, 1987) assumes functional groups with less than 100 cells. Other authors (for instance Squire, 1987) assume minimal functional units in cortical macrocolumns with approximately 30-100 neurons. In addition, memory for whole events is represented in several areas of the brain; however there is no functional redundancy or replication over several areas.

Feldman (1986) studies the neural representation of conceptual knowledge. In his summary, which he calls an "interim neuron doctrine", he points out that the physical stimulus on higher levels of sensory processing is represented gradually more abstractly and that the number of neurons that take part in the representation decreases gradually. Statistical considerations lead to the assumption that about 15-20 cells are involved in a conceptual representation. Clearly, this is a rough estimate; nevertheless, Feldman (1986) and Squire (1987) arrive in principle at the same result.

Recent years have seen a shift of attention to more localist or punctuate representations. Bowers (2009) surveys simulation results in psychology and attributes the popularity of distributed representations to a misunderstanding of how localist models works. Quiroga et al. (2005) describe a neuron that selectively responds to images of the actress Halle Berry as well as the name "Halle Berry". This is seen by some authors as evidence for a high-level, multi-modal, localist representation.

1.4.4 Learning-Based Changes of Receptive Fields

Authors such as Merzenich et al. (1988) point out that neurons in the neocortex may, during the course of a lifetime, change or lose their complex or hypercomplex receptive fields, and cells which had no receptive fields may develop these in response to stimuli. Recruitment learning is the abstraction of such a process in artificial neural networks.

The most important changes in primate neural structures take place during a critical period before and after birth. The structural changes in the early postnatal phase can be summarized as follows (Honavar, 1989, p.18):

1. There is a rapid growth of neural connections, possibly triggered by environmental conditions.
2. There is an activity-dependant purge of connections ("pruning") as a result of a competition between neural links with the goal to transfer inputs into gradually higher brain regions.
3. There is an activity-dependent fine adjustment of neural connections which is Hebbian in nature (Hebb, 1949).

In the current context, structural changes immediately after birth are of limited interest compared to topological modification in later periods, in particular changes which are activity dependent and triggered by sensory inputs. Merzenich et al. (1988) provide a good summary of these processes. Topographic changes in the somatosensory cortex are summarized as follows:

1. Cortical representations are substantially reorganized, if peripheral neural connections are interrupted (this applies to adults rats, cats and primates). Neurons in a broad cortical sector including the original region as well as a broad zone around this field get new sensory inputs after such peripheral deprivation. Neural representations of skin surface areas, which lie close to the cut off, expand into the free cortical region, i.e. free cortical sectors are occupied immediately after the interference (Merzenich et al., 1988, p.41).
2. There are substantial movements in the receptive fields of primates (Merzenich et al., 1988, p.42).
3. In adult primates, tactile stimulation substantially reorganizes and extends cortical representations. If a limited skin region is strongly stimulated, nearly all neurons participating in hand surface representations get new afferent inputs (Merzenich et al., 1988, p.42).
4. The hand surface representation is substantially reorganized after cortical damage (Merzenich et al., 1988, p.43).
5. In adult primates, functional hand surface representation is substantially reorganized after surgical interference (Merzenich et al., 1988, p.43).

Similarly to Merzenich et al., Gilbert & Wiesel (1992) refer to the ability of the brain to adapt to changes as a result of sensory input, this time in the visual system. These sensory changes lead to direct, fast modifications of the receptive field organization; i.e. durable reorganizations over months. This includes enlargements of receptive fields, the reactivation of cortical regions after deprivation and shifts of receptive fields. Gilbert & Wiesel (1992) conclude from their studies that topographic reorganization results from synaptic changes in the cortex and is possibly caused by long, horizontal connections. After Gilbert & Wiesel (1992), these results allow for the possibility of continuous, dynamic changes of the receptive field structure during the normal perception.

In their experiments with cats and primates, Gilbert & Wiesel (1992) interrupted visual inputs by binocular damage to the retina. Dramatic changes were observed after minutes: cortical cells with receptive fields close to the border of the loss area showed a fivefold increase in receptive field size. In addition, there was a small centrifugal shift of receptive field locations in relation to the center of the damage area.

In order to investigate long-term effects compared to short-term changes, Gilbert & Wiesel (1992) observed the reorganization of the receptive field structure over a period of two months after damage. After this time interval, cortical areas with cells which had become inactive by the damage could be excited again by visual inputs. Further large shifts were observed in receptive field location. These changes of field location extended in some experiments to cells which were not directly affected by the damage.

Gilbert & Wiesel (1992) assume subtle horizontal influences, which become effective by sensory deprivation. In particular, the observed short term changes suggest an unexpectedly large plasticity of the adult brain. According to Gilbert & Wiesel (1992), this opens the possibility that dynamic changes of receptive field structure are a component of normal cortical processing.

Since cortical neurons can be changed by experience, it is fair to conclude that the neocortex represents memory contents (Squire, 1987, pp.117). Spinelli & Jensen (1979, quoted after Squire, 1987) trained cats to lift a paw in order to avoid a small shock. This shock was announced by the presentation of horizontal and vertical lines. After 10 weeks of training (8 minutes daily), the area of the sensorimotor cortex with cells responsive to the shock was two to six times larger than the corresponding cortical area of an untrained second paw. Furthermore, 75% of the cells in the trained sensorimotor cortex reacted to tactile and visual stimuli which resembled the conditioned input, while this was only the case in 30% of the cells of the untrained area. In addition, responsive neurons in visual association areas reacted to samples used in the learning phase, while the corresponding cells of untrained cats exhibited the same probability of reaction to all stimuli.

Bear et al. (1987, p.295) point out that spatial changes in the topographic representation do not go beyond one millimeter. There are therefore limitations on topographic reorganization, caused by the stable axonal structure which limits plasticity at the adult age. Nevertheless, it is important to investigate to what extent these changes have functional relevance.

1.4.5 Changes in the Functional Representation of Neural Circuits

Nerve cells can leave functional circuits and become an integral component of other neural groups (Hooper & Moulins, 1989). This functional change correlates with a change of membrane characteristics (Hooper & Moulins, 1989, p.1587) and is partly due to sensory inputs. Hooper & Moulins (1989) studied a neuron with synaptic connections to two different neural circuits. In the case of external stimulation, this neuron fires with neurons of the first group (it originally belonged to), if the second network is inactive, however, it exclusively fires with neurons of the second net as soon as this is active. Hence, the examined neuron becomes a functional component of the second circuit, independent of the activity of neurons in the first group (Hooper & Moulins, 1989, p.1588).

1.4.6 Learning-Based Structural Changes in Single Neurons

There are solid results on the experience-dependent change of the structure of nerve cells. Some of the most important aspects are summarized here:

1. Many nerve cells have several contacts (= synapses) with the same, neighbouring neuron. This is qualitatively different from an increase in synaptic efficiency (Lynch & Granger, 1989). The activation of these synapses is experience dependent. This opens the possibility of a multiplicity of nonlinear interactions between synapses (see Lo & Poo, 1991).
2. Non-local learning (heterosynaptic regulation; see below) is probably of great importance. Pre- and postsynaptic activation do not depend on an individual synapse in isolation, but several contacts may be affected (Singer, 1986).

3. Nerve cells change as a result of experience (Coss & Perkel, 1985; Nottebohm, 1989); this includes the formation of additional dendrites. Components close to the soma are not affected and genetically fixed; parts of the dendrite further away from the soma, however, can change under the influence of sensory inputs (Coss, personal communication). Chang & Greenough (1982, quote after Squire, 1987) report on measurable changes in the dendritic structure after 20-30 hours of training in the labyrinth (1 hour daily).
4. The postsynaptic contact points of synapses ("spines"), which can be found at the dendrites of pyramid cells in the cortex, change as a result on sensory input (Coss & Perkel, 1985). The structural modification is paralleled by a change of the electrical conductivity. This can be regarded as the physiological correlate of the connectionist "weight". In addition, results point to the activity-dependant change of the number of postsynaptic receptors for neurotransmitters (Lynch & Baudry, 1984).

1.4.7 Diffuse Activation and Plasticity at the Adult Age

There are good results on the role of non-specific activation in the nervous system of the adult, which according to some authors makes cortical plasticity possible (Singer, 1987, pp.314ff). Signals from the retina result in changes of the cortical representation if they serve motor control (i.e. trigger motoric actions). The neural feedback from actions can cause a necessary and non-specific activation in conjunction with other influences. A threshold mechanism is assumed which is effective on the postsynaptic side. This mechanism is attributed to voltage-dependant Ca2+ channels in the cell membrane. The non-specific excitation adds to the excitatory input to the cell, and thereby allows cortical plasticity, i.e. the change of the cortical representation.

1.4.8 Heterosynaptic Regulation

Singer (1986) confirms the Hebbian learning rule and extends it. The stabilization or attenuation of a synaptic connection is not based on the strength of the activation, but on the temporal correlation between pre- and postsynaptic activity. The same mechanism also regulates the competition between synaptic contacts. If two neural paths converge, i.e. have contact points with the same postsynaptic neuron, and their activity is uncorrelated, then the afferent connection, which stimulates the postsynaptic cell most effectively, becomes stronger, and the other synapse weakens. This principle applies to afferent, synaptic groups as well: The populations, which stimulate the postsynaptic cell most effectively, or have the strongest connections to begin with, will win and efficiency will increase.

Lo & Poo (1991) observed a clear competitive behavior between directly neighbouring synapses (heterosynaptic inhibition). In developing nervous systems, several nerve fibers normally stimulate a muscle at the same neuromuscular end plate, but there is only one axon even during adulthood. Lo & Poo (1991) stimulated one neuron from a group of neurons which had formed neighbouring synapses. The stimulated synapse increased its efficiency significantly, while the other synapses were suppressed. The effective inhibition of neighbouring synapses requires both pre- and postsynaptic activity. The synchronous stimulation of several synapses, however, causes no heterosynaptic inhibition. Based on these investigations, Po & Loo (1991) propose an extension of the Hebb' learning rule (Hebb, 1949). A synapse is strengthened and/or stabilized by its own activity, suppressed and/or eliminated by the asynchronous activity of other synapses at the same postsynaptic cell (Po & Loo, 1991; p.1021). Such a synaptic surrounding field inhibition could certainly contribute to compact neural representations.

Supported by simulations, Brown et al. (1991) describe a process similar to Po & Loo (1991). The interaction between Hebb-type synapses in the hippocampus causes a cell to react only to a subset of available synapses. The same process can lead to the generation of groups of active synapses at dendrites.

According to Singer (1987), cortical nerve cells can be in two different conditions, a "relay state" and a "plastic state". In the first condition, activation is passed on via sodium-induced activation potentials, without a change of synaptic efficiency. In the second condition, the postsynaptic cell is stimulated by excitatory influences in addition to (cholinergic) inputs and recurrent stimulation. There is thus a cooperating interaction between specific and modulating inputs. This interaction causes a change of synaptic efficiency in the plastic condition.

It is conceivable that these interactions between synapses lead to the formation of "gates"; modeled at the connectionist level by "connection" or "binder units".

Heterosynaptic regulation provides answers to some of the problems raised by purely Hebbian learning: What allows some form of normalization of synaptic plasticity? How can neurons react to groups on inputs only and maintain the response pattern even if synapses decay (Chechik et al., 2002)? Synaptic competition, however, may also arise as a result of spike-time dependent plasticity (STDP). If a pre-synaptic action potential briefly precedes the post-synaptic spike, potentiation may occur. Depression results if the post-synaptic spike shortly precedes the pre-synaptic action potential. Assuming that STDP is weaker than depression in a neuron with net positive input, the excitatory synapses will be weakened on average, resulting in the survival of strong excitatory synapses only (Chechik et al., 2002, p.6).

1.4.9 Fast Learning and the Biological Plausibility of a Declarative Long-Term Memory

Squire & Zola Morgan (1991) describe the function of the hippocampus and adjacent cortical regions during the fast acquisition of declarative knowledge. The impairment of memory function in amnestic patients is highly selective, which can be seen as evidence for a modular organization of the memory system. The ability to acquire new memory contents is a cerebral function, which is different from other perceptive or cognitive processes (Squire & Zola Morgan, 1991; p.1380)). Patients with damage to the hippocampus and adjacent cortical regions have limited memory performance, while other cognitive functions (simple habits, certain forms of learning by conditioning) are not impaired.

According to Squire & Zola Morgan (1991, p.1384), the hippocampus is involved in the formation of long-term memory at the time of acquisition. Distributed cortical activity is translated into stable memory contents; the hippocampus has the function to form conjunctions between events or bind stimulus characteristics, which are represented in different cortical locations. In this sense, the hippocampus is a memory system for simple contents, summaries ("a summary sketch") or for an index. Every time shifting attention leads to a change of the elements in short-term memory, the hippocampus binds contents and forms a memory trace for a whole event.

The hippocampus thereby serves the fast acquisition of new information on facts and events, which can be processed consciously. LTP ("long term potentiation", a relatively long-term change of synaptic efficiency) in the hippocampus can be made responsible for forming conjunctions. However, the hippocampus is only temporarily involved in the formation of long-term memory. Slow synaptic changes can lead to genuinely long-term storage. Memory processes are then possible without participation of the hippocampus.

The hippocampus is also one of a small number of sites for which adult neurogenesis has been confirmed experimentally. This is considered further in the next section.

1.5 Neuro- and Synaptogenesis – New Synapses and Neurons in Adulthood

Neurogenesis is probably the most active and vibrant area in brain research. The "no new neurons" dogma, i.e. that no new neurons are formed in the adult mammalian brain, has been convincingly rejected and the generation of new neurons and synapses in both the hippocampus and the olfactory bulb has been confirmed. Even more important, new neurons are being recruited into functional circuits and the process appears to be highly relevant for theories of learning and memory.

Empirical evidence for the formation of new synapses in the adult cortex of several species was reported two decades ago. Greenough (1987, quote after Honavar, 1989, p.12) describes the process of the formation of new synapses as follows:

> The process of the generation of new synapses in later phases of development and at adult age is similar to the growth of the sensory system in the early postnatal period. In the "plastic" part of the brain, new synapses are produced randomly, but only those synapses which receive reinforcement survive, all others are eliminated. The formation of new synapses in late phases of development and at adult age is a result of the activity of neurons, i.e. information processing or the execution of neuromodulatory functions. The newly formed synapses are located in those cortical regions, in which specific activation led to their generation.

The generation of synapses is triggered by local activation and is spatially constrained. This process is economical on the one side, since only a limited number of synapses must be formed, and contributes to the stability of the system on the other side: The modification of information processing structures is always spatially limited (Honavar, 1989).

1.5.1 Neurogenesis

The studies summarized in earlier sections refer to changes of the neural representation, with no new nerve cells being formed. Neurogenesis has been a very controversial topic in the neurosciences, but significant new evidence has emerged in recent years. Previously, Rakic (1985) stated that neurogenesis plays a limited role in primates and humans at the adult age. The view that neurogenesis is not concluded shortly after birth in mammals received increasing attention over the last twenty years only. Reynolds & Weiss (1992) observed the formation of new nerve cells "in vitro" in the brain (striatum) of adults mice (Barinaga, 1992). At this time, however, the standard view remained that there is no neurogenesis in mammals during adulthood.

It was acknowledged, however, that neurogenesis exists in some full grown vertebrate animals, e.g. fish (Rakic, 1985) and birds (Nottebohm, 1989), and this includes an embedding of the newly formed cells into functional circuits, including the formation of memory contents.

Nottebohm and co-workers offered empirical evidence for a neurogenesis at the adult age in a set of experiments with birds. Patton & Nottebohm (1984) could demonstrate neurogenesis in adult canaries by chemical markers over a period of 30 days. According to Patton & Nottebohm (1984), these new cells are recruited into functional circuits. The recruited nerve cells release action potentials and show slow changes of membrane potentials which are typical for synaptic activity (Patton & Nottebohm, 1984, p.1048). All newly formed cells react to auditory inputs, either at the onset of the signal or at the end of the stimulus.

Canaries are continuously adaptive birds, and acquire new songs every year (Nottebohm, 1989). This acquisition is dependent on the season and is probably steered by hormonal influences. The new nerve cells do not differ from other neurons and like these are replaced after approximately eight months. This is an indication that the new cells are used for memory formation.

Gould et al. suggested that adult neurogenesis may also occur in the human neocortex (e.g., Gould et al., 1999). Others, including Rakic (2002), have questioned these findings, suggesting that the new cells may be glia. It is confirmed now that adult neurogenesis in the mammalian brain leads to the functional integration of new neurons, and the question to what extent this functional integration includes learning has been investigated. A paradigm shift has occurred in the neurosciences which significantly strengthens the biological foundation of recruitment learning.

Leuner et al. (2006) discuss the all important question of a link between adult neurogenesis and learning. Four questions are investigated: (1) Is the number of new neurons positively correlated with learning? (2) Does learning increase the number of new neurons? (3) Are new neurons necessary for learning? (4) How might new neurons participate in learning and memory? Leuner et al. (2006) summarize positive and negative evidence for each question. Due to the relevance for recruitment leaning, the focus here is on the last two questions.

The argument in favor of new neurons as a condition for learning is based on studies using agents to block adult neurogenesis in rats (Leuner et al., 2006, p.219). If a substantial reduction in the number of adult born neurons is achieved by using a blocking agent over a two-week period, then an impaired ability to acquire the trace eyeblink conditioning task was observed. When the population of new neurons was allowed to replenish itself, the ability to acquire memories was restored (Leuner et al., 2006, p.219). Leuner et al. report that there are probably an equal number of studies to date that have failed to demonstrate that newly generated cells are involved in hippocampal learning.

The final point addressed in Leuner et al. (2006), how new neurons participate in learning and memory, is answered as follows (Leuner et al., 2006, p.220):

> … the production of new cells in the DG (dentate gyrus)increases the opportunity for learning in the future by providing more cells that can be recruited into existing circuits (Kempermann, 2002). However, it is possible that newly born cells influence learning processes even before they achieve full maturation. Certain characteristics of synaptic plasticity are enhanced in adult-generated neurons and these characteristics may make them particularly useful for the processing of new associations.

Shors (2008, p.254) describes the process of neurogenesis in the mature hippocampus: New neurons are generated specifically in the dentate gyrus. As they mature, newly generated neurons migrate short distances into the granule layer. Within a few days, new cells develop dendrites and start to form synapses with pyramidal cells in area CA3. Over the course of a few weeks, neurons express molecular markers of immature cells, followed by markers of mature neurons a week later. Within several more weeks, the new cells can produce action potentials and are established as functional neurons in the adult brain (Shors, 2008, p.254).

Toro and Deakin (2007, p.5) describe the process of adult neurogenesis in the olfactory bulb: During neurogenesis in the adult subventricular zone (SVZ), newly formed neurons migrate in chains along the rostral migratory stream to the olfactory bulb. When they reach the olfactory bulb they detach from chains and migrate radially while they differentiate into GABAergic interneurons. Reelin, a protein that regulates neural migration and synaptic plasticity, plays an important role in the detachment of new neurons from chains once they reach the destination before their radial migration and differentiation.

Social dominance has an impact on the survival rate of new neurons:

> Animals that exercise or live in enriched environments produce more cells, whereas those exposed to stress or deprived of sleep produce significantly less … Perhaps more amazing is the effect of social dominance on cell number. Males that are higher in the hierarchy produce more than submissive males, and female rodents exposed to the pheromones of a dominant male produce more cells than when exposed to smells of a submissive male … (Shors, 2008, p.254).

Shors (2008) emphasizes that special types of learning involve the hippocampus, generally learning that requires a mental effort, attention, several trials and has indeed a successful outcome. Shors (2008) concludes that the newly generated neurons do not form "grandmother" cell representations in the narrow sense, but represent "templates for new learning" Shors (2008, p.257). Following Nottebohm (2002), Shors (2008) concludes

> the generation of new neurons serves to replenish a circuit with cells that had yet to experience ''learning'' so that the subject could respond anew ... new cells have the ability to respond anew to a learning experience without the challenge of distinguishing old from new" (Shors, 2008, p.257).

Kempermann et al. (2008) come to a similar conclusion. They believe that adult hippocampal neurogenesis allows the lifelong adaptation of the network to levels of novelty and complexity that an individual encounters. The process is cumulative and optimizes network efficiency which allows avoidance of cross-talk (i.e. catastrophic interference; the detrimental interaction of new information with older information; Kempermann et al., 2008, p.290–295).

1.6 Neuroplasticity and Mental Health: Key Issues for Recruitment Learning

Throughout its existence, psychology has investigated mental disorders to infer from abnormal function how the healthy brain and mind works. In a similar way, neurogenesis and mental disorders are discussed here because it clarifies and reinforces the role that neurogenesis plays in both learning and memory. Neurogenesis is relevant for the models of recruitment learning discussed in later chapters of this book, because it allows for a supply of functionally available units, While models of recruitment learning developed decades ago could not take advantage of recent discoveries, there is a very good chance that future models will be based on adult neurogenesis during normal processing as well as in brains affected by disorder.

1.6.1 Depression and Adult Neurogenesis

After the confirmation of the generation of functional neurons in the adult human brain in areas such as the hippocampus, the interest moved to implications for clinical disorders such as depression and schizophrenia. Magnetic resonance imaging studies have shown that hippocampal volume is decreased in patients with chronic depression, and this is even more evident in patients with posttraumatic stress disorder (PTSD). The loss correlates with the degree of cognitive impairments in hippocampus-related verbal memory tasks (Bremner & Vermetten 2004, in Perera et al., 2008). Antidepressant treatment arrests hippocampal volume loss in patients with major depressive disorder and PTSD, and reverses cognitive deficits (Bremner & Vermetten 2004, in Perera et al., 2008, p.330).

Perera et al. (2008) study the role of neurogenesis in depression as well as the effect of anti-depressant treatment. According to Perera et al. (2008), neurogenesis within the adult hippocampus is suppressed by factors that predispose the patient to major depression and stimulated by anti-depressant interventions, and this means all forms of anti-depressant therapies. That is, the use of newborn neurons essential for effective antidepressant action.

Perera et al. (2008) present a theory of depression which is based on adult neurogenesis. In this theory, newborn hippocampal neurons are necessary for detecting environmental changes and linking emotions with the external context. The stress-induced inhibition of adult neurogenesis decouples emotions from external context, resulting in depression. Persistence of negative mood beyond the duration of the initial stressor is defined as major depression. Anti-depressant induced neurogenesis therefore restores the coupling of mood with the environment, leading to the resolution of depression (Perera et al., 2008, p.326).

In this theory, a positive event in the life of a person suffering from severe depression does not necessarily lead to a mood change because the absence of neurogenesis means that the crucial link between emotion and the positive external environment cannot be made. The person continues to suffer from depression while the original stressor is long gone and the "positive" environment is ineffective. Only therapy (medication, electroconvulsive treatment or psychotherapy) restores neurogenesis and with it the link between context and emotion. The "positive" environment (or rather, the perception of a positive environment) can have an effect and the person recovers from depression. It is tempting to develop the Perera et al. (2008) theory further to the point were neurogenesis allows the perception of the environment "colored with the right emotion" and hence, the maintenance of good health.

In summary, all antidepressant drugs and electroconvulsive therapy stimulate hippocampal neurogenesis (Nasrallah, 2008). Mood stabilizers such as lithium and valproate which are used for the treatment of bipolar disorder promote neurogenesis and regeneration of cortical gray matter (Nasrallah, 2008).

1.6.2 Schizophrenia and Adult Neurogenesis

Neurogenesis is the focus of intense research and a significant part is directed to neuroplasticity, or lack thereof, in patients suffering from severe mental health problems such as schizophrenia (Mei & Xiong, 2007). The DISC1 gene (DIsrupted in SChizophrenia 1) was discovered in large Scottish and Finnish families and constitutes a risk factor for schizophrenia, schizoaffective and affective disorder (Kempermann et al., 2008, p.293). DISC1 is expressed in the hippocampus and may have diverse functions in neuronal migration and differentiation, neurite outgrowth, and synaptic plasticity. Among other things, DISC1 controls the integration of new neurons in the course of adult hippocampal neurogenesis. Mice with reduced DISC1 levels in just a few new brain cells showed accelerated integration of those neurons into the existing brain circuitry (Duan et al., 2007). The abnormally shaped neurons also positioned themselves improperly. DISC1-deficient newborn neurons were increasingly excitable and developed a greater number of synapses with other brain cells. Neurons also grow a lot more dendritic processes and fire at a faster rate. Synapse formation is faster as well (Duan et al., 2007). Many of these findings have led to the view that schizophrenia may be an adult neuro-developmental disorder.

Schizophrenia is associated with a loss of cortical and subcortical gray matter and new generation antipsychotic drugs can stimulate the growth of neurons and synapses. This includes areas of the brain traditionally of interest to recruitment learning models such as the hippocampus. Studies with subjects experiencing a first-episode psychosis demonstrated a progressive loss of cortical and subcortical gray matter in schizophrenia associated with acute psychotic episodes (Nasrallah, 2008, Keshavan & Hogarty, 1999). In childhood-onset schizophrenia, the cortical volume loss was estimated at 1%-3% per year during the first five years (in Alzheimer's disease, the annual volume loss is about 5%, Nasrallah, 2008). Affected brain regions are the frontal and temporal cortices, the hippocampus, the amygdala and the thalamus. The shrinkage includes a reduction in dendrite length by a half and a decrease in the number and size of dendritic extensions (Nasrallah, 2008).

The low level of connectivity between neurons in the brain of patients suffering from schizophrenia has been investigated by use of computer simulation (Calitoiu et al., 2005). In this model, both very high levels of connectivity in the piriform cortex (due to lack of pruning) and very low levels of connectivity (due to excessive pruning) are investigated. Both increasing and lowering connectivity levels decreases the level of chaos (in a technical sense) in the system with excessive synaptic pruning having the higher impact.

The picture is more complicated in schizophrenia than in depression since "first generation antipsychotic" drugs such as haloperidol do not consistently stimulate neural growth in the adult brain. "Second generation antipsychotic" drugs such as olanzapine have shown the following effects in patients suffering from schizophrenia (cited after Nasrallah, 2008): Increased neurogenesis in the hippocampus (in one study this required administration of the drug for 28 days) and an increased proliferation in the prefrontal cortex. Nasrallah (2008) believes a paradigm shift is taking place: an impairment of adult neurogenesis is being considered as an important mechanism for psychosis.

The following chapters introduce various recruitment learning models, originally designed without adult neurogensis in mind and solely based on the functional recruitment of units. However, the solutions that these models provide in conjunction with very recent results from neurobiology, can lead to a new class of models that enjoy a very high level of biological plausibility.

Chapter 2
One-Shot Learning – Specialization and Generalization

Joachim Diederich

2.1 Introduction

Most work in connectionist learning, and indeed in the present volume, is devoted to induction from examples, but instruction is as important in symbolic artificial intelligence as it is in nature. This chapter begins our exploration of recruitment learning by examining a connectionist learning system capable of rapidly integrating a new fact into an existing knowledge base in response to a single instruction expressed in a formal description language. Such instructions are transformed into an input for a connectionist knowledge representation system, which in turn changes the network in order to integrate new knowledge. In subsequent chapters we shall consider models based explicitly on neurally plausible circuits, with operations defined at the level of the individual neuron or assemblies of neurons. Here the representation is more abstract, in principle replacing the neural assembly by a single node – demonstrating both the power of recruitment learning as a knowledge acquisition strategy, and the potential alignment between connectionist systems and the established agenda of symbolic artificial intelligence.

Integration is important when a single new fact causes changes in several parts of the knowledge base; it is an adjustment which cannot easily be realized with inductive techniques alone. The new, integrated knowledge can be used in conjunction with prior knowledge. As before, learning is performed through recruitment, the conversion of network units from a pool of free units into units which carry meaningful information, i.e. to represent generic concepts.

The objective of this chapter is to demonstrate how high-level learning techniques such as "learning by being told" can be realized in connectionist systems. One-shot learning methods, which require a single training instance only, have been explored in symbolic artificial intelligence but there have been limited attempts to realize instruction in connectionist networks. The next sections describe a "learning by being told" technique for connectionist knowledge-based systems.

Similarity-based learning, or "learning from examples," is the original domain of learning in neural networks. Competitive, reinforcement and backpropagation learning are all inductive techniques as they require the presentation of an ensemble of input vectors. One of the disadvantages of these methods is that they make rather strong assumptions about the nature of the environment, e.g. the availability of a teaching or reinforcement input and structured data sets. One-shot learning, in

J. Diederich, C. Günay, and J.M. Hogan: Recruitment Learning, SCI 303, pp. 37–56.
springerlink.com

contrast, is sometimes necessary and has advantages, for example when only a single new fact is available and has to be integrated.

The instruction used here has two parts: a goal-concept which is already part of the connectionist representation system and a training example which consists of a set of attribute/value pairs, represented by units in the network. The instruction is expressed in a symbolic description language and entered into the system. On the network level, the presentation of the input triggers recruitment learning, i.e. the acquisition of a network unit from the pool of free units and the integration of this unit within the network of concept units. The recruitment learning process guarantees that the newly recruited unit is integrated immediately below the goal-concept; a new concept unit is generated this way. The newly recruited unit is the immediate sub-concept of the goal-concept and the immediate super-concept of former sub-concepts of the goal-concept.

It is important to note that the integration process requires changes in the network of concept units, i.e. knowledge is not only added to the network, but the network of knowledge units is changed in order to allow real integration.

In summary, the approach has two objectives: 1) to demonstrate "learning by being told" in a neural network and 2) to propose a solution to the integration problem in recruitment learning, i.e. the integration of a newly allocated unit in a richly structured connectionist semantic network without any programming or external modification of the network structure.

2.1.1 Knowledge-Intensive Learning in Symbolic Artificial Intelligence

"One shot" learning techniques are an important part of symbolic artificial intelligence. The term explanation-based generalization (EBG) was introduced by Mitchell et al. (1986) and refers in artificial intelligence to knowledge-intensive learning methods which require the presentation of only a single example, using domain knowledge to constrain any search for a possible generalization. This involves generating a new chunk of knowledge which describes a set of features including the properties of the training example. EBG is a two-step learning method: First, construct an explanation of why an example fits a particular goal concept and find those features of the training example which are relevant to satisfy this goal concept. Secondly, search for sufficient features of the example to build the general concept definition, the goal of the method. EBG, however, requires a strong domain theory, a pre-condition which is unrealistic for real-world applications. Consequently, efforts were made to use weaker domain models.

It is important to note that the recruitment learning systems described here do not realize analytical generalization in the sense that EBG does, but rather shares some aspects like input requirements and the use of built-in knowledge. EBG requires unification and deduction which is hard to realize in connectionist networks without variable binding.

2.1.2 An Example for the Learning Process

The following is a rather informal "real-world" example for the kind of learning realized in the subsequent sections. A person sees a car and forms the hypothesis that this car is a "BMW." The hypothesis is verified by comparing the features of the perceived car with typical features of "BMWs." The person realizes that this is a special kind of "BMW" with additional features which are not owned by "BMWs" in general. The new knowledge about this special kind of "BMW" is added to the general knowledge about cars from this manufacturer.

BMW is the goal-concept. BMW 520 and BMW 730 are two subclasses of BMW and are part of the built-in knowledge. BMW 520 owns the attribute "number of front lights" with value "four," BMW 730 owns the attribute "price" with the value "expensive." The just-seen BMW looks rather expensive and has four front lights. This perception and the "best match" BMW/just-seen car triggers learning. A new class is formed, e.g. "BMW upper class," which is an immediate subclass of "BMW" and the immediate parent of BMW 520 and BMW 730.

2.1.3 The Organization of the Chapter

This chapter is organized as follows. First, the integration of a newly recruited unit in a connectionist semantic network is described, essential parts of the recruitment learning algorithm are outlined, including a formal description, and results of a test run of the system are shown. Next, an extended architecture for the recruitment of binder units, i.e. units that connect objects, attributes, and values is introduced, including empirical results for this architecture. Finally, advantages and shortcomings of the proposed models are discussed.

2.2 Forms of Recruitment Learning in Connectionist Ontologies

There are several possible applications for recruitment learning as part of connectionist knowledge representation systems. Oscillatory neural networks are not considered here; Part II of this monograph is devoted to these types of neural networks.

a) Concept Binding

Shastri (1988) used a partitioned random graph to build basic knowledge structures. It consists of units representing concepts and instances, attributes and values of attributes. The central idea is to use binder units to connect concepts and instances with attributes and values, and to make these structures accessible for retrieval and question answering. A typical binder unit has three sites where input lines come in and an activation function which requires input from at least two sites for a positive output.

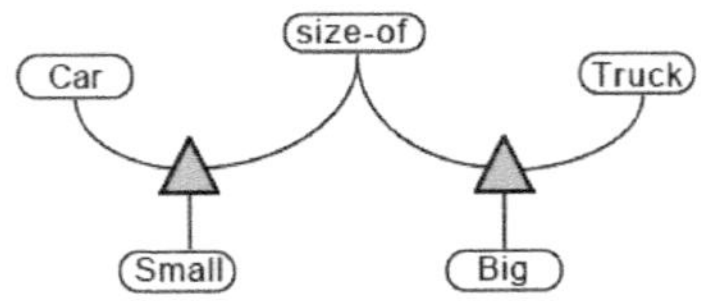

Fig. 2.1. Concept "Car" owns the attribute "size-of" with value "Small;" concept "Truck" owns the attribute "size-of" with value "Big."

For example, an agent sees an interesting red pyramidal object and recruits a structure equivalent to: polyhedron(B1), color(B1, red), top(B1, pointed). During classification, the instance "polyhedron"[1] can be retrieved by clamping the units "color" and "red" on ("clamping on" means a unit has high output for a given number of updates). On the other hand, it is possible to retrieve the value "red" by clamping "polyhedron" and "color" on during property retrieval.

b) Concept Formation from Several Instances
Concept formation from several instances includes presenting multiple examples to a network and to analyze similarities among these examples. Concept formation takes place if the number of instances exceeds a certain threshold and if high similarity among the examples is determined. One or more concept units is/are then recruited and weighted connections between these concept units and common features of the training examples are established. For example, the agent sees several instances $B_i \in (B_1, B_2 \ldots B_n)$ with top(B_i, pointed) and forms a concept "pyramid" with an attribute top(pyramid, pointed).

c) Higher-Level Concept Formation
Higher-level concept formation means the generation of a new super-concept triggered by the analysis of a number of concepts which are already part of the connectionist ontology. Learning is similarity-based. A super-concept is built and integrated into the existing organization of the knowledge network, and the analyzed concept units get positive weighted connections to the new super-concept. This process requires not only a positive weight change between a new knowledge structure and the already existing network of committed units, but also the "unlearning" of weights, e.g. between the analyzed concept units and the former super-concept.

For instance, an agent knows object(C_i) where "object" and C_i are concepts and $C_i \in$ (block, pyramid, cylinder) with side(C_i, plane). The agent forms a new concept "polyhedron" which is an "object" with the attribute side(polyhedron, plane). Furthermore, the agent deletes for all C_i: object(C_i) and asserts for all C_i: polyhedron(C_i).

[1] "The Original Sin_in the theory of polyhedra goes back to Euclid, and through Kepler, Poinsot, Cauchy and many others ... [in that] at each stage ... the writers failed to define what are the 'polyhedra' ..." Grünbaum (1994, p.43).

d) Simple Specialization
Specialization means adding a new chunk of knowledge as a specialization of an existing concept to a knowledge network, based on the analysis of a single example. The new concept unit must establish weights to the specialized concept and to the features which define the newly acquired concept. No modification of the network of committed units is required. In our example, the agent knows side(polyhedron, plane) and sees an instance B_i of "polyhedron," i.e. polyhedron(B_i) with top(B_i, pointed). Consequently, the agent forms the concept "pyramid" and asserts polyhedron(pyramid) with top(pyramid, pointed). It is also possible that the new concept "pyramid" gets additional attributes inferred from domain knowledge, i.e. super-concepts of "polyhedron", but this is not part of the example.

e) Specialization by Integration
Specialization by integration requires not only the enrichment of a network by a more specialized concept, but also the modification of parts of the knowledge structure. This is necessary when the new, specialized concept has sub-concepts which must cut their connections to the former super-concept and must establish connections to the newly recruited concept. This case is similar to higher-level concept formation, but requires the specification of a goal-concept as part of the input, as well as a training example which consists of a set of attribute/value pairs. For instance, an agent knows object (C_i) where "object" and C_i are concepts and C_i $\in$ (block, cylinder). The agent sees an instance B_i of "object" with side(B_i, plane) and forms a new concept "polyhedron" which is an "object" with the attribute side(polyhedron, plane). Furthermore, the agent deletes for all C_i: object(C_i), i.e. connections between all C_i and "object" are cut in the connectionist system, and asserts for all C_i: polyhedron(C_i).

This modification of existing network structures is a non-trivial process in a massively parallel connectionist system. Some part of the network must remain unchanged while other parts undergo modifications. All decisions and modifications must be done the connectionist way; there is no top-level interpretation and no programming of network structures. The following sections of this chapter describe a specialization by integration process which is triggered by an instruction to the system.

2.3 The Integration of a Newly Recruited Unit in a Connectionist Ontology

2.3.1 The Architecture of the Network

The network used in this study has five modules; each module is an n-layer network with mutual excitatory connections between neighboring layers (in both directions) and inhibitory connections within layers. This architecture is similar to the interactive activation model in McClelland & Rumelhart (1981). Each layer is a specialization of a "winner take all" (WTA) network (Feldman & Ballard 1982). Competition among units in a layer results in the strong activation of a winner unit, but several units might be active simultaneously if they receive strong excitation. Initially, there are small random weights in this network: weights within a layer are inhibitory [-1, -10] (weights are integers [-1000, 1000], in general), weights to the layer above are initially excitatory [1, 10], as well as weights to the layer below [1, 100]. Recruitment learning can change random initial connections between layers to strong positive weights. The total network has four "spaces," i.e. three network modules with the architecture described above, and an additional single space containing a set of units without internal organization (the instance space). In detail, there is a space for the representation of structured objects, a space for the representation of attributes, a space for the representation of values of attributes and the instance space mentioned above. All representations are localist; there is a single unit for the representation of each structured object, attribute, value or instance.

On the network level, there are two classes of units in each space: concept units and relay units. Concept units may have one of two states, "free" or "committed." A concept unit in the "free" state is called a free unit. Recruitment learning transforms free units into committed concept units.

Committed units build a hierarchy, i.e. committed object, attribute and value units are embedded in a multi-layer network and form a straight-forward spreading activation network. Initially, there is at least one layer with relay and free units between each layer with committed concept units, i.e. committed concept units in different layers are connected by relay units.

Units are not only connected with units in the same space, but also with units in other spaces. Object, attribute and value units are connected by binder units. A binder unit has three "sites" where symmetric input-lines come in, and has an activation function which requires input from at least two sites for a positive output (See Chapter 1 for details on the architecture of units in structured connectionist networks). There is one single binder unit for each connection between an object, attribute and value.

Relay units connect those concept units which have no direct connections, i.e. which are separated by at least one layer. The output of relay units is weaker than the output of concept units (< 250 compared to < 1000 for concept units).

2.3.2 The Specialization of a Concept in a Connectionist Semantic Network

The objective of this recruitment learning process is to create a generic concept as a specialization of an existing concept as described in the instruction, and to integrate the new concept in the hierarchy of structured object units embedded in the multi-layer network. In the following, we restrict the recruitment to the object space.

Integration of a new concept unit includes the following processes:

1. The newly recruited concept unit develops an excitatory link to the to-be specialized concept unit;
2. The new concept unit gets excitatory links from lower-level concepts which are already part of the network of committed units and which are sub-concepts of the to-be specialized concept;
3. Sub-classes of the newly recruited concept unit develop null valued or small random connections to their former super-concept (now specialized by the newly recruited unit); and
4. The new concept unit gets connections to and from attribute and value units by the recruitment of binder units.

Note that the four points above define a recruitment process which requires changes in the network of committed units. These changes must be done in a connectionist way: there is no top-level interpreter and units communicate by non-symbolic messages only. Weight change is strictly local.

We first specify the pre-conditions for the recruitment process and will explain the "learning by being told" procedure afterwards.

2.3.2.1 The Input

The input has two parts: 1. a goal-concept, which is the to-be-specialized concept and part of the hierarchy of committed concept units. The goal-concept unit is clamped on during learning. 2. A training example, which is an instance of the goal-concept and consists of a set of attribute/value pairs, represented by units in the attribute and value space.

An input to the system is a statement in a symbolic description language, such as

(DEFEXAMPLE Block_1 `Block `((SIZE SMALL)(COLOR RED)))

where Block_1 is the name of the training example which consists of the attributes "size" and "color" and the corresponding values "small" and "red." "Block" denotes the goal-concept. An alternative method for presenting the input is a system-guided interview, i.e. the user is asked to name goal-concept and features of the training-example.

The instruction is then compiled into a script file which can be used by the Rochester Connectionist Simulator (Goddard et al. 1987), which is used for this study, and results finally in the presentation of an input to the network, i.e. units representing concepts which are specified in the instruction are clamped on. The goal-concept gets a very high potential and output, while features of the training examples are continuously activated with the maximum potential and output (1000).

2.3.2.2 The "Learning by Being Told" Procedure

The input is presented to the network, i.e. the goal-concept unit and attribute/value units describing the training example are clamped on. Learning is triggered by clamping on the goal-concept unit with an extremely high activation value, which is out of the range for activation values without external modification. The goal-concept unit sends activations in two directions: in the direction of the top-level layer of the network which results in the (limited) activation of super-concepts of the goal-concept, and in direction of the layer below which causes competition among the free units in this layer (for reasons of simplicity we assume that the layer below the goal-concept units consists of free and relay units only and that sub-concepts of the goal-concept are immediately below this layer).

Attribute and value units representing the training example activate their binder units, and the corresponding concept units in the object space receive activation. The condition that the training example is an instance of the goal-concept is important at this point. Sub-concepts of the goal-concept which own some of the attributes/values of the training example, and which might own attributes and values not shared by the goal-concept, will receive activation from attribute and value units. Note that the excitation of these concept units results only in the activation of one single branch of the hierarchy of concept units while the other branch receives inhibition from the activated concept units (the WTA effect) and has no activation at all.

During the next time steps there will be a winner unit in the layer below the goal-concept unit as a result of the competition among the free units. The winner unit among the free units establishes excitatory weights to the goal-concept unit in the layer above and to activated units in the layer below by use of the strictly local weight change rule described below. At this point during learning, the sub-classes of the goal-concept have positive connections to the winner unit in the layer above and to the goal-concept unit via the relay units (See Figure 2.3).

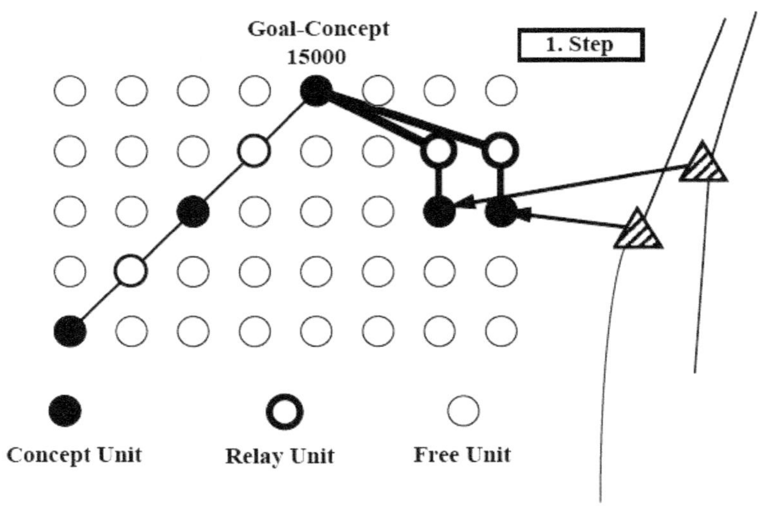

Fig. 2.2. The initial network before recruitment learning. Thick lines show connections which are affected by weight change.

During the third and final step, the "cleanup" phase, the goal-concept has to decrease incoming weights from its former sub-classes in order to get full integration of the new concept unit (the winner among the free units). The special characteristics of the weight change rule and the transmission characteristics of the relay units are important at this point. The goal-concept will increase the incoming weight from the new concept unit because both units have high activation values. Relay units connecting the goal-concept unit with its sub-classes shut off because there is a winner unit in the same layer. As the relay units decay (or become inactive), the former sub-classes of the goal-concept have no path to the goal-concept anymore, but they do have connections to the new concept unit. The final state of the network is shown in Figure 2.4.

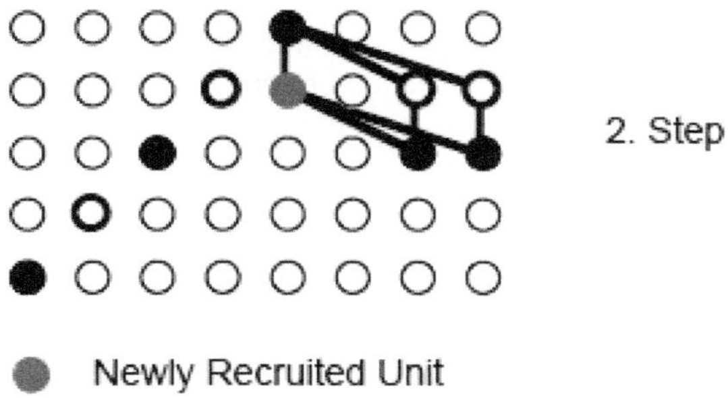

Fig. 2.3. Weight change between the winner unit and the goal-concept, and sub-concepts of the goal-concept.

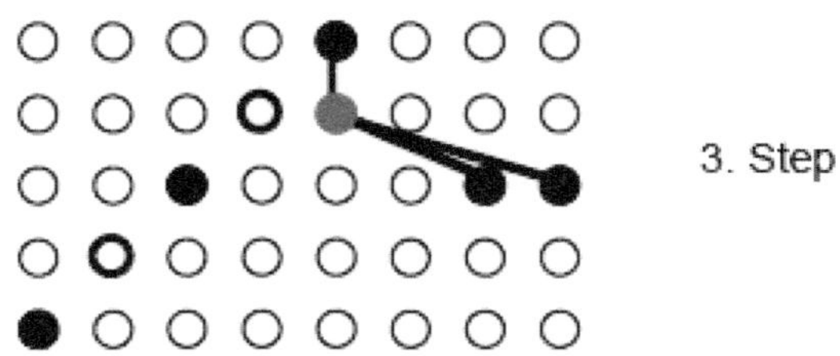

Fig. 2.4. The state of the network after cleanup: Sub-concepts of the goal-concept have links to the newly recruited unit.

2.3.3 Simulation Results: Integrating a Unit in a Connectionist Semantic Network

Simulations are done with a 400 unit object space, a 100 unit attribute and value space, a 10 unit instance space and 10 binder units. In total, the network has 620 units and 45600 links as part of the multi-layer networks, i.e. links from and to binders are excluded. Fair asynchronous update is chosen to prevent oscillations due to interactions between spaces. During each update, 50% of all units are randomly selected for firing. Each unit has fired at least once in 234 updates.

Potentials, outputs and weights are integers as described above. All excitatory sites in all units calculate the weighted sum of inputs. Exceptions are described in section 3.4. Potential and output of a unit is simply the sum of all sites. Units in the object space start with random potential and output [0, 100], binder units start with an initial output 1 and attribute/value units with an initial output 0. The distribution of initial weights and the weight change rule are described above.

Figures 2.5 – 2.7 show the competition among free units in the layer immediately below the goal-concept during the first 15 updates. The goal-concept is clamped on with an output of 15000. Figure 2.5 shows potentials of the first 4 free units in this layer, figure 2.6 shows potentials of free units 5 to 7 and figure 2.7 shows free units 8 to 10. For reasons of simplicity, there are only two objects in the object space which have connections to attribute and value units (via binder units): the initial subclasses of the goal-concept have one feature each; both features are part of the training example.

Figures 2.5 to 2.7 also show a clear "winner takes all" behavior, due to lateral inhibition. Free unit 7 is fully activated after 15 updates, all other free units on this layer have a zero potential and output. This pattern is stable. Figures 2.8 and 2.9 show competition in the same layer as the goal-concept! The very strong activation of the goal-concept, together with feedback from the layer above and below the goal concept causes competition among units in the same layer as the goal-concept. The full activation of an additional unit in this layer would obviously result in false, additional recruitments (i.e. cross-talk). The behavior of units in the goal-concept layer is important and shown in Figure 2.8 and 2.9. These figures demonstrate that all units - with the exception of the goal-concept - have zero or negative potentials. One unit (free unit 5) has a negative potential due to strong inhibition from the goal-concept. This does not affect weight-change and spurious recruitments are not possible.

2.3.4 A Formal Description of the Method

The system is specified on two related levels: 1. on the conceptual level where generic concepts denote classes of individuals with common features and individuals represent instances of these concepts. 2. On the network level, units are organized in sets called "spaces" and interact via weighted links. Because we are using a localist approach, in most cases conceptual entities are represented by a single unit. However, it is important to distinguish between the conceptual level and the network realization, i.e. the mapping from a conceptual description to a connectionist realization. In contrast to similar approaches, such as Shastri (1988), the focus of attention is on the enlargement of the network through recruitment learning.

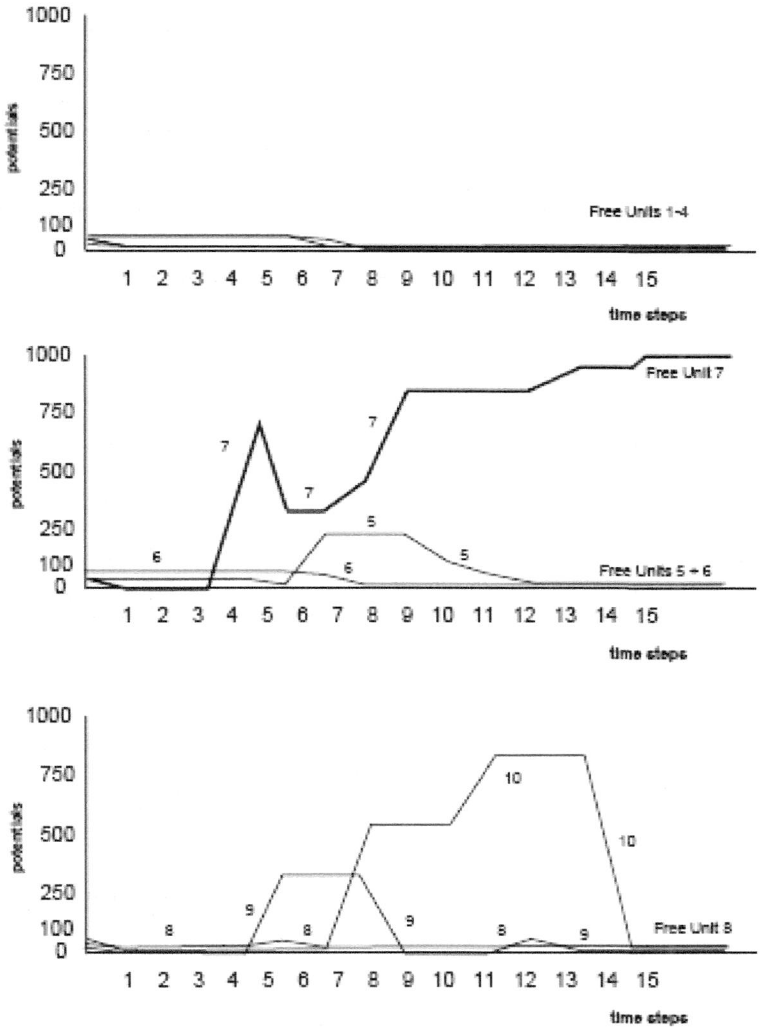

Fig. 2.5 – 2.7. Potentials of free object units in the layer below the goal-concept unit during the first 15 time steps.

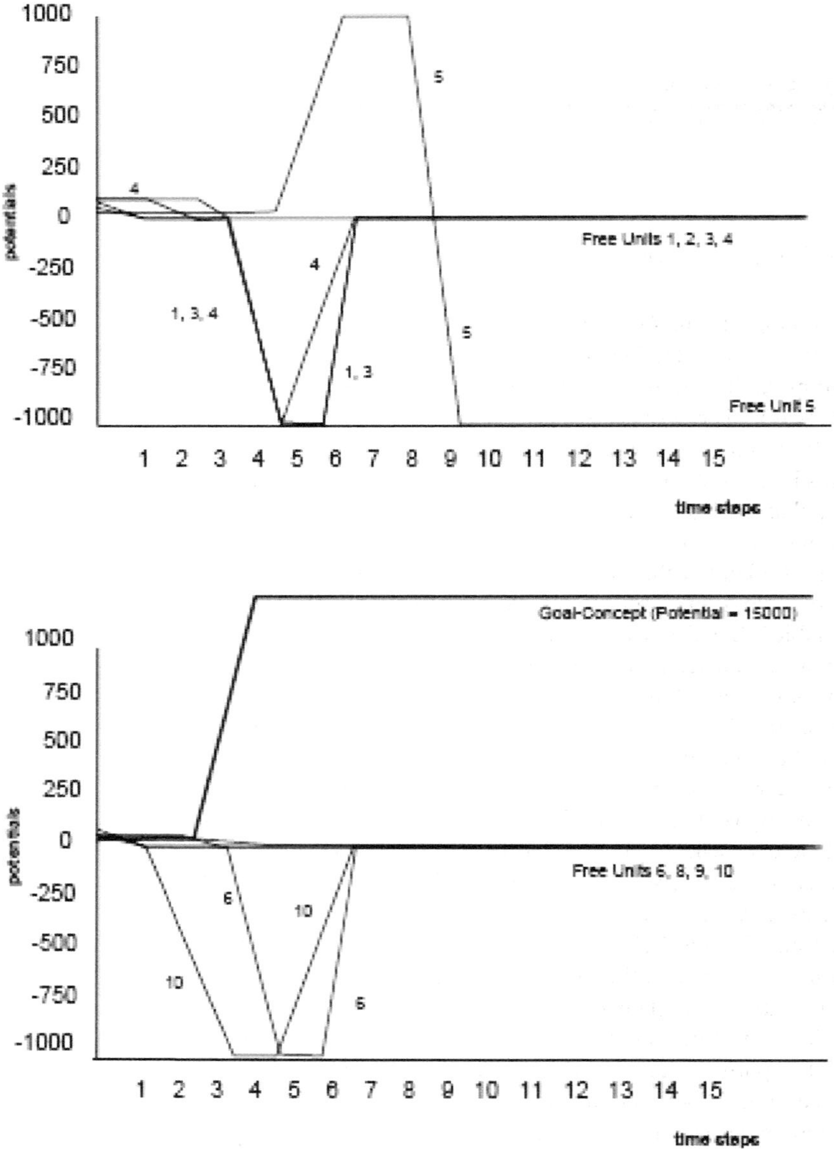

Fig. 2.8 – 2.9. Potentials of free units in the same layer as the goal-concept.

The following notation is used throughout this section: capital letters (O, A, ...F) denote sets. Lower case letters denote individual members of sets, e.g. $o \in O$, $a \in A$.

2.3.4.1 The Conceptual Level

On the conceptual level the model is a n-tuple $W = \{O, A, V, I, F, \lambda, \alpha, \leftarrow\}$, where O is the set of all objects in a domain, A is the set of all attributes, V is the set of all possible values of all elements in A (features), I is the set of all instances, F is the set of all free, uncommitted entities which may become committed during learning, λ is a mapping from O to the power set of A, α is a mapping from A to the power set of V, and $\leftarrow$ is a partial ordering defined on O, A and V. Note that not only elements in O are generic concepts, but elements in A and V too.

O, A and V have a partial ordering, organized as a taxonomic hierarchy. $o_{gc} \in O$ is the goal-concept. o_{gc} is part of the input and is continuously clamped *on*.

Free entities have no attributes and values.

In our system, $i_e \in I$ denotes the training example. Here, i_e must be an instance of the goal-concept o_{gc}. If the example i_e is represented by a unit in the instance space, it will therefore inherit all attributes and values from o_{gc} as described above. Moreover, i_e might have attributes and values not shared by o_{gc}.

Recruitment learning means the transformation of free entities into committed entities: $f_j \in F \wedge \neg(f_j \in O) \Longrightarrow \neg(f_j \in F) \wedge f_j \in O$. The operator ==> refers to the process of transforming a free entity into a committed entity $\in O$.

Let $F_b \in F$ be the set of all free entities which are available for recruitment, and $f_j \in F_b$ is selected to become the newly formed conceptual entity (on the network level, F_b denotes to the set of free units on the layer immediately below the goal-concept unit, and f_j is the winner free unit). Let $O_{\underline{sc}} \in O$ be the set of all subclasses of $o_{\underline{gc}}$, i.e. $\forall o_{\underline{j}}, o_{\underline{j}} \in O_{\underline{sc}}, o_{\underline{gc}} \leftarrow o_{\underline{j}}$. We define a function

$$r(s, o_{gc}, i_e)$$

which accepts o_{gc} and i_e as input, and s is the conceptual state of the network, including the taxonomic hierarchy of objects and their attribute/value structure. The function r changes s, i.e. forms the new conceptual entity f_j (i.e. $f_j \in F \wedge \neg(f_j \in O) \Longrightarrow \neg(f_j \in F) \wedge f_j \in O$) and asserts $o_{gc} \leftarrow f_j$ and $f_j \leftarrow o_i$. Furthermore, the relations $o_{gc} \leftarrow o_i$ are deleted.

The following section describes the realization of function r on the network level.

2.3.4.2 The Network Level

On the network level the model is a tuple of elements $M = \{U, \omega\}$. U is the set of all units in the network and is ω the set of all weights.

Members of U are partitioned into six classes of elements which make up U. U^o denotes the class of all object units: $U^o{}_i$ represents $o_i \in O$. U^a is the class of all attribute units with $U^a{}_i$ representing $a_i \in A$. U^v is the class of all feature units with $U^v{}_i$ representing $v_i \in V$. Furthermore, U^i is the class of all instance units in the attribute space with $U^i{}_i$ representing $i_i \in I$. U^f consists of free units with $U^f{}_i$ representing $f_i \in F$. The sixth and final class of units is U^b, the set of binder units.

2.3.4.3 Unit Types and Functions

On the network level, there are two classes of units (with exception of the instance space): concept units and relay units. Relay units have a special architecture. There are three "sites," i.e. locations of separate computation on arriving input: "up," "down" and "inhibition." "Up" is the site where input from the layer above arrives, "down" is the layer where input from lower-level units arrives and "inhibit" is the site where input from units within the layer arrives. Relay units have a special "site function" for the "inhibition" site. The site function computes the weighted sum of inputs from this site, but the unit goes off if there is a winner unit among the free units in the same layer (i.e. strong positive input from a unit in the same layer). The inhibition is strong enough to absorb the input from a possible goal-concept unit in the layer above and possible sub-classes of a goal-concept in the layer below the relay unit.

Concept and free units have the same structure as relay units, i.e. two sites where input from the layer above and below arrives and one site where inhibitory links from units within the layer arrive (no special site function for the "inhibit" site). In addition there is an "interaction" site where links from other spaces enter.

More formally, the site-value x is simple the weighted sum of inputs:

$$\text{site-value}_x = \sum_{i=0}^{N-1} i_i\, w_i$$

with exception of the "inhibition" site of relay units, where

$$\text{site-value}_{inh} = \text{if}\left(i_i == \text{max\_input}\right) \text{ then } 0 \text{ else } \sum_{i=0}^{N-1} i_i\, w_i$$

and max_input is the maximum input of 1000. The output and potential of units is simply the sum of all sites

$$\begin{aligned} \text{output} = \text{potential} = {} & \text{if } \left(\left(\text{sum} = \sum_{i=0}^{N-1} \text{site-value}_i\right) > \text{max\_sum}\right) \text{ then max\_sum} \\ & \text{else if (sum} < 0) \text{ then } 0 \\ & \text{else sum} \end{aligned}$$

again with the exception of relay units, where

$$\text{output}_{rel} = \text{potential}_{rel} = \begin{array}{l} \text{if (site-value}_{inh} == 0 \text{) then } 0 \\ \text{else if } \left(\left(\text{sum} = \sum_{i=0}^{N-1} \text{site-value}_i\right) > \text{max\_sum}\right) \text{ then max\_sum else sum} \end{array}$$

site-value_{inh} is the value of the inhibition site of relay-units and max_sum is the maximum potential and output of relay-units.

In general, output and potentials are set to 1000 if the sum of sites is greater than 1000 and output and potentials are set to 0 if the sum of sites is smaller than 0.

2.3.4.4 Weight Change

Each weight w_i is an integer $\in$ [-1000, 1000]. Positive associated units have weights $w_i > 0$, negative associated units have weights $w_i < 0$, a zero weight means no association.

The following weight change rule is applied (inputs and potentials are integers):

$$\Delta w_{ij} = \begin{array}{l} \text{if}\left(i_i >= 0 \wedge w_{ij} < p_j \wedge w_{ij} < i_i \wedge w_{ij} < \max\right) \text{ then } +1 \\ \text{else if}\left(w_{ij} > \min\right) \text{ then } -1 \end{array}$$

where i_i is the i_{th} input to the receiving unit, p_j is the potential of the receiving unit, max is an upper limit for weights (= 990), and min is a lower limit for weights (= -990).

The simulations were done with this modified version:

$$\Delta w_{ij} = \begin{array}{l} \text{if}\left(i_i >= 250 \wedge w_{ij} < p_j \wedge w_{ij} < i_i \wedge w_{ij} < \max\right) \text{ then } +10 \\ \text{if}\left(i_i >= 100 \wedge w_{ij} < p_j \wedge w_{ij} < i_i \wedge w_{ij} < \max\right) \text{ then } +1 \text{ else if}\left(w_{ij} > \min\right) \text{ then } -1 \end{array}$$

The increase of weights depends on the receiving unit's potential and on the amount of the received input. Furthermore, the upper limit for this increase is the lower value of the potential p_j and the amount of transmitted activation i_i.

2.3.4.5 The Recruitment of the Newly Integrated Unit

Recruitment is based on the layered organization of the underlying "winner takes all" networks. Free concept units are forced into competition by the strong output of the goal-concept in the layer above. The full activation of a winner free unit makes weight change to the goal-concept unit in the layer above, and to subclasses of the goal-concept unit in the layer below possible by Hebbian learning only. The "winner takes all" behavior of the free units is crucial in this context.

Competition depends on three factors: (1) on the initial distribution of weights, (2) the initial distribution of potentials and outputs, and (3) on the number of free units in the layer below the goal-concept. The free unit in the layer below the goal-concept with the strongest positive connection to the unit representing o_{gc} receives the strongest input from the goal-concept (in comparison with other free units), and therefore has the potential to absorb inhibitory inputs from competing free units. Because of this inhibitory influence, the potential and output of the goal-concept has to be greater than the number of free units times the maximum potential/output. In other words, the WTA network below the goal-concept converges and generates a clear winner if

$$p_{gc} > \#free\_units\ max\_potential$$

where p_{gc} is the potential of the goal-concept, #free_units is the number of free units in the layer below the goal-concept, and max_potential is the maximum potential of free units (i.e. 1000). A potential of 15000 was sufficient in the simulations, because the number of free units in the layer below the goal-concept was 10. The additional potential helped to speed-up convergence. Note that potential == output for all units.

2.4 The Recruitment of Binder Units

This section provides a brief description of the recruitment of binder units. The newly recruited unit in the object space did not get connections to attribute and value units yet. This section gives a brief description of the dynamic generation of these connections.

The network structure outlined in section 3 has to be changed in order to allow the recruitment of binder units. This changed network has four "spaces," i.e. two network modules with the architecture described in section 3, plus the additional single space with a set of units without internal organization (the instance space) and the so-called "binder space." In detail, there is a space for the representation of structured objects, a single space for the representation of both, attributes and values, the instance space mentioned above, and the binder space. In other words, the most important modification of the network structure is the integration of the attribute and value space in one module, now called attribute space. This attribute space includes a hierarchy of generic attributes, where all leaf nodes are values of attributes.

Attribute and value units have connections to object units by binder units. Binder units, in the same way as concept units, are either "free" or "committed." The recruitment learning process transforms free binder units into committed binder units. Every non-relay unit in the object space has connections to every potential binder unit and vice versa. Binder units, however, are connected to a subset of all attribute/value units only. This particular connection pattern is described below.

2.4.1 The Organization of the Object and Attribute Space

This section describes the organization of the object and attribute space in more detail. The object space includes the connectionist inheritance hierarchy, i.e. units in this hierarchy are representing classes of objects which have the same attribute/value structure. This conceptual hierarchy is embedded in the multi-layer network and builds a coalition of units with strong excitatory connections. Paths between these units, i.e. connections over an arbitrary number of relay units, are representing "superclass/subclass" relations. These relations are implicit in the structure of the connectionist inheritance hierarchy in the object space.

It is important to note that committed units in the attribute space represent a different, implicit structure. Again, the attribute space is a multi-layer network and attributes and values are embedded in this network. The attribute space, however, has an implicit "a kind of" hierarchy. For example, "Small" and "Big" are values of the attribute "Size." They have positive weighted connections, over an arbitrary number of relay units, to "Size," which itself has a positive weighted connection to "physical-property." In other words, "Small" and "Big" are a kind of "Size" and "Size" is a kind of "Physical-Property." Activation of "Small" and/or "Big" will result in an activation of "Size" and an activation of "Size" will result in an excitation of "Physical-Property" (unless the relay units are disabled).

2.4.2 The Architecture of the Binder Space

The next space is the binder space. This module consists of several subsets of binder units. Each of these subsets is a WTA network. Binder units in a WTA network compete whenever activation flows into the corresponding net, and there will be one single binder unit active as a result of this competition. This binder unit becomes recruited and makes the connection between the newly learned concept unit in the object space as well as an attribute unit and a value unit in the attribute space.

The crucial question is: How are binder units connected to units in the object and attribute space in order to allow the "appropriate" recruitment of binder units; i.e. semantically correct associations between activated attributes, values and the newly recruited concept without crosstalk?

The answer to the questions above is based on the organization of so-called receptive areas for each subset of binder units. Each WTA network in the binder space has a receptive area in the attribute space, i.e. a set of attribute and value units which are connected to binder units in this subset. A receptive area is basically a diagonal or vertical line of units through the attribute space. For example, each of the following number is representing a unit in the attribute space (for reasons of simplicity we are ignoring relay units here):

1 2 3 4 5 6 7 8 9 10
11 12 13 14 15 16 17 18 19 20
21 22 23 24 25 26 27 28 29
...
71 72 73 74
81 82 83
91 92 93

The first subset of binder units (first WTA network) has a receptive area of one single unit: 91. The second subset of binder units has a receptive area of two units: 81 and 92. The third subset of binder units has a receptive area of three units: 71, 82 and 93 and so on: each receptive area is a diagonal line of units through the attribute space, actually in "both" directions. In other words, units are part of several receptive areas. "91," for example, is one receptive area for its own, but also part of the receptive area which includes 91, 82, 73, 64 ... 10.

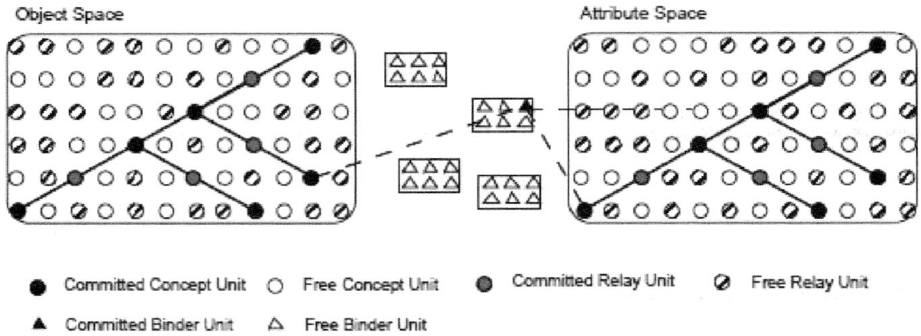

Fig. 2.10. The organization of the object and attribute space.

2.4.3 *The Recruitment of Binder Units*

Learning is triggered by presenting an input to the network and is performed in two separate learning phases. During the first step, the goal-concept unit and units representing the training example are activated as described in section 3. In the second step, the goal-concept unit is turned off.

Recruitment in the object space is done in phase 1 and is identical to the processes described in section 3. The goal-concept, which is also part of the input, is clamped on the object space. Activation flows to the layer below the goal-concept and causes competition among the free units in this layer. This results in the activation of a single winner unit. Furthermore, attribute and value units in the attribute space activate additional units in the object space.

Activated attribute and value units send activation to the corresponding WTA binder unit networks; causing competition among the binder units. Competition among binder units results in a single winner unit in those binder subnets which have connections to activated units in the attribute space. These activated binder units make the connection between the winner unit in the object space and attribute and values in the attribute space.

The generation of these connections, i.e. recruitment, can be done by local weight change only. The activated new concept unit in the object space gets strong excitatory connections to all activated binder units and these binders units get strong connections to activated attribute/value units. Note that the organization of receptive areas in the attribute space guarantees that the appropriate bindings are done, i.e. semantically illegal bindings are not possible. The built-in "a kind of" network of committed attribute and value units allows only bindings between attributes and values with an implicit "a kind of" relationship, due to the organization of the receptive areas. Each receptive area includes only those units which represent conceptual entities with implicit "a kind of" relations.

2.4.4 *Simulation Results: The Recruitment of Binder Units*

This section describes experimental results with this kind of architecture. Simulations are done with a 400 unit attribute space and a 200 unit object space. Asynchronous update is chosen to prevent oscillations due to interactions between spaces. Potentials, outputs and weights are integers [-1000. 1000]. Unit functions in binder units compute the product of the sum of sites and a constant (10 in the simulation described here). Units in the object space start with random output [0, 1000], binder units start with an initial output 1 and attribute/value units with an initial output 0.

Slow weight change is used for connections between binder units and attribute/value units (± 10) and extremely slow weight change is used for connections between binder units and object units (+1 if both units are active for 5 updates). The network becomes unstable otherwise. The initial weights in the object and attribute space are described above. Initial weights between object and binder units and between binder units and attribute/value units are chosen to be small, random, excitatory values in the range [1, 10].

Figure 2.11 shows the competition among units in the object space and the activation of a winner unit. The potentials of 5 free units in the layer immediately below the goal-concept during the first 15 steps of a test run are shown. The goal-concept is clamped on with output 25000. There is only one attribute and one value unit active in the attribute space. Both units are in the same receptive area and clamped on with output 250.

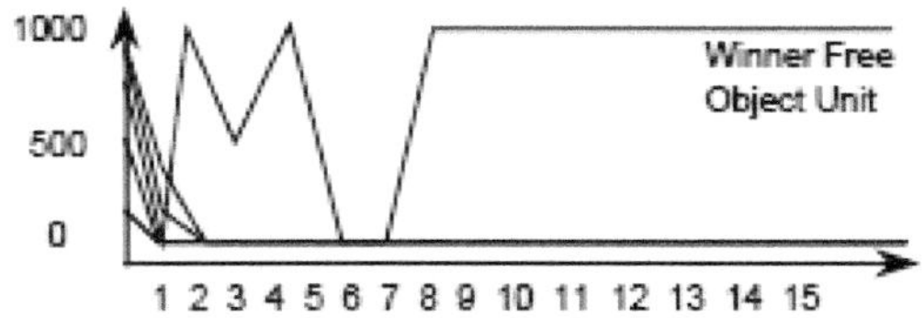

Fig. 2.11. Potentials of free object units in the layer below the goal-concept unit in the object space during the first 15 time steps.

Figure 2.12 shows the potentials of three binder units during the first 25 updates of the network. Each unit is a "winner" unit in its WTA network of binder units. Learning causes the activation of these three binder units only, all other binder units are inactive.

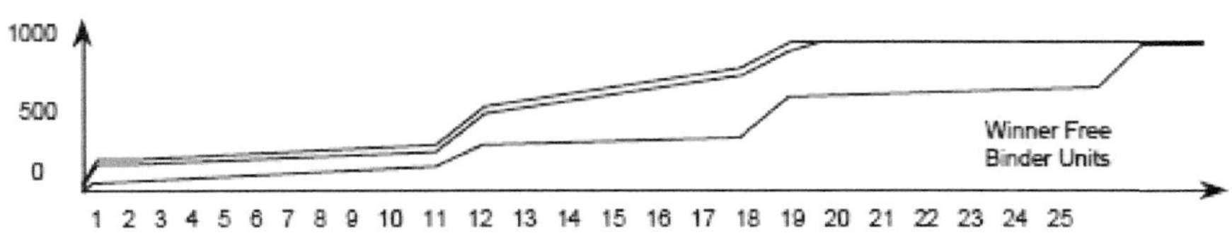

Fig. 2.12. Potentials of 3 binder units; each unit is a winner in a WTA sub network during the first 25 updates.

2.5 Discussion

A connectionist learning system is described which transforms an instruction expressed in a description language into an input for a connectionist knowledge representation system and which changes the network in order to integrate new knowledge. This realizes a high-level learning procedure in a connectionist network which can augment learning-by-examples methods that are widely explored in the neural network area.

However, the approach has several shortcomings. For example, it is not possible to learn a specialization of the just-learned concept. More than one layer of relay units has to be used in order to allow this additional flexibility, and unit structure must be changed. This modification, however, would allow the incremental generation of the network of committed units. Competition among the free units must be realized in a layer with some distance to the former sub-classes of the goal-concept and the goal-concept itself, in order to allow further specialization and generalization of the newly learned concept.

Recruitment of binder units was briefly outlined and it was shown how structuring the attribute space avoids semantically illegal bindings. On the other side, there is no mechanism which guarantees that the newly recruited object unit gets only those connections to attributes and values which are sufficient for recognition. This would allow taking full advantage of the connectionist inheritance hierarchy in the object space. Future research has to provide a mechanism which recruits these particular connections only.

In summary, it was shown how "learning by being told" can be realized in a structured connectionist network, and a solution for the integration of units in connectionist semantic networks was proposed. Furthermore, it was demonstrated how the introduction of structure facilitates the recruitment of connections between units in a conceptual hierarchy and units representing features. The outlined solution is not complete, but avoids semantically incorrect recruitments.

In subsequent chapters, we shall examine the representation and recruitment of conceptual knowledge of this type in networks developed under neurally plausible scarcity constraints.

Acknowledgement

This chapter is an updated version of: Diederich, J.: Instruction and High-Level Learning in Connectionist Networks. Connection Science 1 (1989) 2, 161-180. Text and figures have been re-used with permission from the publisher Taylor & Francis Group: http://www.informaworld.com who is gratefully acknowledged.

Chapter 3
Connectivity and Candidate Structures

James M. Hogan

3.1 Introduction

We have earlier considered the foundations of recruitment learning, an approach grounded in the understanding of neural organisation and psychological function which emerged during the second half of the twentieth century. While the revolutionary findings in respect of adult neurogenesis explored in section 1.5 have deep implications for recruitment learning, we focus initially on Arbib's earlier contention that "... a certain critical degree of structural complexity is required of a network before it can become self-modifying ... in a way that we could consider intelligent". This chapter begins a computational exploration of these ideas, adopting the strategy of converging constraints as pioneered by Feldman and Valiant to help delineate the tasks which might be realised through the recruitment of general purpose circuits.

The convenient abstraction of *committed* and *free* units introduced in section 1.1 allows a ready distinction between functional circuitry and those units yet to be allocated to a functional task. In contrast to Chapter 2, however, we shall be concerned with the application of these ideas in a neurally plausible setting, and with the exploitation of function supported within nascent neural circuits, having available neither the abstractions just considered nor the local specialisation characteristic of extensive training[1]. Such a 'library' of circuitry might correspond to that available in an early postnatal phase, or indeed, emerge on a smaller scale within neural populations supplemented through adult neurogenesis. We proceed by generating random networks under constraints of structure and biologically plausible connectivity levels, yielding an initial random knowledge base, K_0, which may be subsequently exploited to facilitate learning.

In the latter sections of this chapter, we make these notions of functional structure more concrete, examining the probability of existence of candidate subnetworks known to support the representation of a number of well-known

[1] See the famous 'Shape from Shading' study of Lehky and Sejnowski (1988) for an example due to backpropagation.

J. Diederich, C. Günay, and J.M. Hogan: Recruitment Learning, SCI 303, pp. 57–81.
springerlink.com

Boolean concepts. In the following chapter, we look more carefully at the capacity of the realised structures to support complex knowledge, and at our ability to recruit them as part of more complex circuits.

The use of random links in connectionist systems dates from the perceptron models of Rosenblatt from the early 1960s, and reappeared in the structurally adaptive *cascade correlation* networks of Fahlman and Lebierre (1990) at the height of the backpropagation era. However their use as a technique for modelling of sparse, cortically inspired connectionist networks appears to date from Feldman's classical study of dynamic connections in neural networks (Feldman, 1982). Feldman examined conditions under which dynamic connections could be established between sets of neurons as a means of rapidly acquiring unfamiliar knowledge. These ideas were examined both in the context of interconnections across a layered architecture, and in the mutually reinforcing activation of a stable cell assembly.

The present work builds in some respects upon Feldman's but also has similarities to models introduced by Valiant in successive contributions since the late 1980s (Valiant, 1988, 1994, 2005). In essence, Valiant combines the data representation of the cell assembly with the probabilistic link arguments of the random interconnection network. His focus is then to find a previously uncommitted assembly or *frontier* to receive projections from existing, committed representations, and to allocate some fraction of these nodes to represent some combination of existing knowledge. Like Feldman, Valiant regards random connectivity principally as a mechanism for concept formation, realising structure as a consequence of intersecting sets of (possibly weighted) activation. The problem thus becomes one of probabilistic linkage under sparsity constraints rather than one of network formation. Indeed, Valiant is explicit in the adoption of a neural *tabula rasa*, arguing for the primacy of learning in the combination of knowledge arising from sensory and cortical feature detectors, and in the latter's construction. In this respect, the approaches considered in the next two chapters differ markedly from the examination of one-shot learning of Chapter 2 and from the spiking neuron studies of Part II, although there is no reason in principle that these ideas cannot be unified.

Nevertheless, random connectivity lies here at the core of knowledge representation, populating the initial knowledge base described above. To investigate the viability of the approach, we determine the likelihood of existence within our networks of neural structures suitable for the representation of particular Boolean concepts, such *candidate structures* forming a library of random domain knowledge whose extent is necessarily linked to the statistical properties of the network. In principle, each initial knowledge base is created through random sampling (with replacement) from some class of deterministic networks, with the network 'dynamics' being selected at 'compile' time, and subsequent modifications localised to one or more recruitment layers. The development of these models, and the conditions governing network realisation, are considered in the following sections.

3.1.1 Structured Random Networks

While functional recruitment may readily be examined within a genuinely random graph, one in which the connection probability is uniform across possible node pairs, there is evidence that neural structure in general, and cortical organisation in particular, optimises connectivity under the volume constraints imposed by cranial capacity. Murre and Sturdy (1995) noted that both fully connected and randomly connected topologies are extremely inefficient in the absence of a defined packing arrangement, and a clearer morphological picture of neural connections has been provided by the optimisation studies of Kaiser and Hilgetag reviewed in Chapter 1.

For connectionist systems, this discussion serves to highlight the ongoing tension between top-down and bottom-up design, the former allowing substantial pre-specification of the system, the latter relying upon learning algorithms of various kinds to *select* through weight adjustment those aspects of the architecture required. The intermediate framework of "partially structured connectionism" (Feldman et al, 1988) provides a rational compromise between the two extremes, coupling systems level architectural design with unstructured local networks which may be recruited in order to realise an overall system task. Both the present work and the interconnection and propositional networks of Feldman and Valiant lie comfortably within this grouping, but our approach is in one sense more dependent upon pre-wired (albeit *randomly* pre-wired) knowledge than the others considered.

As our focus is upon computations exploiting or operating within cortical circuits, we shall usually impose some overarching layered structure, allowing recurrent connections when required to support sequence processing. Such 'structural innateness' may also include differences in connectivity regimes between subsystems, supporting bias in link distributions and their associated constraints. Typically we will work with a K-layer feedforward network, each layer consisting of N nodes. The relationship between nodes in different layers will be determined by a single Bernoulli trial, where the success probability may be fixed, or may depend in some way on the separation between them. Grounding of this parameter in actual neural connectivity levels is considered in the following section.

3.2 Models of Cortical Connectivity

Alignment of network sparsity with anatomically observed levels of neural connectivity is more complex than may initially be apparent. Direct counting of neurons and synapses at the desired resolution has not proven feasible, and estimates have traditionally been based on population counts after sectioning of representative tissue. Current work as part of the *Connectome Project* (Pfister et al, 2010), using high-throughput, high-resolution microsocopy (See for example Jeong et al, 2009) offers the promise of significant improvements in the accuracy of local population estimates, and of connectivity regimes specialised to functional regions. For models such as those discussed in this chapter, these technical developments may usher in a revolution as significant as that which may emerge from the confirmation of adult neurogenesis.

At prsesent, however, local connectivity remains speculative, and one must work with models informed by the available observations. Our approach is consistent with the modular structure preferred by Murre and Sturdy (1995), and the reviews and optimization studies of Kaiser and Hilgetag (2006). In particular, these studies offer strong evidence for relatively dense local neural circuitry coupled with sparser long range connections, consistent with an imperative to minimise steps in the processing chain while respecting physical constraints.

Hogan (2000) presents an extensive discussion of these issues, exploring a range of anatomical studies and the assumptions under which connectivity estimates were obtained. The essential principle is that some estimate of the connectivity associated with each neuron may be inferred from gross measures of connection volume and density. Stevens (1989), for example, relies upon two parameters which are approximately constant across selected mammalian species:

- The volume density of cortical synapses ($r = 0.6 \times 10^9$ synapses per mm^3); and
- The number of neurons underlying a given area of cortical surface ($p = 1.48 \times 10^5$ neurons per mm^2).

Assuming cortical homogeneity[2], it is possible to fit an expression for the the mean number of synapses per cortical neuron, $q(n)$, based on the relative size $s = n/n_0$ relative to some reference brain of size n_0:

$$q(n) = q(n_0)s^{0.09}. \tag{3.1}$$

Assuming, following Stevens that "...each neuron is connected to an *almost-constant* number of other neurons irrespective of brain size" (emphasis added), an indicative value may be obtained through simple manipulations involving the cortical volume V and surface area A (where Q and n are respectively the total number of cortical synapses and neurons):

$$q(n) = \frac{Q}{n} = \frac{rV}{pA} = \frac{rT}{p}. \tag{3.2}$$

The mean thickness $T = V/A$ is then chosen arbitrarily within the accepted range, Stevens using a reference value of $T = 1mm$ to give $q(n_0) = 4.05 \times 10^3$ synapses per neuron[3]. Noting the value of $p = 1.48 \times 10^5 mm^{-2}$, the Stevens analysis leads to the conclusion that the cortical network is sparse, with each

[2] Increases in cortical network size across species here result from the addition of structurally similar modules.

[3] He reports 4.12×10^3.

neuron likely to receive direct synaptic input from fewer than 3% of its near neighbours. By implication, an even sparser long range connectivity must operate, but these conclusions are far from universally accepted, and as we shall see the Stevens estimate lies at the lower end of those available.

Murre and Sturdy (1995) undertake a similar analysis, but lie closer to the spirit of the present work, allowing randomness to operate within a series of encompassing models, each imposing constraints on the network structure. Most of these models support stratification of the neurons, with connectivity between strata modelled through a connectivity fraction f, similar in principle to the branching factor employed by Feldman. Here f represents the fraction of the *entire* target population to which a particular neuron may be connected, and thus may be *extremely* small for large networks.

Murre and Sturdy are principally concerned with the scalability of each packing arrangement with network size (as reflected in the number of neurons). None of the stratified models are plausible, offering only limited packing advantages over full connectivity. In contrast, the modular arrangement, which aligns with the present approach, balances dense local connectivity with relatively sparse links across modules. Here, volume scales far more slowly than the almost cubic behaviour exhibited by the other regimes:

$$V_{mod} = O(mN^{3/2}), \tag{3.3}$$

where m is the number of modules supported. We note that the modular network is far superior in accounting for the observable properties of the cerebral cortex. While there are some significant technical differences between this modular network and the models developed below, we nevertheless concur with the authors that this type of connectivity pattern is the most plausible and also likely to prove the most useful in supporting sophisticated processing.

The more recent studies of Kaiser and Hilgetag (2006) offer striking confirmation of these earlier ideas, and a more formal consideration of the connectivity regimes through the traditional graph theoretic measure of the clustering coefficient. As recounted in Chapter 1, these authors report clustering coefficients around three times the comparable random network value, and evidence of tightly bounded shortest paths. Significantly, Kaiser and Hilgetag suggest that long range projections account for as little as 10% of neural connectivity. While it remains difficult to ensure consistency within these estimates, and still more difficult to apply them directly to the layered network structures considered below, it is clear that these figures support the view that cortical connectivity follows a small world network regime (Watts and Strogatz, 1998). We shall return to this issue in the analysis below.

Valiant (1994) also considered models of this kind as a means of establishing the structural plausibility of algorithms presented earlier in his monograph, with graph models "... based on the availability both of dense local connections [and] sparser global connections". Valiant's central concern is that the expected number of synapses on each *local* neuroid be sufficient to support *vicinal* linkages – overlapping projections from distinct assemblies or node sets to a common region to support a new relation – with high confidence. Local connection probability is here determined from the expected number of local synapses onto each neuron within a cubic millimetre of cortex, and is inversely proportional to the volume occupied by each dendritic tree. In this fashion, Valiant is able to develop Poisson-based expressions for the expected number of neuroids with which a given neuroid synapses *at least* some threshold number of times – in effect the size of the set on which the node has a substantial *influence* or effective weight. Using both laminar and columnar organisation of the local areas, he is able to demonstrate that his algorithms are sustainable through a "... small but sufficient fraction of the [target neuroids] ..." receiving synapses at far above the average rate.

3.3 The Connectivity Model

In what follows we shall model layered networks supporting strong local connectivity within some radius R of the neuron, while limiting long range connectivity outside this region through some decay of the success probability α_0. Some difficulty arises here as a result of the substantial role of recurrent excitation in cortical circuits (Douglas et al, 1994), but the extent to which this dominance is reflected in the local topologies at a columnar or lesser scale is debatable. Our approach is pragmatic, choosing feedforward or recurrent systems according to computational need, applying connectivity estimates with reference to the structure chosen. Additional feedforward models were discussed in Hogan (2000), but were of limited utility and are not considered here.

More formally, we examine a structured network of K layers, $0, 1, \ldots, K-1$, each of N nodes, for a total of

$$M = KN \tag{3.4}$$

nodes in the network. Usually N is large, while K is tightly restricted – in a cortical context being limited to around 6. We consider connections between arbitrary nodes i in layer k (which we shall write as the ordered pair (k, i)), and j in layer $k + \nu$ $((k + \nu, j))$. To avoid double-counting, we allow links only for $\nu \geq 0$, and impose a feedforward structure between the input layer 0 and output layer $K - 1$, allowing in principle the special case of lateral and self-recurrent connections, but prohibiting genuine feedback links.

In the absence of a readily identifiable length scale for this architecture, we define a 'Manhattan' node separation ρ, by the following relation (for $\nu \geq 0$)

$$\rho_{\nu ij} = \rho((k + \nu, j), (k, i)) = \nu + |j - i|. \tag{3.5}$$

Following in some respects the anatomical work of Sholl (1956), we assume a relatively high probability of connection within a *local radius* R of each unit, with decay of probabilities for nodes outside this region. The appropriateness of the decay regime is dependent in part upon the consequent value of α. In its most extreme form, we may use a power law decay with each unit increase in node separation beyond R, or alternatively one may postulate decays based on concentric regions out from each node, the probabilities to be governed by ranges defined by real or integer multiples of R.

We consider here the former case, for which a power law decay applies for radii greater than R, and for which local connectivity will be marginally higher than for the concentric ring approach:

$$Pr((k,i) \to (k+\nu,j)) = \begin{cases} \alpha_0, & \rho \leq R \\ \alpha_{\rho-R} = \alpha_0^{\rho-R}, & \rho = R+1,\ldots,N+K-2 \end{cases}, \tag{3.6}$$

where appropriate writing α_ρ to represent either expression. Again, we allow for self loops and lateral connections, and at this point assume that they follow the same regime as the feedforward links.

Connections from a node (k,i) to other nodes are again modelled as a series of independent Bernoulli trials with success probabilities α_ρ based upon the node separation. The outcome of each of these trials is a Bernoulli distributed random variable with generating function

$$\omega_\rho(s) = (\alpha_\rho s + \beta_\rho), \tag{3.7}$$

where $\beta_\rho = 1 - \alpha_\rho$ is the failure probability. We follow a procedure similar to that used above to develop the generating function for n_{ki}, the number of connections emanating from a particular node.

We consider connections from a (non-central) node (k,i) to nodes in the same or subsequent layers. It is readily apparent that for small R (relative to N), the Bernoulli trials form equivalent pairs (until the end of a layer is reached) and may be treated as repeated trials. Figure 3.1 shows an example network layer in which $N = 9$ and $R = 2$. We consider possible connections from the source node (say $(k,2)$). It is readily seen that the following trials (with associated success probabilities) are possible within layer k:

$$\begin{array}{cc} (k,2) \to (k,2) & (\alpha_0) \\ (k,2) \to (k,1),\ (k,2) \to (k,3) & (\alpha_0) \\ (k,2) \to (k,0),\ (k,2) \to (k,4) & (\alpha_0) \\ (k,2) \to (k,5) & (\alpha_1) \\ (k,2) \to (k,6) & (\alpha_2) \\ (k,2) \to (k,7) & (\alpha_3) \\ (k,2) \to (k,8) & (\alpha_4). \end{array} \tag{3.8}$$

The situation is similar when the analysis is extended to the following layer, excepting that the separation has increased by one, so that the probabilities are correspondingly lower, and another α_ρ is introduced:

$$\begin{array}{lr}
(k,2) \to (k+1,2) & (\alpha_0) \\
(k,2) \to (k+1,1),\ (k,2) \to (k+1,3) & (\alpha_0) \\
(k,2) \to (k+1,0),\ (k,2) \to (k+1,4) & (\alpha_1) \\
(k,2) \to (k+1,5) & (\alpha_2) \\
(k,2) \to (k+1,6) & (\alpha_3) \\
(k,2) \to (k+1,7) & (\alpha_4) \\
(k,2) \to (k+1,8) & (\alpha_5).
\end{array} \tag{3.9}$$

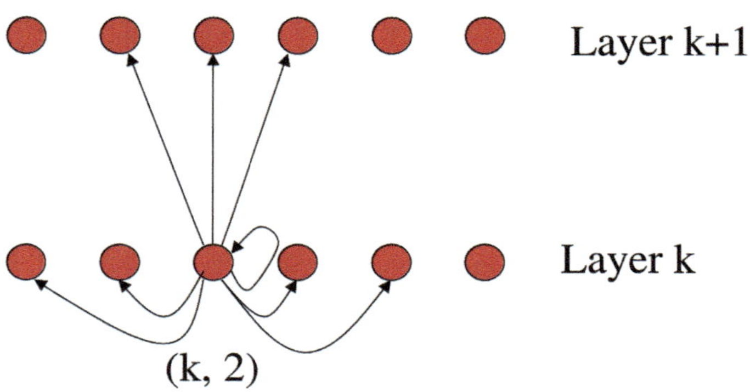

Fig. 3.1. Local connections from $(k,2)$ within layers k & k+1: $N=9$, $R=2$

Detailed consideration of the resulting end effects leads to an extremely complex expression for the generating function, largely through the appearance of unmatched Bernoulli trials of relatively low success probability. Provided R is of moderate size relative to N and expressions of high-degree in α may be neglected safely, we may effectively ignore the end effects, in essence folding the architecture around a cylinder and allowing wrap-around connections up to a limit of l.

Under these assumptions, the number of connections n_{ki} emanating from the node (k,i) has the generating function

$$A(s) = \omega_0^{\gamma_{Rm}} \omega_1^{2m+2} \ldots \omega_{l-R}^{2m+2}, \tag{3.10}$$

where

$$\gamma_{Rm} = 2R(m+1) - m^2 + 1, \tag{3.11}$$

and m is the number of layers legally traversable from layer k. Taking logs, we obtain

$$\ln A(s) = \gamma_{Rm} \ln \omega_0 + (2m+2)[\ln \omega_1 + \ldots + \ln \omega_{l-R}]. \tag{3.12}$$

For small α_ρ, we write

$$\begin{aligned} \ln \omega_\rho &= \ln(\alpha_\rho s + \beta_\rho) \\ &\approx \alpha_\rho(s-1). \end{aligned} \tag{3.13}$$

Using this result and noting that the decay amongst the α_ρ's leads to a geometric series in $\alpha = \alpha_0$, we obtain (for large l)

$$\ln A(s) \approx -\lambda_{Rm}(1-s), \tag{3.14}$$

where

$$\lambda_{Rm} = \left[\gamma_{Rm}\alpha + \tfrac{(2m+2)\alpha^2}{1-\alpha}\right]. \tag{3.15}$$

Thus

$$A(s) \approx e^{-\lambda_{Rm}(1-s)}, \tag{3.16}$$

the generating function of a Poisson distribution, so that n_{ki}, the number of connections emanating from (k,i) is approximately Poisson distributed and we may write

$$Pr(n_{ki} = n) = e^{-\lambda_{Rm}} \frac{\lambda_{Rm}^n}{n!}. \tag{3.17}$$

The number of connections from layer k, n_k, is a random variable, the sum of the N independent random variables $n_{ki}, i = 0, \ldots, N-1$, having generating function

$$A_k(s) = e^{-N\lambda_{Rm}(1-s)} \tag{3.18}$$

and hence the Poisson distribution

$$Pr(n_k = n) = e^{-N\lambda_{Rm}} \frac{(N\lambda_{Rm})^n}{n!}. \tag{3.19}$$

The number of connections within the entire network, C, is in consequence a random variable, the sum of the K independent random variables n_k, with generating function

$$A_C(s) = e^{-N\lambda_R(1-s)}, \tag{3.20}$$

where

$$\begin{aligned} \lambda_R &= \textstyle\sum_{m=0}^{K-1} \lambda_{Rm} \\ &= \gamma_R\alpha + \tfrac{K(K+1)\alpha^2}{1-\alpha}, \end{aligned} \tag{3.21}$$

and

$$\gamma_R = RK(K+1) + K - \frac{K(K-1)(2K-1)}{6}. \tag{3.22}$$

Thus, C has the Poisson distribution

$$Pr(C = n) = e^{-N\lambda_R} \frac{(N\lambda_R)^n}{n!}, \tag{3.23}$$

so that

$$E[C] = N\lambda_R. \tag{3.24}$$

For parameter estimation we equate the expected number of connections with the apparent connectivity, leading to a quadratic equation for estimation of α. Substituting again for $K = 6$ and re-arranging, we obtain

$$7\left[\tfrac{6R}{N} - \tfrac{13}{N}\right]\alpha^2 - \left[7\left(\tfrac{6R}{N} - \tfrac{7}{N}\right) + 36f\right]\alpha + 36f = 0. \tag{3.25}$$

In normal circumstances, with large N and moderate R, we may safely neglect terms of $O(1/N)$ against those involving $\xi = R/N$. If $R \rightarrow 1$, then such assumptions fail, and moreover, $\alpha \rightarrow 1$ in consequence. A lower limit for ξ, below which α must be taken as unity, is given by

$$\xi_{min} = \frac{6f}{7}, \tag{3.26}$$

and the quadratic may be recast to give:

$$\alpha^2 - (1 + \frac{\xi_{min}}{\xi})\alpha + \frac{\xi_{min}}{\xi} = 0. \tag{3.27}$$

The solution for α is readily found to be

$$\alpha = \frac{\xi_{min}}{\xi}. \tag{3.28}$$

Figure 3.2 shows values of α as a function of ξ for $f = 0.03$ and 0.16, which we shall take as the extremes of our admissible connectivity range, although higher values may be plausible. Appropriate values of α for a given parameter set may be obtained by inspection.

Provided R is chosen to be a significant fraction of N, and N remains large[4], further analysis may be simplified by disregarding possible connections with probability of $O(\alpha^2)$ – certainly for $f = 0.03$, and for all except small values of ξ even when $f = 0.16$. We shall find it convenient to make this assumption when dealing with local networks – in particular when considering the existence of small candidate structures – to allow simplification of some tiresome counting arguments. However, the extent to which higher powers of α may be neglected depends critically upon the value of ξ chosen, and each network must be considered on its merits. While we naturally exclude direct identification between our networks and a specific cortical region, models of

[4] No attempt has been made to identify a formal lower limit, but there is certainly no difficulty for $N \geq 100$.

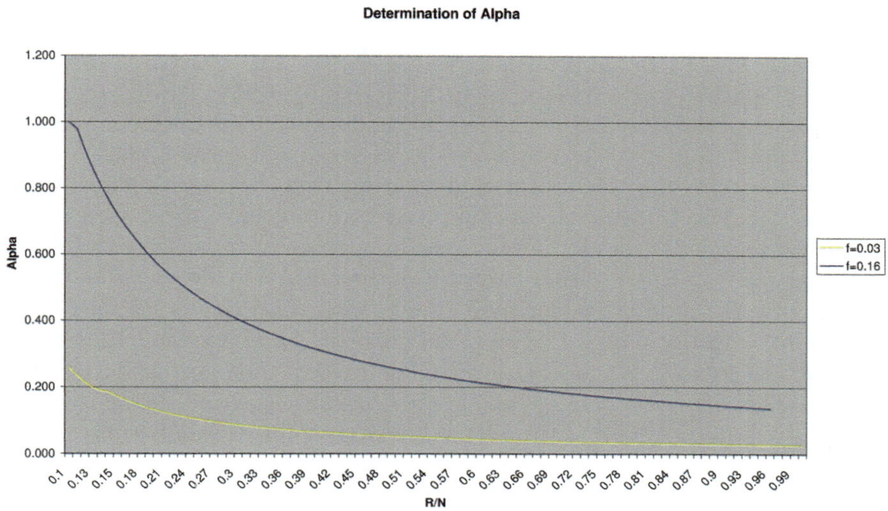

Fig. 3.2. Variation in α w.r.t ξ for the selected values of f.

connectivity may ultimately help elucidate more precise limitations on biologically realisable computations, and it is useful to consider our assumptions in this regard. At first glance the assumption of a feedforward network rules out any significant contribution to the debate, but the introduction of recurrent linkages serves only to *increase* the connection density near the source node, and the present model is at worst a loose lower bound on the available linkages.

The question is whether 'difficult' cortical computations occur within small regions of extreme local connectivity which are balanced (in the global view) by other, perhaps much larger, sparse domains. One approach, and one emerging from our assumptions, holds that local arbourisation of significant complexity is possible provided that long-range connections are correspondingly penalised. In this view, long range connections may provide input and output channels for the more sophisticated local *decision* circuitry.

It is worthwhile at this point to examine these regimes using the machinery of the clustering coefficient as employed in the cortical analyses of Kaiser and Hilgetag (2006). Direct comparison, as noted above, is difficult due to the constraints imposed by the layered architecture. Nevertheless, we may observe that a typical node (k, i) in a network layer may be expected to provide projections to around

$$Z = 2\alpha R \tag{3.29}$$

nodes in each subsequent layer under the usual assumptions of large R, large N and moderate ξ. The union of these node sets thus constitutes the neighbourhood of (k, i), and approximately $\frac{K(K-1)}{2}Z^2$ connections are possible in the neighbourhood of a node in layer 1, for example. It is readily seen that the node clustering coefficient reduces to the local connection probability, α, each potential link as before being the outcome of a Bernoulli trial.

The corresponding result for a random graph over the nodes defined is well approximated by f, the apparent connectivity fraction introduced earlier. Reference to Figure 3.2 shows the relationship between α and selected values of f for varying ξ. There is clear alignment of the present models with the small world network assumptions for moderate ξ, with clustering coefficient ratios increasing with decreasing ξ. The result reported by Kaiser and Hilgetag of 0.49 for Macaque circuits offers some confidence that the information processing capacity of the networks realised in the subsequent sections remains within reasonable bounds, and that the relatively rich structures available may indeed serve as a basis for the acquisition of complex knowledge.

Nevertheless, connectivity in these networks *does* remain sparse, and sparse connectivity dictates that some limitations on local structure are inevitable, with consequent restrictions on the functional architectures realised. These issues are considered in the following sections.

3.4 Candidate Structures for Boolean Concepts

In this section we consider the realisation of local arbourisation sufficient to support 'intelligent processing' in the sense discussed by Arbib. More specifically, we shall be concerned with structure capable of realising the elemental building blocks of more sophisticated computations. It is natural to begin with the representation of Boolean concepts of modest complexity, this approach having the additional virtue of honouring Boole's vision of formal logic as the fundamental *Laws of Thought.*

Realisation of Boolean concepts requires that nodes exist within the network which receive connections from an appropriate subset of the network inputs. Fundamental to Boolean logic are the notions of *conjunction* (logical AND across the specified inputs) and *disjunction* (the logical inclusive OR), and local connectivity sufficient to represent one such n-degree concept is perfectly adequate to represent the other. Differences in functionality then arise from the relationship between the (randomly chosen) weight and threshold values, interactions naturally governed by the underlying distributions, and admitting a wide range of Boolean concepts in which some classes may be far more probable than others. More complex structures may then result from suitable associations between these basic nodes, whether randomly occurring or realised through learning.

3.4.1 Limitations on Node In-Degree

Assessing the likelihood of existence of an *n-degree* node is a straightforward exercise if the problem is restricted to connections into a single layer. Figure 3.3 shows a set of n_0 input nodes making random connections into a layer of width $W = 2R + 1$, the central region of a much larger layer of size N. We examine the probability of existence of an n-degree node for various values of N, ξ and α.

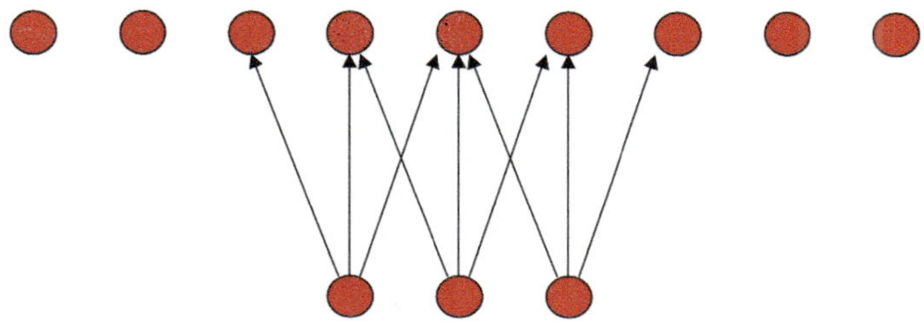

Fig. 3.3. Connections from n_0 input nodes to the central portion of a forward layer.

As before, the existence of a connection may be regarded as a Bernoulli trial of success probability α and failure probability $\beta = 1 - \alpha$. Assuming independence, the existence of an n-degree node is in turn a Bernoulli trial, this time with success probability $\alpha_n = \alpha^n$. With W trials, the probability of existence of *at least one* such node is given by the expression:

$$Pr(n) = 1 - \beta_n^W \approx 1 - \exp(-W\alpha^n). \tag{3.30}$$

Figures 3.4 and 3.5 show the alarming decrease in existence probability as n increases for the given network layer size. Patently, high-degree conjunctions (say for $n \geq 6$) cannot be realised directly within the networks proposed without a vast number of units being present in each layer, and concepts of this type necessarily rely upon combinations within one or more hidden layers. However, such concepts remain linearly separable, and to demonstrate that even sparse random networks may support more complex Boolean concepts, we shall consider two distinct architectures for *XOR*, and a closely related architecture for the well-known *Monk I* problem (Thrun, 1991).

The problem of realising complex structure under sparsity constraints leads naturally to consideration of additional resources through neurogenesis, and perhaps some unspoken feeling that incremental growth of this nature might provide a panacea for information representation. Unfortunately, uncommitted nodes in the present context are not a scarce resource, and we have argued earlier that the structures being examined here might emerge in the neuron-rich environment of early post-natal life. Neural prior knowledge may arise initially through chance, but recruited circuits are subject to the same synaptic modifications as any other cortical neurons, and one must expect reinforcement of functional relationships and the decline of others. Inclusion of neurons through adult neurogenesis is thus qualitatively different from the development of a random knowledge base as considered here.

Even discounting these distinctions, undirected inclusion of free neurons into the 'candidate pool' would offer only a very marginal improvement in the overall probability of recruitment. Cortical modellers in particular have little cause for optimism: in reviewing the experimental evidence for adult neurogenesis, Gould (2007) notes that cortical neurogenesis appears to take place at a rate less than *one thirtieth* of the comparable figures for hippocampus and dentate gyrus. Moreover, we may make some cautious inferences about the role of neurogenesis in functional circuits from the studies reviewed in Chapter 1. While one cannot rule out the possibility that 'fresh' neurons might be integrated into structures of the kind examined here, evidence that neurogenesis is implicated in restorative processes in depression and perhaps in the processes of short term memory suggests a more focused, organising role during learning, perhaps akin to the binding units of earlier models. We shall return to these issues in the subsequent chapter.

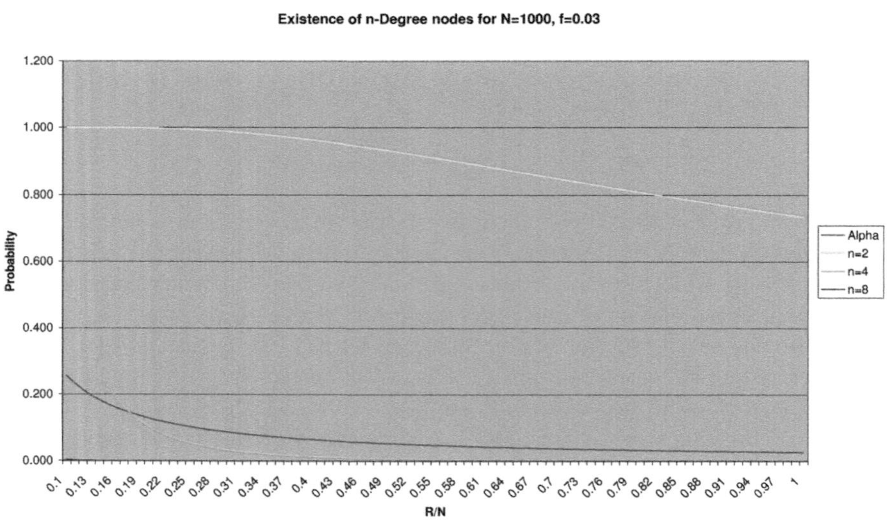

Fig. 3.4. Existence probability of n-degree nodes for increasing n, for $f = 0.03$ and $N = 1000$.

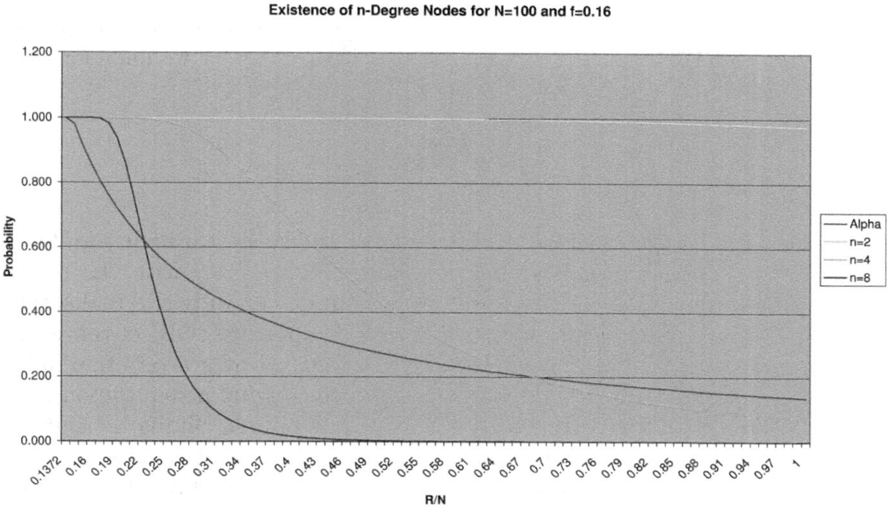

Fig. 3.5. Existence probability of n-degree nodes for increasing n, for $f = 0.16$ and $N = 100$.

3.4.2 Example Problem Architectures

We consider initially an architecture for the *XOR* problem characterised by a single hidden unit (hereafter the *single HU* or $HU1$ architecture) and short-cut connections from the inputs to the single output unit (Figure 3.6). For our purposes, we focus upon these adjacent inputs and examine the probability that this pair will fail to be connected to *at least one* candidate structure. For clarity, we restrict the analysis to a three-layer network – the two input nodes in layer 0, with N nodes in each of layers 1 and 2 – noting that this provides a loose upper bound for the six-layer result.

Let us define a *reinforcing doublet* as a node in a forward layer which receives an input from each of the two input units. The existence of such doublets may be modelled as a series of Bernoulli trials, each with success probability $\alpha_r = \alpha^2$, and failure probability $\beta_r = 1 - \alpha_r$. Then r_1 layer-1 doublets exist with probability:

$$\binom{N}{r_1} {\alpha_r}^{r_1} {\beta_r}^{N-r_1}, \tag{3.31}$$

and a similar expression arises for r_2 layer-2 doublets.

Potential $HU1$ architectures then require a connection between one of the r_1 layer-1 doublets, and one of the r_2 layer-2 doublets. There are N^2 possible connections between the layers, each of which may be regarded as a Bernoulli trial of success probability α. From this set of potential links, there are $r_1 r_2$

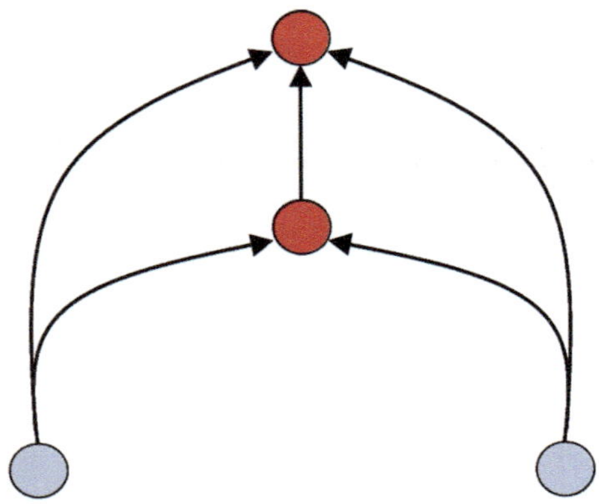

Fig. 3.6. The *HU*1 or shortcut architecture for the *XOR* problem.

links which are *marked* as connections between doublets. If we assume that l of the N^2 possible links are realised, then the two input units are connected to $r \leq l$ candidate architectures with probability

$$\binom{N}{r_1}\binom{N}{r_2}\binom{r_1 r_2}{r}\binom{N^2 - r_1 r_2}{l - r}\alpha^l {\alpha_r}^{r_1+r_2}\beta^{N^2-l}{\beta_r}^{2N-(r_1+r_2)}, \tag{3.32}$$

where we have used an hypergeometric distribution for the hidden-output link. This expression is readily shown to be a probability distribution using an identity on binomial co-efficients. The input units fail to be connected to at least one candidate architecture with probability:

$$\sum_{r_1=0}^{N}\sum_{r_2=0}^{N}\sum_{l=0}^{N^2-r_1 r_2}\binom{N}{r_1}{\alpha_r}^{r_1}{\beta_r}^{N-r_1}\binom{N}{r_2}{\alpha_r}^{r_2}{\beta_r}^{N-r_2}\binom{N^2-r_1 r_2}{l}\alpha^l\beta^{N^2-l}. \tag{3.33}$$

For large N, and moderate values of α, we may use the Poisson approximation:

$$\sum_{r_1=0}^{N}\sum_{r_2=0}^{N} e^{-\lambda_r}\frac{{\lambda_r}^{r_1}}{r_1!}e^{-\lambda_r}\frac{{\lambda_r}^{r_2}}{r_2!}\beta^{r_1 r_2}, \tag{3.34}$$

where $\lambda_r = N\alpha_r$.

In order to construct an *XOR* network under the distance-based probability regime, it is necessary to ensure that forward nodes are reachable by *all* of their input nodes from previous layers. Neglecting probabilities of $O(\alpha^2)$ at the level of an individual connection, this requirement means that each possible forward node must be within radius R of its possible input nodes. This somewhat cumbersome requirement is greatly simplified by the choice of adjacent input nodes, and this difference in 'reach' may be safely neglected. Moreover, as in the previous section we effectively consider a localised sub-layer of width $W = 2R + 1$, situated within a much larger layer of size N, with R and W recoverable from the stated value of ξ.

3.4.2.1 Results for the $HU1$ Architecture

Results for representative values of ξ and f using equation 3.34 are shown in table 3.1, for network layer sizes of $N = 1000, 5000$, and 10000.

Table 3.1. Probability that the input units fail to be connected to at least one HU1 architecture for selected values of N. Values of the parameters $\xi = R/N$, f and α are shown in columns 1-3, and the table is ordered according to increasing α.

ξ	f	α	$N = 1000$	$N = 5000$	$N = 10000$
0.25	0.03	0.11	$7.90E-2$	$7.55E-13$	$1.15E-26$
0.1	0.03	0.25	$1.63E-5$	$1.44E-27$	$1.03E-54$
0.4	0.16	0.34	$1.37E-40$	0.00	0.00
0.2	0.16	0.69	$3.93E-83$	0.00	0.00

The figures presented – chosen to represent the effect of variations in the size of the local, favoured region (i.e. $\rho \leq R$), and variations in global connectivity – suggest that recruitment of XOR structures may be viable in biologically plausible networks of quite moderate size. It is apparent that for $N \geq 5000$, the input pair is virtually certain to be connected to at least one candidate architecture – regardless of the values of ξ and f – and even smaller layer sizes may prove viable if a greater number of layers is employed.

At this point we must emphasise that the role of a network of layer size $N = 10000$ will certainly be far more extensive than the computation of a simple Boolean function of two of the available inputs. The results of this and later sections of this chapter serve only to demonstrate that such computations may be possible within some localised subregion of the network[5], and we assume it likely that numerous parallel computations will take place within the remaining circuitry.

3.4.2.2 The *Monk I* Problem

The limitations of such high-degree hidden units may be demonstrated by considering a similar architecture for the *Monk I* problem. The three

[5] In table 3.1 the *available* network layer size for $N = 10000$ ranges between 1000 and 4000.

Monk's Problems (Thrun, 1991) are a set of machine learning tasks of varying degrees of difficulty developed to facilitate comparison between a wide variety of training algorithms. The monks exist in an artificial robot domain, and each monk is characterised by six different attributes: `head_shape` and `body_shape`, which each take on a value from the set {round,square,octagonal}; `is_smiling`, {yes,no}; `holding`, {sword, balloon, flag}; `jacket_color`, {red,yellow,green,blue}; and `has_tie`, {yes,no}. The tasks describe certain Boolean expressions over these attributes (see Thrun (1991) for full details). Here we are concerned only with the simplest of the problems, *Monk I*, which is specified by the expression:

```
(head_shape=body_shape) or (jacket_color=red).
```

Numerous architectures are possible for the *Monk's* problems, but a sparsely coded input leads naturally to a $17-1-1$ structure. Consideration of the nature of the *Monk I* problem and elimination of irrelevant attributes allows reduction to a $7-1-1$ configuration, and we present existence results for both architectures. Figure 3.7 shows both the complete and reduced candidates.

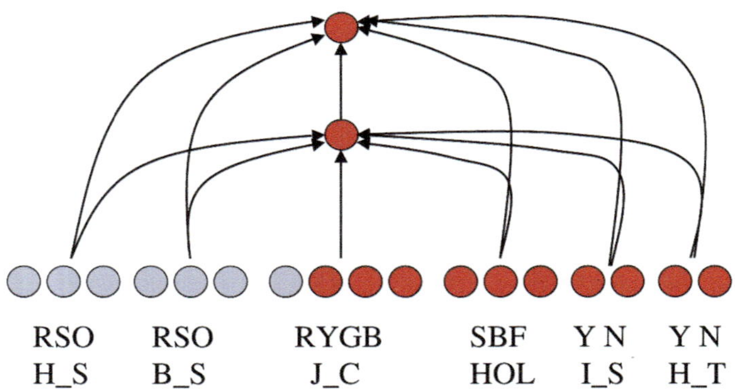

Fig. 3.7. The $17-1-1$ complete and $7-1-1$ reduced shortcut architectures for the *Monk I* problem.

3.4.2.3 Results for the *Monk I* Problem

Table 3.2 shows the probability that a set of chosen input nodes fail to be connected to at least one of the $(7-1-1)$ and $(17-1-1)$ *Monk I* candidate architectures. The three layer sparse network, highly suitable for the representation of the $HU1\,XOR$ structure, is limited in its ability to harbour the $17-1-1$ architecture, and even the $7-1-1$ *Monk I* architecture is troublesome for $f=0.03$. It is apparent that representation of problems as complex as *Monk I* requires the higher connection probability arising from the Murre and Sturdy result, or some alternative architecture based upon the use of a greater number of hidden units, or perhaps a greater number of layers. While we consider such an architecture for the *XOR* problem in Section 3.4.3.1, there exists a highly plausible alternative, similar in principle to the random-chunking network arguments advanced by Feldman (1982).

Table 3.2. Probability that the input units fail to be connected to at least one $7-1-1$ and $17-1-1$ candidate architecture for the *Monk I* problem for $N = 10000$ and $N = 50000$ respectively. Values of the parameters $\xi = R/N$, f and α are shown in columns 1-3, and the table is ordered with respect to increasing α.

ξ	f	α	$7-1-1$	$17-1-1$
0.25	0.03	0.11	1.00	1.00
0.1	0.03	0.25	1.00	1.00
0.4	0.16	0.34	$6.19E-01$	1.00
0.2	0.16	0.69	$1.11E-89$	$3.02E-16$

3.4.2.4 Input Redundancy

Suppose that instead of a single unit, each of the input components is represented by a small *assembly* of k units, each having a similar activation for a given input pattern. Then the desired candidate structure may appear if *at least one* of the assembly units achieves a connection to appropriate units in the forward layers.

This approach is similar to that taken by Valiant, ensuring representation of Boolean concepts through overlapping connections of sufficient density from the constituent regions. The situation is conveniently modelled through the use of an effective *multiple-trial* probability for each of the inputs. We consider now a hidden unit requiring in-degree n, for a k-redundant input layer. Then the existence of the appropriate connectivity may be considered a Bernoulli trial with success probability:

$$\alpha_{n,k} = (1 - \beta^k)^n, \tag{3.35}$$

where as usual $\beta = 1 - \alpha$.

To illustrate the effect of this approach, we consider again the in-degree studies of Section 3.4.1, this time adopting a local connection probability of $\alpha_{n,k}$, rather than α^n as before. The existence probability is then given by a modified version of equation 3.30:

$$Pr(n;k) = 1 - \beta_{n,k}^W. \tag{3.36}$$

Figures 3.8 and 3.9 show the marked improvement in the complexity of local nodes that may be supported if the network allows assembly-based inputs.

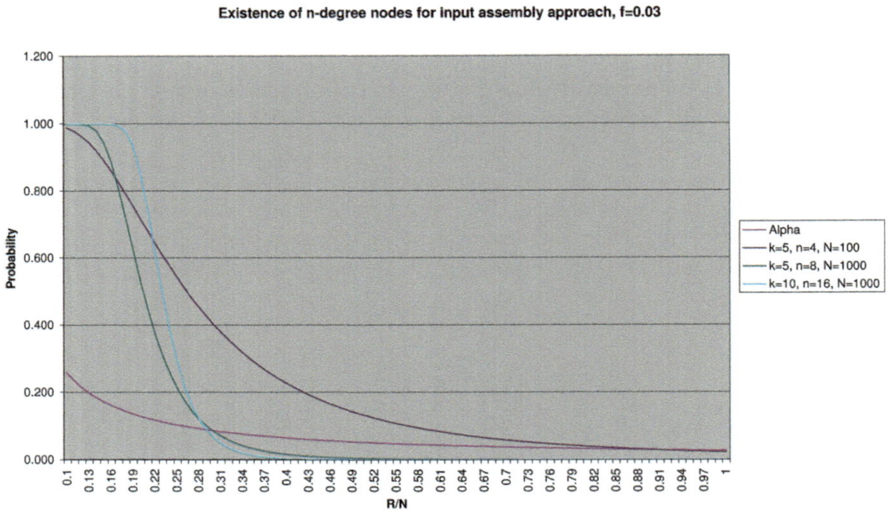

Fig. 3.8. Existence probability of n-degree nodes under the k-redundancy assumptions, for $f = 0.03$.

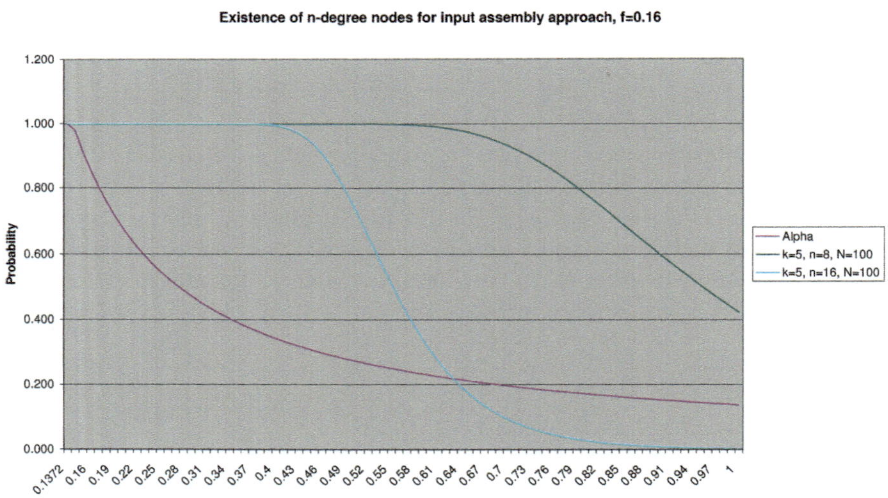

Fig. 3.9. Existence probability of n-degree nodes under the k-redundancy assumptions, for $f = 0.16$.

We note in particular that for $f = 0.16$, nodes of extremely high degree may be realised, and this is reflected in the results of table 3.3, showing the existence probabilities for the $17-1-1$ and $7-1-1$ *Monk I* architectures, under the input assembly approach. Note that in these calculations, we have retained the local probability α for the *internal* network connections, suggesting that a concept network based upon compact internal representations may be possible even for dense inputs, without the necessity for the complex dynamics associated with a full-blown interacting cell assembly model. Patently, even smaller network architectures may be viable for larger input assemblies, the choice of $k = 5$ being extremely modest by neural standards.

Table 3.3. Probability that the input units fail to be connected to at least one $7-1-1$ and $17-1-1$ candidate architecture for the *Monk I* problem under the input assembly assumption ($k = 5$). Network layer sizes are $N = 5000$ $N = 10000$ respectively. As before, values of the parameters $\xi = R/N$, f and α are shown in columns 1-3, and the table is ordered with respect to increasing α.

ξ	f	α	$7-1-1$	$17-1-1$
0.25	0.03	0.11	$3.49E-1$	1.00
0.1	0.03	0.25	$3.75E-50$	$4.91E-9$
0.4	0.16	0.34	0.00	0.00
0.2	0.16	0.69	0.00	0.00

3.4.3 Probability of Connection in Multipartite Random Graphs

It is useful to consider at this point the related work of Gunay, which builds directly upon the earlier approach of Valiant and extends his neuroidal model to support spiking neurons (see Part II of this volume). The results presented here involve random projections from assemblies in a designated source area A_s to images in a destination area A_d. As in the analysis of candidate structures, we wish to identify some region of the destination area which may be expected to receive projections from each of the source ensembles, thus providing a robust basis for representing some function of the concepts they represent. From the outset, we assume a parameter r known as the replication factor, which reflects the number of nodes necessary to represent a concept reliably in a neural region – a quantity directly analagous to the input redundancy. Disjoint source ensembles $\tilde{x}$ and $\tilde{y}$ are drawn from the same or distinct source areas, and project to the destinaton area, where the intersection of these projections defines an ensemble $\tilde{z}$. Our goal is to determine conditions under which this new ensemble may be expected to have size approximately r_d – the replication factor for the destination area – and thus be recruited reliably. Formal derivations of the necessary probabilities are given in the Appendix, and the two cases treated by Gunay are shown respectively in figures A.1 and A.2. Here we give an outline of the approach.

Let us denote by $p_{s,d}$ the probability that a *neuroid* in A_s is connected to a neuroid in A_d. We are concerned with the probability p^* that a neuroid from the destination area A_d will be a member of the recruitment set $\tilde{z}$, the intersection of the *frontiers* or images of $\tilde{x}$ and $\tilde{y}$ within A_d, and wish to relate these two probabilities in order to determine conditions favourable to stable recruitment.

More formally, we require:

$$\varepsilon(|\tilde{z}|) = N_d \, p^* \simeq r_d \,, \tag{3.37}$$

where N_d is the number of neuroids in the destination area. The probability that a neuroid in A_d is not connected to a neuroid in A_s is

$$\bar{p}_{s,d} = 1 - p_{s,d} \,,$$

so that the probability that a neuroid in A_d is not connected to *any* of the neuroids of an ensemble of size r_s in A_s is

$$(1 - p_{s,d})^{r_s} \,.$$

In turn, the probability that a neuroid in A_d is connected to at least one of the neuroids of each of the two ensembles of size r_s in A_s, and is thus a member of $\tilde{z}$, is

$$p^* = (1 - (1 - p_{s,d})^{r_s})^2, \tag{3.38}$$

which is readily approximated to second order by

$$p^* = r_s^2 \, p_{s,d}^2. \tag{3.39}$$

The value of $p_{s,d}$ required is then obtained by substitution in equation 3.37, yielding

$$p_{s,d} = \frac{1}{r_s} \sqrt{\frac{r_d}{N_d}}. \tag{3.40}$$

This result may be generalised to the case of disjoint source areas by using the expression of equation 3.40 to calculate independently the probability of connections from each ensemble.

Let

$$p_{s_1,d} = \frac{1}{r_{s_1}} \sqrt{\frac{r_d}{N_d}} \,,$$

and

$$p_{s_2,d} = \frac{1}{r_{s_2}} \sqrt{\frac{r_d}{N_d}} \,.$$

The probability p^* that a unit in A_d lies in the recruited set as a result of projections from disjoint areas A_{s_1} and A_{s_2} is given by the product of the independent probabilities:

$$p^* = r_{s_1}\, p_{s_1,d}\, r_{s_2}\, p_{s_2,d}.$$

The results of this section provide both a more general statement of the problems being considered and a exposition of the trade-offs between the robustness of the concept ensemble and the utility of the layered structures more traditional in connectionist models. Recruitment of $\tilde{z}$ – the intersection of the frontier of two source ensembles – is readily seen as a generalised version of the reinforcing doublets and other local structures considered in the present chapter, promising robustness in return for a greater commitment of neural resources. The area projection probability $p_{s,d}$ is seen to turn on the ratio of the replication factors, with the familiar $\frac{1}{\sqrt{N}}$ behaviour returning with balanced ensembles. In contrast, the layered architecture trades some local robustness for ease of selection, and as we shall see in the following chapter, some control of input saturation.

3.4.3.1 The *HU2 XOR* Architecture

As is noted in Section 3.4.1 and is further apparent from the results for the $17-1-1$ *Monk I* architecture shown in table 3.2, realisation of architectures requiring high-degree hidden nodes may be difficult within sparse networks, and more probable architectures may be constructed over a number of succeeding layers, or through the use of a greater number of units in a single hidden layer. In this section we consider the well-known double hidden unit (hereafter *HU*2) architecture for the *XOR* problem, developing existence results in a manner similar to the previous analysis for the shortcut approach.

As before, each candidate hidden unit may be regarded as a reinforcing doublet with existence probability $\alpha_r = \alpha^2$, and h such nodes are realised within a layer of size N with probability

$$\binom{N}{h} {\alpha_r}^h\, {\beta_r}^{N-h}. \tag{3.41}$$

In a similar fashion, a particular output node will receive connections from l of the N nodes in the hidden layer with probability

$$\binom{N}{l} \alpha^l\, \beta^{N-l}, \tag{3.42}$$

and using the previous hypergeometric approach, we see that our selected output node receives r linkages from among the h *marked* hidden layers with probability

$$P(r;h,l) = \binom{N}{h} \alpha_r{}^h \, \beta_r{}^{N-h} \binom{h}{r} \binom{N-h}{l-r} \alpha^l \beta^{N-l}. \tag{3.43}$$

Clearly, $P(r \geq 2; h, l) = 0$ for $h, l < 2$, and $P(r)$ is obtained by summation over both h and l. The probability that at least one candidate may be realised with our chosen output unit is then

$$P(r \geq 2) = 1 - P(r \leq 1). \tag{3.44}$$

Under the usual Poisson approximations, this leads to the expression

$$1 - \sum_{h=0}^{N} \sum_{l=0}^{N-h} e^{-\lambda_r} \frac{\lambda_r{}^h}{h!} e^{-\lambda} \frac{\lambda^l}{l!} \beta^h - \sum_{h=1}^{N} \sum_{m=0}^{N-h} e^{-\lambda_r} \frac{\lambda_r{}^h}{h!} e^{-\lambda} \frac{\lambda^m}{m!} h\alpha\beta^{h-1}, \tag{3.45}$$

where $\lambda_r = N\alpha_r$ and $\lambda = (N-h)\alpha$.

Table 3.4 shows the probability that such a structure will fail to exist between two defined input nodes and one single output node for selected values of the parameters α and N. Note in particular that layer sizes of around 500 suffice at the 16% connectivity level, while even the 3% level may be expected to support such structures for layer sizes of several thousand or more. This approach may be straightforwardly extended to include architectures involving a larger number of hidden units – which may in principle support more complex Boolean expressions – but as before the node in-degree will prove the most demanding limitation.

Table 3.4. Probability that the *HU*2 architecture fails to exist between defined input and output nodes for selected values of N. Values of the parameters $\xi = R/N$, f and α are shown in columns 1-3, and the table is ordered according to increasing α.

ξ	f	α	$N = 500$	$N = 1000$
0.25	0.03	0.11	$9.56E-1$	$8.56E-1$
0.1	0.03	0.25	$5.37E-1$	$1.81E-1$
0.4	0.16	0.34	$2.49E-6$	$7.17E-13$
0.2	0.16	0.69	$1.95E-27$	$1.13E-55$

3.5 Discussion

The lengthy and often laborious analyses of this section are an important prelude to recruitment, demonstrating that our networks may, with high probability, harbour candidate structures suitable for the representation of non-trivial Boolean concepts. While these probabilistic arguments owe a good deal to the earlier work of Feldman and Valiant, it should nevertheless be noted that the approach differs significantly in its focus on the recruitment of larger subnetworks – confronting directly the problem of realising more complex knowledge.

The principal lesson of this chapter lies in the advantage of assembly-based representations or their analogues at all levels of the functional circuit. While we may reason about the probability of individual connections, robust structures may be realised only through a pool of chances large enough to absorb the inevitable failures. While we allow a layered model, we should be under no illusions that candidate structures will be neatly disjoint and localised, as in the illustration of Figure 3.6. Weight distributions and learning schemes must thus be chosen carefully if competing inputs – through connections unwanted for one target concept – are to be managed without compromising their role in the representation of other targets.

These issues dominate the following chapter, which develops these ideas first at the local unit level before extending the analysis to encompass entire networks and the deleterious effect of saturation if weight magitudes are allowed to become excessive. Following a detailed look at learnability and updates of varying degrees of plausibility, we proceed to actual recruitment of randomly occurring structures in support of Boolean concept learning.

Chapter 4
Representation and Recruitment

James M. Hogan

4.1 Introduction

In the preceding chapter we were more concerned with the neural foundations of knowledge representation and the existence of structures capable of supporting non-trivial concepts than with actual recruitment of functional circuits, which so far has been limited to the case of one-shot learning (Chapter 2). This chapter makes explicit the linkages between structural plausibility, knowledge representation and recruitment, exploring the interactions between them and ultimately demonstrating the success of our approach when applied to a number of standard benchmark problems. Later in Part I of the book, in Chapter 5, we come somewhat closer to the biological context in which the ideas were developed, recruiting sparse random networks of this type to encode the spatial relations semantics of simple images.

It must be emphasised that the candidate structures of the previous chapter provide only scaffolding to support knowledge representation, and that capacity is critically dependent upon the weight distributions in operation as individual nodes are generated. The problem of representation of Boolean knowledge through linear threshold gates – the Linear Threshold Unit (LTU) or Threshold Logic Unit (TLU) to the neural network community – has a lengthy history which spans several disciplines, each of which has provided insights of some use in the present context. In particular, both theoretical computer science – through 'Cover's Argument' (Cover, 1965) – and Electrical Engineering – through the careful analyses of Muroga (1971) – provide a cogent reminder of the capacity limitations of a single unit, rather brutally puncturing any delusions that universal concept representation might be possible in a node of sufficient in-degree.

Here we are less concerned with these limitations at the individual node than with their compounding effect across a 'neural' population. The networks have the same form as those defined in Chapter 3, but we now enforce a separation between one or more fixed network layers (known subsequently as

J. Diederich, C. Günay, and J.M. Hogan: Recruitment Learning, SCI 303, pp. 83–135.
springerlink.com

the *random black box*) in which weights and thresholds remain fixed, and a single output or concept layer which must ultimately host one or more Boolean targets. In keeping with the candidate library approach, weight change is permitted only upon projections between these two areas.

The role of the network is then to approximate successfully the behaviour of some target function $t(x)$ by a network hypothesis $h(x)$, through recruitment of existing network knowledge. Each node of the black box which projects into the concept layer acts as a basis function for $h(x)$, and such nodes are termed *feeder* or *basis* units, $f_i(x)$, where f denotes the node-level transfer function. Many choices are possible for f, but here we shall work with either the linear threshold or step function:

$$u(y) = \begin{cases} 1, & y \geq 0 \\ 0, & y < 0 \end{cases}, \tag{4.1}$$

the *sigmoid* or *logistic* function:

$$\sigma(y) = 1/(1 + e^{-\lambda y}), \tag{4.2}$$

where the steepness parameter λ is taken to be equal to 1; or more usually, the rounded sigmoid:

$$\sigma_R(y) = \begin{cases} 1, & \sigma(y) \geq 0.5 \\ 0, & \sigma(y) < 0.5 \end{cases}, \tag{4.3}$$

which permits some control of input saturation while preserving a binary response.

Formally then, the network hypothesis has the form:

$$h(x) = f(\sum_{i=0}^{F-1} \nu_i f_i(x)), \tag{4.4}$$

and we see that the activation of the hypothesis unit may be viewed as a linear combination of functions over (some subspace of) the input patterns, a view we shall find more convenient than direct consideration of the input vector. The nature of this basis is crucial in determining the representation capacity of the network, and in facilitating learning within an acceptable timeframe. The effect is analogous in some respects to the feature mappings commonly discussed in kernel methods, although the effective dimension of the function space may at times become disturbingly low. Much of the following analysis is thus concerned with mechanisms for influencing the feature basis, so as to support a richer family of targets.

Learning in connectionist systems of this type suffers from an irreconcilable tension between power, efficiency and biological plausibility, and the simulations of the latter sections of this chapter involve a number of delicate trade-offs between them. *All* of the computational experiments of this chapter respect biologically plausible levels of connectivity and weight precision, but they may differ markedly in the plausibility of the associated learning rules. On the one hand, we are concerned with the limits of Boolean knowledge acquisition in a system abstracted carefully from natural computation. Here, we wish to identify some practical bounds on neural learnability and thus remain as faithful as possible to the Hebbian ideal, returning to first principles in assembling the mathematical form of the weight update. On the other hand, we take a more pragmatic view, regarding the randomly generated knowledge base of our system as a resource to be harvested in the cause of rapid learning, and we shall lose little sleep if the final updates owe more to computational learning theory than to neurobiology.

Yet like Feldman, Valiant and their many predecessors, we do not rule out the possibility of a perfect synthesis of these approaches, with neural mechanisms identified for some of the more powerful, computationally motivated updates. If in this respect we are retracing the history of connectionism, we nevertheless do so without recourse to backpropagation of error. Indeed, by relying upon existing knowledge to provide the network hidden units, we do not so much solve the credit assignment problem for multi-layer networks as avoid it altogether, although we pay a representational price for this agility.

All learning methods considered in this chapter owe some basic debt to the Hebbian hypothesis – in the sense that they rely upon a strengthening of the connections from inputs which fire whenever the unit itself fires, or at least when the unit *should* fire if a particular target function is to be achieved. It is tempting, in the light of this basic reliance upon Hebbian correlation, to consider the 'itermediate' alternative of reinforcement learning, which is thought to have a neural analogue in the action of the neurotransmitter dopamine in reward-based associative learning[1]. However, it remains unclear whether reinforcement learning is sufficient to support the acquisition of concepts which rely on successful processing of negative examples. Evidence reviewed by Hyman et al (2006) suggests that the method offers only limited advantages over unsupervised learning in the present context, with over-generalisation a significant risk.

Consequently, we shall usually follow Feldman (1982) and Valiant (1984) in arguing that some supervised update mechanism is necessary if complex knowledge is to be encoded, but with weight change made in accordance with the correlation between the network constituents and some desired target Evidently, we have departed from the unsupervised purity of Hebb's original idea, yet compromise is unavoidable, particularly if the target concept is not easily distinguishable from a set of similar competitors.

This issue - discussed below in Section 4.3.1 - is especially troublesome when complex knowledge is realised directly, by aggregation of elemental relationships through high degree conjunction and disjunction. Our architecture imposes limitations on the direct realisation of 'feature rich' conceptual

[1] See Hyman et al (2006) for a recent review, of these ideas in the context of drug addiction.

knowledge through one or two layers, and there is an implicit focus on construction and realisation of concept hierarchies over multiple layers rather than through a single pool of hidden units as in backpropagation networks. The essence of representation is here the composition of low degree Boolean functions across successive layers, allowing the formation of intermediate feature units. As the central issue in recruitment from the feeder basis is then to *learn rapidly when irrelevant attributes abound*, it is unsurprising that the Littlestone Winnow rules should have a prominent role in our success with benchmark problems.

We begin this chapter by examining in some detail the representation of Boolean concepts at a single unit receiving up to n input connections (Section 4.2). After concluding this section by analysing the effect of random connectivity on the expected input activation, we consider the special case of non-negative weights and thresholds, the basis of later recruitment of k-DNF expressions (Section 4.2.1). Section 4.3 examines these effects on a network scale, with the phase volume analysis of learning and generalisation in random networks introduced in Section 4.3.1, and used to analyse correlation-based updates in Section 4.4.4.

Recruitment learning is introduced initially through consideration of the Hebbian hypothesis, and we offer a number of mathematical formulations for this principle, leading ultimately to the convergent Hebbian update of Section 4.4.1. A second, less plausible class of learning rules is explored in Section 4.5. While these updates do not rely upon backpropagation of error across several layers of the network, an error signal is nevertheless employed to control the adjustment of individual weights. Thus, we hold that their existence owes more to the perceptron update of Rosenblatt (1961) and the Probably Approximately Correct (PAC) framework of Valiant (1984) than to Hebb's ideas. Ultimately, the discussion will lead us toward the optimal mistake bound Winnow rule of Littlestone (1988), and this is explored in Section 4.5.2.

These rules are subsequently employed in a variety of applications, sections 4.7.2 to 4.7.9 addressing the recruitment of Boolean problems of progressively increasing difficulty, culminating in benchmark problems from the UCI Machine Learning Repository (Asuncion and Newman, 2007). Section 4.8 presents a short discussion of the potential role of adult neurogenesis in recruitment learning models of this nature, with a particular focus on the opportunities in competitive learning systems. The chapter then concludes with some discussion of generalisation performance and of the application of our approach within more elaborate domains.

4.2 Representation at a Single Unit

The mathematical foundations of our system lie in the well-developed theory of Boolean concepts defined over a domain of bit vectors of some fixed dimension, here denoted n. A non-sparse n-input Boolean concept is based upon an input space of $b = 2^n$ patterns, and in consequence the vector or bit-string specifying the concept is also of this dimension. Let us denote the set of bit-strings of length n by:

$$B^n = \{0, 1\}^n, \tag{4.5}$$

so that each concept $c \in C_n$ is a mapping

$$c : B^n \to \{0, 1\}. \tag{4.6}$$

As is well known, there are $|C_n| = 2^b$ possible concepts of this type, and in consequence, the concept space may be spanned only by a set of basis concepts of dimension at least b, the most obvious choice being the 'Euclidean' basis E^b. So, for example, the set of 2-input Boolean concepts may be generated by

$$E^4 = \{1000, 0100, 0010, 0001\},$$

where the vectors are shown as bit-strings for the sake of brevity.

Here we are less concerned with the space of Boolean functions *per se* than with their representation by neural analogues such as the LTU or the perceptron, and with the capacity of networks of such units to realise more complex Boolean relations. For clarity, we shall work initially with the sharp edges of the hard-limiting LTU, although we shall utilise other activation functions as appropriate. Similarly, we shall at times distinguish the input dimension n from the 'input' dimension of the feeder basis, $b = 2^n$, although the actual dimension of the function space may be a small fraction of this figure.

Classification through a linear threshold unit is dependent upon the placement of the associated hyperplane

$$\sum_j w_{ij} x_j - T_i = 0, \tag{4.7}$$

which partitions the input space B^n into two disjoint classes. The resulting split between *concept* and *non-concept* examples is usually termed a *dichotomy*, and the capacity of the unit is characterised by the extent to which it may implement an arbitrary dichotomy amongst a set of points drawn from the pattern space.

More formally, the representation capacity of a classifier may be expressed through its Vapnik-Chervonenkis dimension (see for example Natarajan (1991)), the cardinality of the largest subset of the domain for which all possible dichotomies may be realised by the classifier. Such a subset is then said to be *shattered* by the classifier, and Cover (1965) showed through an inductive argument that a linear threshold node is unable to shatter the subset if the number of sample points exceeds the number of trainable weights[2].

[2] The paper is often informally cited as "Cover's Argument".

Nodes of the type described (with n inputs excluding the threshold) thus have VC dimension $n+1$, and it is immediately apparent that as n increases, the number of linearly separable dichotomies, $N(n)$, is a rapidly decreasing fraction of the 2^{2^n} possible partitions. For large n, Muroga (1971) reports

$$\log_2 \frac{N(n)}{|C_n|} \approx kn^2 - 2^n, \tag{4.8}$$

confirming the relative paucity of separable concepts.

While VC dimension bounds the range of dichotomies possible at an individual unit, realisation of a particular concept or set of concepts relies on achieving an appropriate balance between the activation (controlled by the weight distribution) and the subtractive threshold. Even if parameter selection is deterministic, some concepts are necessarily easier to represent than others, occupying a greater phase volume (see for example Gordon and Peretto (1990), Van den Broeck and Kawai (1990)) in parameter space, and these effects are especially important in the case of random networks.

Weights in the present work are initialised under some discrete probability distribution $\Omega(W)$ over the set

$$\{-W_S, \ldots, -W_1, W_1, \ldots, W_S\},$$

where the values are (usually) held to be equally spaced and of separation Δ, so that $\pm W_i = \pm i\Delta$. Weights are thus restricted to a limited bandwidth of $1 + \log_2 S$ bits, although we may work solely with excitatory weights where appropriate, and the bandwidth of synaptic connections is kept to no more than a few bits.

Unit thresholds are similarly restricted, and will be selected according to a distribution $\Theta(T)$ of similar functional form to $\Omega(W)$, acting over the set

$$\{-T_m, \ldots, 0, \ldots, T_m\},$$

although we frequently work with only the non-negative values. Without wishing to limit the choice of Ω and Θ, threshold and weight settings will usually be drawn from the domain under a uniform distribution.

Gordon and Peretto (1990) considered the distribution of Boolean concepts for *fully-connected* two and three-input nodes, randomly sampling real-valued weights and thresholds according to a uniform distribution[3] over symmetric intervals $[-W_S, W_S]$ and $[-T_m, T_m]$ respectively. Through straightforward symmetry arguments, the authors were able to partition the available

[3] Normal distributions for the weights were considered, but the results were surprisingly insensitive to the choice of distribution. While the authors also analysed a more general network of H hidden layers, each containing N units, we are at this point concerned with their preliminary work with a single unit.

relations into families of equi-probable concepts[4]. For restricted cases it is possible to derive analytic expressions for the family probabilities in terms of the ratio of the maximum weight and threshold values:

$$r = T_m/W_S. \tag{4.9}$$

Broadly speaking, the effects can be partitioned into three regimes:

- Small r, say $r < 0.5$: a region dominated by simple concepts based on a small number of strongly weighted inputs;
- Intermediate r, say $0.5 \leq r \leq 2$: in which the probability of representing a given concept family is governed by its characteristic split between positive and negative examples; and
- Large r, $r > 2$: which favours those concepts for which there are a small number of positive examples – there being a reduced likelihood that the magnitude of the activation will exceed that of the threshold.

In the limit as $r \to \infty$, only those families consisting of 'unbalanced' concepts converge to a finite probability – a consequence of the increasing likelihood that the weight magnitude will fail to exceed the threshold magnitude for *any* input pattern.

Hogan (2000) used Monte Carlo simulations to extend these results to cover higher input dimensions and a number of alternative weight and threshold regimes, the principal goal being the investigation of higher order conjunctions and disjunctions across the inputs. Asymptotic behaviour similar to that described above was observed for $n = 3$ and $n = 4$, and may be safely predicted for higher degree units, although these may have low existence probabilities under the regimes of the previous chapter. This analysis differed further in its use of binary rather than bipolar inputs, thereby eliminating an additional symmetry relation and causing some concept families to fracture into sets of lower cardinality.

Figure 4.1 shows the results of Monte Carlo simulation with symmetric weights and thresholds and binary inputs for $n = 3$. Fracture of concept families is clearly evident, and here we observe 13 rather than the 6 families identified with bipolar inputs. While the resulting distribution provides some probabilistic sifting among diverse concepts, there is a relatively fine gradation across class probabilities, with strong correlations among the families making it difficult to favour one such sub-class at the expense of others.

Of most concern, however, is the possibility that random generation of weights and thresholds will produce an excessive number of constant and 'single-input' concept units, with potentially dire consequences for representation. The problem is compounded by the effects of sparse connectivity, and the likelihood that many of the potential linkages into each neuron will fail to exist, and the effect of these deletions on the distribution of activation.

[4] The probability of implementing a given concept is then obtained by dividing the family probability by its cardinality.

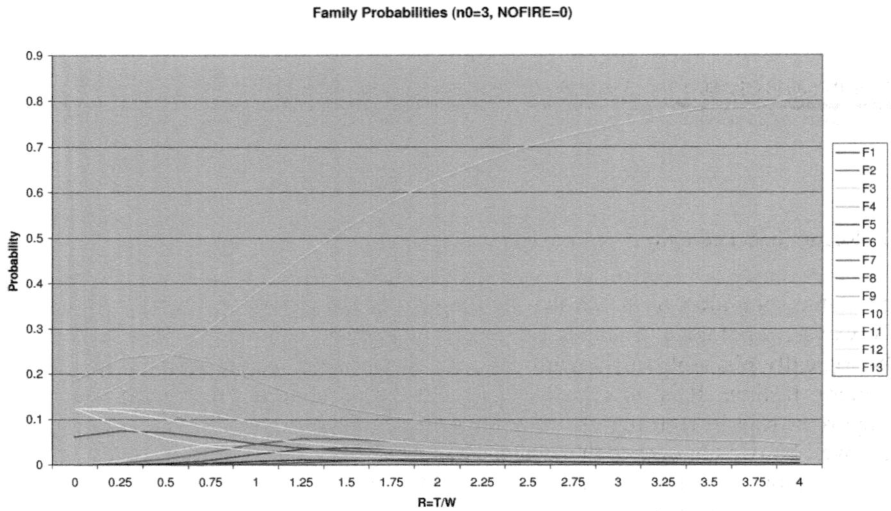

Fig. 4.1. Variations in *family* probability w.r.t. r for $n_0 = 3$, allowing symmetric weight and threshold ranges, but using binary inputs.

Consider now a single unit with weights drawn uniformly from $[-W_S, W_S]$ as before, excepting that non-existent (or zero weight) connections are assumed to have probability $\beta = 1 - \alpha$. Discarding the usual threshold T_i, the input activation y_i may then be written:

$$y_i = \sum_j w_{ij} x_j, \tag{4.10}$$

with expectation $E[y_i] = 0$, and variance

$$Var(y_i) = \frac{\alpha n W_S^2}{3}, \tag{4.11}$$

where as before n is the in-degree. The corresponding analysis for full connectivity yields

$$Var(y_i) = \frac{n W_S^2}{3}, \tag{4.12}$$

and so the net effect of sparseness is to scale the input activation, effectively contracting the weight range by a factor $\sqrt{\alpha}$, and increasing the value of r.

Unsurprisingly, a reduction in the number of realised input connections has the effect of focusing the distribution toward low input activation, and when coupled with either a narrow weight band (i.e. small W_S), or a large threshold range, the concept distribution is necessarily limited. From equation 4.11, it is clear that the activation variance is linear in both n and α, but quadratic in W_S, indicating that the deleterious effects of sparse connectivity may be less pronounced for a larger input dimension. Yet while an unbalanced concept such as the n-input conjunction or disjunction is favoured relative to more balanced concepts as r increases, it is unlikely that the computation will be realisable at a single node, due to the difficulty of generating one of sufficient in-degree.

In practice, sparse connectivity necessarily favours relations over some small subset of the potential inputs, and the most promising strategy is to tailor our approach toward a class which supports the hierarchical construction of conceptual knowledge. We thus assert that our primary concern will be the representation of Boolean knowledge for which an equivalent disjunctive normal form expression may be readily found, and we shall take a particular interest in weight and threshold distribution choices which may facilitate the association or binding of disparate input through conjunctions and disjunctions. In the following sections we shall explore the concept families which arise under this *class-specific* distributional bias.

4.2.1 Non-Negative Weights and Thresholds

Restriction of the weights and thresholds to non-negative ranges rather than the earlier symmetric intervals leads to a pronounced thinning of the concept distribution, it being impossible to implement variable complements with non-negative weights.

Table 4.1. Families of equi-probable Boolean concepts for the two-input linear threshold unit, allowing only non-negative weights and thresholds, and binary inputs.

Family	Description	Cardinality	Example
F_{21}^{+}	Low output	2	x_1x_2; 0
F_{22}^{+}	Determined by one input	2	x_1, x_2
F_{23}^{+}	High output ($r > 0$)	1	$x_1 + x_2$
F_{24}^{+}	High output ($r = 0$)	1	1

Tables 4.1 and 4.2 show the concept family partitions respectively for $n = 2$ and $n = 3$, and variations in family probabilities with respect to r may be seen in figures 4.2 and 4.3. Of principal interest here is the differentiation between the previously equi-probable disjunction and conjunction concepts, the former being favoured for small r and the latter for large r. In contrast to the symmetric ranges, non-negative weights and thresholds provide a clear

Table 4.2. Families of equi-probable Boolean concepts for the three-input linear threshold unit, allowing only non-negative weights and thresholds, and binary inputs.

Family	Description	Cardinality	Example
F_{31}^{+}	2, 3-conjunctions	7	$x_1x_2x_3$; x_2x_3; $(x_1+x_2)x_3$
F_{32}^{+}	Constant zero	1	0
F_{33}^{+}	Majority	1	$MAJ(x_1,x_2,x_3)$
F_{34}^{+}	Single dominant	6	x_1, $x_1+x_2x_3$
F_{35}^{+}	2-disjunctions	1	x_1+x_2
F_{36}^{+}	3-disjunction	1	$x_1+x_2+x_3$
F_{37}^{+}	Constant one ($r=0$)	1	1

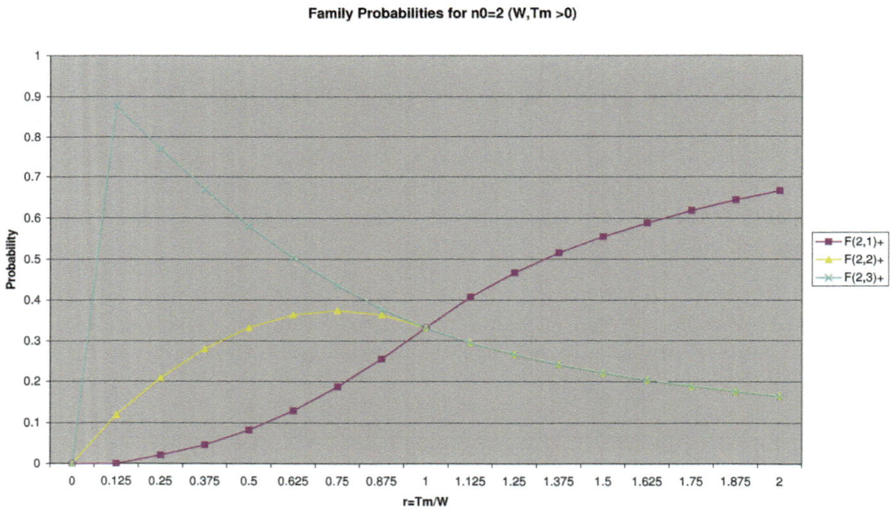

Fig. 4.2. Variations in *family* probability w.r.t. r for $n_0=2$, allowing only non-negative weights and threshold, and binary inputs.

separation between concept families, and some opportunity for the control of unwanted network side-effects. In particular, we see that a specialisation to DNF-based representations need not require substantial weight precision, and may indeed be enhanced through the use of low variance – even fixed – weights coupled with appropriate threshold distributions. Under such a regime, the value of r directly determines the range of realisable conjunctions, and the knowledge base may be tailored toward a particular class of concepts – say k-DNF for tightly bounded k – while retaining the assumptions of

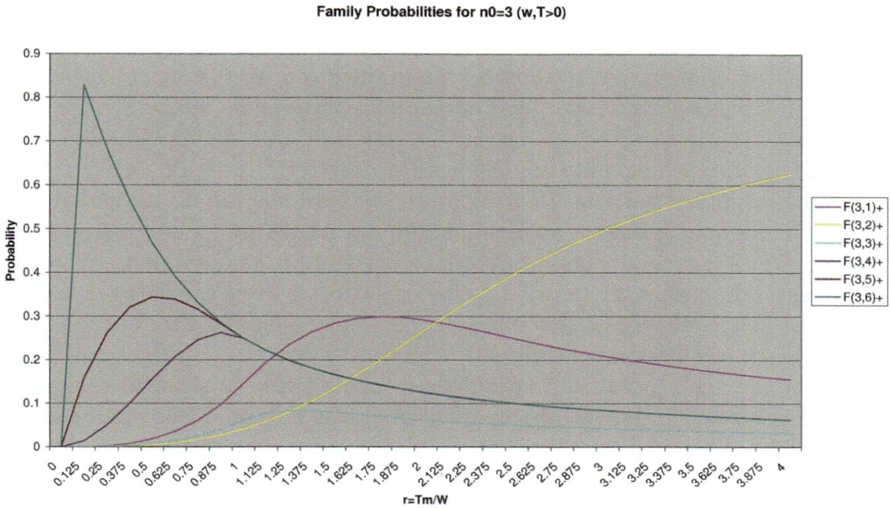

Fig. 4.3. Variations in *family* probability w.r.t. r for $n_0 = 3$, allowing only non-negative weights and threshold, and binary inputs.

random connectivity and threshold selection. From this point, it is a natural simplification for r to be fixed as an integer, and for each threshold to be drawn from the set

$$\{W, \ldots, (r-1)W, rW\},$$

where $W > 0$ is a suitably chosen weight value, and we shall adopt this course throughout our simulations.

This approach introduces its own concerns in regard to input saturation. In the absence of inhibitory input weights, only the unit threshold may balance excessive activation. If the unit exists in a forward layer of the black box, one may expect around $\alpha\xi N$ excitatory inputs from the preceding layer (of size N), and the problem may become especially acute, with the layer dominated by constantly firing units. Unfortunately, saturation of this kind has a cumulative effect, with units in subsequent layers served by (again excitatory) inputs with a firing probability far higher than the 0.5 typical of the input layer. While the problem may be partially moderated by replacing the threshold unit $u(y)$ by the real-valued sigmoid $\sigma(y)$ or its integer-valued cousin, careful control of the connection probability and radius remains necessary. Composition of these effects across populations of interconnected units, and the nature of the concept distributions supported, are considered in the following sections.

4.3 Representation at the Network Scale

In making the transition from individual unit to the network scale we are unavoidably confronting the history of connectionism, seeking to overcome the limitations of the single layer perceptron without recourse to backpropagation of error. In principle, the introduction of only a single hidden layer allows the representation of any Boolean concept (Hertz et al, 1991), or real-valued continuous function (Hornik et al, 1989) of the inputs, provided that sufficient hidden units are available and suitable transfer functions are employed. Yet existence results of this nature are of little consequence if we are unable to select, generate or modify the basis provided by the network hidden units, so as to encompass the subspace hosting the classification problem.

In the present approach, hidden layers are fixed at the time of network generation, each separating hyperplane in effect being positioned randomly. Available bases – intentionally or otherwise – favour some Boolean concepts over others – particularly those which may require a fine balance between a number of decision surfaces.

Suppose for example that a task is (minimally) dependent upon the realisation of H hidden unit dichotomies, the most probable of which occurs with probability $\alpha_h \leq \epsilon$, where $0 < \epsilon \ll 1$. Thus, the representation probability for the task is bounded by an expression of $O(\epsilon^H)$, and resource constraints rapidly intervene, with layer sizes of $O(\epsilon^{-H})$ or greater required if we are to have any reasonable expectation of recruiting the target. In particular, we cannot expect to address concept classes for which H grows exponentially with the input dimension.

More fundamentally, the approach described by equation 4.4 involves the approximation of the target by a *threshold function* over the feeder basis, with each of the feeders in turn a (composition of) threshold function(s) over (some subset of) the network inputs. Evidently, the feeder basis will seldom be of the exponential dimension sufficient to span the concept space, and we must rely upon so-called sparse approximations, in which the hypothesis is defined over polynomially many basis functions. Fortunately, approximations of this type may be established under quite general conditions[5], and we shall in principle employ a basis composed of low degree conjunctions over the inputs.

Formally, we define a set of 2^n conjunctions over n Boolean inputs by the relations

$$\wedge_\omega(x) = \begin{cases} 1, \; x_j = 1, \; \forall j \in \{k | \omega_k = 1\} \\ 0, \; otherwise \end{cases}, \tag{4.13}$$

where $\omega \in B^n$ is the index vector, and the conjunction is vacuously satisfied for $\omega = 0$. Thus, each $\wedge_i$ implements a conjunction over the components of x corresponding to those for which ω is non-zero. The set is readily seen to be

[5] Technically, the requirement is that the Boolean concept have bounded generalised spectral norm, and that the basis be linearly independent. See Siu et al (1995) and Hogan (2000) for details.

linearly independent, and in consequence forms a basis for the concept space – and one for which many of the set have a high probability of existence. We note in particular that the basis incorporates many of the low-degree localised concepts which one would expect to be formed through random connections from a set of input nodes.

In principle our networks may approximate any Boolean concept of polynomially bounded generalised spectral norm, but in practice their *realisable* capacity will depend critically on the extent to which such representations may be recast in terms of low-degree (and hence more probable) basis functions. One attractive notion is that a decomposition reliant upon at least some high degree conjunctions may be realised through an additional layer supporting conjunctions over the more probable nodes of $2, 3$ and 4 inputs.

Yet a low existence probability for appropriate basis concepts may not be the only barrier to the acquisition of complex knowledge. Of equal concern is the extent to which both the target and its constituents may be isolated from similarly probable but erroneous alternatives. These issues are considered in the next section.

4.3.1 Concept Entropy and Phase Volume

We have earlier considered the difficulties posed by concepts which may be represented only through a fine balance between a number of decision surfaces – noting in particular the low probability of existence of such concepts under the random network assumptions. Yet such problems may be difficult even with the advantage of backpropagation of error, with training times substantially greater than for less complex tasks of comparable input dimension – a fact readily illustrated by training simple networks on the *XOR* and 2-input conjunction problems.

Such phenomena are readily explicable within the framework of statistical mechanics, in which each range of architectures – subject to variations in connectivity, threshold and connection weights – prescribes some entropy distribution in accordance with the number of configurations which support each realisable concept. Thus, while a search of the available configurations through gradient descent is guided in a sense in which the network generation of the present study is not, the 'target volume' embodying those configurations which implement tasks such as *XOR* remains relatively small, and potentially difficult to access. Learning, regardless of the algorithm employed, is profoundly influenced by this prior entropy, and the task is rendered substantially less difficult if the concept is favoured by the distribution.

Van den Broeck and Kawai (1990) examined a feedforward network of 2-input Boolean gates, with each gate receiving distinct input signals drawn from among the inputs ($n_0 = 4, \ldots, 7$) and gates in preceding layers. The concept implemented at a given gate is randomly selected (under a uniform distribution) from a restricted set comprising *ZERO, AND, OR, XOR* and their respective negations, with the number of gates in the network ranging between 2 and 200. Learnability is then governed by the phase volume associated with the concept – the number of network configurations which implement the given task. Thus, noting that within their system the largest phase volumes are those associated with the constant output concepts (which we have denoted *ZERO* and *ONE*), they remark that such high entropy concepts may be easy to learn – in the sense that a large fraction of the *accessible* network configurations are consistent with such concepts.

However, the effect of network training on the entropic landscape is markedly more complex, as supervised learning corresponds to a "[search] for a configuration ... compatible with the teaching examples". As the "accessible phase volume is reduced, ... a new hierarchy of ... Boolean function[s] is established [within] this reduced phase space". Thus, it "is not the entropy *per se* that dictates the [learnability] of a Boolean function, but rather the entropy of a given function compared to that of its "neighbours"".

More formally, we have seen that each Boolean concept over n input variables may be characterised by a single binary vector of dimension $b = 2^n$, the components of this vector corresponding to the concept response for each of the b input patterns of dimension n. In this context, we may define the *Hamming distance* between two concepts f and g by the expression

$$H(f, g) = \sum_{p=0}^{b-1} |f(x_p) - g(x_p)|, \qquad (4.14)$$

which is readily seen to be the number of bits of disagreement between f and g over the example space.

Each pattern drawn from the 2^n possible examples of the concept has the effect, if successfully learnt, of fixing one of the corresponding output values of the network hypothesis. As a result, concepts for which the Hamming distance is large i.e. those which have a *higher* probability of disagreement with the target on a particular example, will be eliminated after relatively few training examples have been presented. The effect of supervised learning of l examples is then to limit the phase space to a Hamming neighbourhood of the target – those

$$V = \sum_{d=0}^{b-l} \binom{b-l}{d}$$

concepts consistent with the training examples. If the target exists at a local entropic maximum within the Hamming space, we see that relatively few training examples will be required to eliminate those (distant) concepts of comparable entropy. Thus, with respect to the other $V - 1$ consistent hypotheses, the target may dominate the available phase volume, resulting in a high probability of perfect generalisation. Conversely, generalisation is likely to be poor for concepts which do not satisfy this criterion, the target being in such cases but one of a number of probable hypotheses consistent with the training examples.

Van den Broeck and Kawai illustrate this point through concepts 0 (*ZERO*), 27030 (*PARITY*), and 32768 (4-input conjunction), each drawn from the set of Boolean concepts over 4 inputs. Under their assumed gate distribution, the parity is acquired straightforwardly as it constitutes a local maximum in the Hamming space. In contrast, the ostensibly simple *AND* problem proves difficult, its nearest neighbour being *ZERO*, the concept of largest phase volume under their regime.

Issues of this nature arise in the present work in two distinct ways: directly – through initialisation of the connection weights and thresholds of the concept layer, favouring some class of threshold functions over the network feeders – and rather more subtly – through the distribution of concepts realised within the network black box layer(s).

Parameter selections for the concept layer are typically biased toward the realisation of disjunctions over a conjunction basis. Indeed, were the target to be a disjunction of moderate degree over some fraction of the *network* inputs, it is difficult to see how successful generalisation could take place, the relevant phase space being densely populated with other disjunctions of comparable and higher degree.

However, the concept layer will normally face a recruitment task (through disjunction) over feeders implementing some range of conjunctions over a subset of the network inputs, and one must expect a relatively low firing probability for such nodes with respect to the training examples. In consequence, concepts within our random knowledge base may exhibit deceptively low inter-concept Hamming distances, with agreement over a large number of examples on which none of a particular subset of feeders fires. As membership of the initial disjunction may be assumed, and a zero output is used to indicate the `NO_FIRE` state, examples for which a particular feeder is silent are of little consequence in determining that feeder's role in the final network hypothesis.

Under these assumptions a feeder node f_j will form part of the final network hypothesis only if there is a high probability that the target t fires on those training examples which cause f_j to fire. Let us define a real-valued distance between f_1 and f_2, $tr_{f^+}(f_1, f_2)$, to be a restriction of the Hamming calculation to those training patterns (for t) which are positive examples of the concept f, where the bit-error is normalised with respect to the number of such patterns. If recruitment is limited to those feeders f_j for which

$$tr_{f_j^+}(f_j, t) \ll 1, \tag{4.15}$$

the primary risk of generalisation error lies in the selection of nodes which fire more frequently than the target over the unseen input patterns, resulting in a strong likelihood of numerous false positive responses.

Suppose now that there exists among the feeders a node implementing the concept f, which forms part of the DNF expression defining the target. Recruitment of such a node into the basis set for the network hypothesis is highly desirable, but the process may be compromised by the collateral recruitment of nodes f_j proximal to both t *and* f – in the sense that both $tr_{f^+}(f, f_j)$ and $tr_{f_j^+}(f_j, t)$ are small[6] – which fire more frequently than t over the generalisation set.

Exposure to risk from this source may be substantially limited if f is constrained to exist at an entropic maximum with respect to concepts proximal under the relation $tr_{f^+}(f, \cdot)$. Within our framework, this may be accomplished only through distributional bias at the conjunction layer, and we shall attempt to tailor the feeder layer so as to provide conjunctions which may reasonably form part of the target DNF. The extent to which we are successful in optimising generalisation performance cannot be discussed without further specification of the learning rules and targets. These issues are considered in turn in the remaining sections of the chapter.

4.4 Hebbian Learning

As noted in the introduction, most of the rules of this chapter may be sourced ultimately to Hebb's famous 1949 hypothesis, which we give again here for convenience:

> "When an axon of the cell A is near enough to excite cell B or repeatedly or persistently takes part in firing it, some growth or metabolic change takes place in both cells such that A's efficiency, as one of the cells firing B, is increased"
> (Hebb (1949), p.62.)

The simplest mathematical statement of this idea is given by the additive update

$$w_{ij} := w_{ij} + \zeta x_i x_j^{(i)}, \tag{4.16}$$

where $x_j^{(i)} \in \{0, 1\}$ denotes the jth component of the input values, $x_i \in \{0, 1\}$ the unit output, and $\zeta \in (0, 1)$ is the learning rate. It is usual also to assume that the weights have bounded magnitude, whether through truncation or as a result of some normalisation procedure. As has been noted by numerous authors, for example (Shouval and Perrone, 1995), this formulation accurately depicts Hebbian strengthening but suffers from the problem of weight saturation, there being no mechanism provided through which weight growth may be limited. In consequence, under repeated presentation of the pattern set, it is expected that all weights will reach their maximum value.

Thus, the Hebbian update is more commonly given with a corresponding penalty for disagreements between the given input and the unit output, which we present below using a contingency table.

---

[6] In consequence, there is a high probability that $tr_{f^+}(f_j, t)$ is also small.

Table 4.3. The contingency table for a common model of Hebb's principle.

	Input Value	
Output Value	$x_j = 0$	$x_j = 1$
$x_i = 0$	$w_{ij} := w_{ij} + \zeta$	$w_{ij} := w_{ij} - \zeta$
$x_i = 1$	$w_{ij} := w_{ij} - \zeta$	$w_{ij} := w_{ij} + \zeta$

Supervised learning is realised naturally by exploiting correlations between the input and a 'target' unit. Changes to the update formulae are trivial, and the approach is neurally plausible if we assume that the concept unit receives strong excitatory input whenever the target is firing, and correspondingly strong inhibitory input when the target is not. Such behaviour requires that the concept unit receive connections from both an oracle and its complement, in the latter case via some inhibitory interneuron.

Formally, we need only write some target value $t_i(x)$ instead of the output x_i, and we shall usually give the rules in this latter form. Some brevity may be achieved if we transform the inputs and target to their corresponding $\{-1, 1\}$ representation, respectively denoted ξ_j and τ_i. In this case the rule has the form of equation 4.16, and we write

$$w_{ij} := w_{ij} + \zeta \tau_i \xi_j^{(i)}. \tag{4.17}$$

This form appears frequently in the associative memory literature, with network weights developed as a linear combination of the outer product of the target and pattern vectors (see for example Hagan et al (1996)).

While the modified form of equation 4.17 ensures that some penalty is exacted for disagreements between inputs and target, it does not remove altogether the problem of weight growth and a variety of normalisation and filtering approaches have been employed[7]. Among the more common ideas (Hagan et al, 1996), is the introduction of a weight decay parameter $0 < \gamma < 1$ such that

$$w_{ij} := (1 - \gamma) w_{ij} + \zeta \tau_i \xi_j^{(i)}. \tag{4.18}$$

Thus, the parameter γ may be used to control the importance attached to new patterns presented to the network. Large values of γ will ensure a strong weighting toward more recent vectors while a small value of γ preserves existing knowledge in much the same way as equation 4.17.

Yet neither of the updates considered so far offers stable convergence of each weight value to some limit under repeated presentation of the pattern set. To some extent this problem may be alleviated through the use of a very low learning rate (Shouval and Perrone, 1995), providing a kind of 'pseudo-convergence' under which network classifications remain correct over one or more epochs – so that tolerance-based stopping criteria may be satisfied – only

[7] See Shouval and Perrone (1995) for a short review.

to become erroneous again if training is continued for a number of epochs. As this behaviour is undesirable in the present context, we must examine more closely our knowledge of Hebb's principle.

4.4.1 Neural Foundations

Levy, Colbert, and Desmond (1990) consider the Hebbian update from the standpoint of modern neuroscience, giving particular prominence to the processes known as Long Term Potentiation (LTP), Long Term Depression (LTD) and synaptic habituation (SH). We shall begin by considering updates consistent with the experimental evidence surrounding LTP and LTD.

After reviewing a number of electrophysiological studies, Levy et al. make the following observations:

1. Modification through *either* of LTP and LTD requires the existence of "powerful net postsynaptic excitation", with weak presynaptic activity producing no consistent long term change in synaptic efficacy;
2. Postsynaptic excitation translates into a "permissive postsynaptic [state]" in a highly non-linear manner.
3. If the pre-requisite of strong post-synaptic excitation is satisfied, then the nature of the modification is governed by the level of pre-synaptic activity.
4. The degree of potentiation and depression is bounded such that synaptic strengths ultimately converge to some limiting value.

In this light, the contingency table of the previous section may be redrawn as shown in table 4.4, where as before y_i indicates the input sum or postsynaptic excitation and $f(\cdot)$ the nonlinear activation function. For clarity, we shall initially take f to be the threshold function $u(y_i)$, and as usual we shall assume non-negative network activations.

Table 4.4. The contingency table for the convergent Hebbian rule.

Postsynaptic Activity	Presynaptic Activity: x_j	
$x_i = f(y_i)$	0	1
0	no change	no change
1	$w_{ij} \Downarrow$	$w_{ij} \Uparrow$

Levy et al. then postulate an update of the form

$$w_{ij}(t+1) := w_{ij}(t) + \Delta w_{ij}^{+}(t, t+1), \tag{4.19}$$

where it is assumed that there exists no more than one synaptic contact between neurons j and i. Three alternative formulations are then considered:

(i) $\Delta w_{ij}^{+} = f(y_i(t))g(x_j(t-\eta)) - c_1$,
(ii) $\Delta w_{ij}^{+} = f(y_i(t))(x_j(t-\eta) - c_1)$, and
(iii) $\Delta w_{ij}^{+} = f(y_i(t))(x_j(t-\eta) - w_{ij}(t))$,

where c_1 is a positive constant, η is a small delay to model membrane effects and $g(\cdot)$ is a non-negative, non-decreasing function of presynaptic activation. Alternative (i) allows spontaneous decay independent of both pre and post-synaptic activity in violation of condition 1[8]. Alternative (ii) correctly models the lower quadrants of table 4.4, but places no bound upon potentiation or depression under constant input. Thus alternative (iii) is selected, providing an important refinement over the updates previously discussed as it ensures asymptotic convergence of the weight to some value in the interval $(0, \max\{x_j\})$[9]. In the present work, the update is simplified and converted to the corresponding supervised form

$$w_{ij} := w_{ij} + \zeta t_i(f_j - w_{ij}), \tag{4.20}$$

where the input value is now the jth feeder unit $f_j(x)$.

Formally, both t_i and the f_j may be viewed as random processes, and equation 4.20 may be recast in the form

$$E[\Delta w_{ij}] = \zeta E[t_i f_j] - \zeta E[t_i w_{ij}], \tag{4.21}$$

where the expectation is understood to be taken with respect to the input distribution. Under network training conditions, assumptions of ergodicity and stationarity[10] are readily justifiable, and the expectation of the weight change in equation 4.20 vanishes as the number of presentations becomes infinite (Levy et al, 1990). Thus, we see that the network weights encode first order statistics of the inputs[11], with convergence to the ratio

$$w_{ij} = \frac{E[t_i f_j]}{E[t_i]}, \tag{4.22}$$

which must always lie in the interval $[0, 1]$.

---

[8] Note the similarity to the decay law of equation 4.18.

[9] Levy et al. give the presynaptic term as $cx_j(t-\eta)$, where c is a constant to ensure dimensional consistency with the weight term $w_{ij}^{+}(t)$. In our discussion this constant is taken to be unity.

[10] See for example Yaglom (1973).

[11] A number of other Hebb-like updates, notably the linear rule of Linsker and the stabilised update of Oja, cause the weight vector to approach the first principal component of the input, although in the former case the weight magnitude is unbounded. See Shouval and Perrone (1995) for detailed discussion of these issues.

4.4.2 Synaptic Habituation

Thus far we have avoided specifying updates of synaptic efficacy when there is little postsynaptic activity. Levy et al. argue that some short term changes may occur as a result of the mechanism of synaptic habituation when $f(y_i) \approx 0$, with adjustments facilitated through the following complementary processes:

1. Repeated activation of *weakly* excitatory inputs *depresses* synaptic strength, a phenomenon which does not appear for repeated activation of strongly excitatory inputs.
2. The synapse spontaneously recovers from the effects of this process, with recovery taking place in the absence of significant pre-synaptic activation.

In particular, the authors assert that the synaptic depression of point 1 cannot persist, as the conditions outlined in point 2 should be regarded as the default system state.

These conditions may be expressed in an update of the form

$$\Delta w_{ij}^{-} = [1 - f(y_i(t))](x_j(t-\eta) - w_{ij}^{-}(t)), \tag{4.23}$$

where w_{ij}^{-} represents an inhibitory weighting to be subtracted from w_{ij}, this decomposition ensuring that potentiation under the conditions of point 2 cannot exceed that needed to recover from the depression due to the conditions of point 1. Thus, the contingencies of the previous section may in principle be revised to take the form shown in table 4.5.

Table 4.5. The contingency table for the convergent Hebbian rule, as modified to incorporate the effects of synaptic habituation.

Postsynaptic Activity	Presynaptic Activity: x_j	
$x_i = f(y_i)$	0	1
0	$w_{ij} \Uparrow$	$w_{ij} \Downarrow$
1	$w_{ij} \Downarrow$	$w_{ij} \Uparrow$

Levy et al. do not resolve the question of interaction between these two distinct processes, an interaction potentially complex in spite of their apparent similarity. At this point we are faced with some conflict between reducing the effect of any input which fires when the target does not (the upper right quadrant of table 4.5) and the unjustified permanence that a combined update in the style of equation 4.17 may give to the effects of synaptic habituation. Moreover, such an update may lead to excessive potentiation through the recovery mechanism.

Fortunately, the problem of excessive input activation may be addressed through adjustment of the unit threshold, and some update mechanism is desirable if concept learning is to be facilitated under the assumption of bounded weight magnitudes. This approach is considered in the following section.

4.4.3 Thresholds and the Perceptron Update

In the present work we have adopted a threshold update drawn from the perceptron rule of Rosenblatt (1961), providing direct linkage between the update and classification errors made by the unit. Rosenblatt's landmark rule tailors its response according to the performance of the unit on the pattern just presented. If the pattern is correctly classified, then weight adjustment is neither necessary nor desirable, for fear of disrupting the present correct selection. If the pattern is misclassified, then the update is governed by the type of error reported:

1 False positive: Unit weights adjusted to increase the likelihood that the decision surface will exclude the current pattern vector from the concept set;
2 False negative: Unit weights adjusted to increase the likelihood that the decision surface will include the current pattern vector within the concept set.

The perceptron rule thus differs markedly from those of the preceding sections in prescribing a specific, error-limiting response for each of the possible categories of pattern. Responses for false positives and false negatives involve a rotation of the weight vector through, respectively, subtraction or addition of the current pattern vector, thus decreasing or increasing its inner product with the weight. Such an update ensures convergence to correct classification of *all* patterns presented if such a weight set exists[12].

In this light, the subtractive threshold T_i is an extra input weight associated with the unit and an update implementing the strategy outlined above is given by the expression

$$T_i := T_i + \mu(h_i(x) - t_i(x)), \tag{4.24}$$

where as usual h and t denote the network hypothesis and target value on pattern x, and μ is a learning rate normally chosen around 0.1. Thus, if the network hypothesis fires when the target does not, the subtractive threshold is increased so as to inhibit firing; conversely, an erroneously inactive hypothesis results in some decrease of the threshold.

In each of the correlation-based rules, the weight update proceeds quite independently from the threshold update, with the latter adjustments

[12] The theorem requires that all patterns be of bounded magnitude, and in the absence of backpropagation of error, the target must be a linearly separable function of its inputs. See Minsky and Papert (1969) or Hagan, Demuth, and Beale (1996) for details.

influenced by the global weight sum for each pattern. While weight convergence may be guaranteed for the rule of equation 4.20, no such result exists for the threshold and its value may oscillate in accord with the patterns presented if no linearly separable representation exists across the network feeders – although this effect may be partially damped by both weight convergence and a low threshold learning rate.

While longer term changes in unit activity levels may be mediated through variations in the effective membrane threshold (Levy et al, 1990), there is little evidence to suggest that an update of the form proposed exists for biological neurons. Our justification is thus purely computational, although we draw some comfort from the relatively minor adjustments that have been required in most of the simulations undertaken. Such an update is a foretaste of the less plausible approaches of Section 4.5, but we must first give due attention to the limitations of the rules we have considered thus far.

4.4.4 Analysis of the Correlation Approaches

We have earlier considered the effect on generalisation performance of the relationship between the target and its nearest neighbours within a conceptual Hamming space. Central to this argument is the effect of partial learning, in essence a refinement of the network hypothesis to the point that disagreement with the target is limited to $b - l$ of the available $b = 2^n$ concept bits. Such learning may greatly affect the entropic landscape, with the result that the desired generalisation may be implemented by a far smaller fraction of the remaining *consistent*[13] configurations than some more probable alternative.

Correlation-based learning rules such as those described may exacerbate such problems when faced with an input (i.e. a feeder unit $f_j(x)$) which, while commonly firing in concert with the target, also exhibits extra responses that we would regard as false positives[14]. Let us as before define the Hamming distance between two concepts f and g through the expression

$$H(f, g) = \sum_{p=0}^{b-1} |f(x_p) - g(x_p)|, \qquad (4.25)$$

which we recast as the sum

$$H_+(f, g) + H_-(f, g), \qquad (4.26)$$

where H_+ and H_- are restrictions of the expression of equation 4.25 to respectively the positive and negative examples of the concept f.

[13] That is, consistent with the target over the l examples that have already been learned.

[14] Similar problems may appear with false negatives if an appropriate update is employed.

The fraction of false positives resulting from some feeder unit $f_j(x)$ with respect to the ith component of the target $t_i(x)$ is then given by

$$\epsilon_+(t_i, f_j) = \frac{H_-(t_i, f_j)}{b - \#\{t_i\}}, \tag{4.27}$$

where $\#\{t_i\}$ is the number of positive examples of the target. In a similar fashion, we may quantify the fraction of false negatives through the expression

$$\epsilon_-(t_i, f_j) = \frac{H_+(t_i, f_j)}{\#\{t_i\}}. \tag{4.28}$$

Under repeated presentation of the pattern set, the rule of equation 4.20 ensures that

$$w_{ij} \to (1 - \epsilon_-), \tag{4.29}$$

and the effect of such a feeder upon the concept representation depends upon both the fraction of false positives as indicated by ϵ_+ and upon the penalty exacted for each such disagreement. While it is difficult, without a detailed concept probability distribution, to place precise bounds upon the disruption caused to the hypothesis, we may nevertheless make some qualitative progress with the problem through the introduction of an 'influence' function

$$I(t_i, f_j) = (1 - \epsilon_-(t_i, f_j))\epsilon_+(t_i, f_j). \tag{4.30}$$

Evidently, $I(t_i, f_j)$ has its maximal value of one when $\epsilon_- = 0$ and $\epsilon_+ = 1$, but a plethora of such units is unlikely.

Of more concern is the situation in which a large number of units exhibit non-zero ϵ_+, say some fraction ϕ of the F feeder units. Then the expected input activation y_i will exceed the threshold whenever

$$E[\epsilon_{ij+}] > \frac{T}{\phi F(1 - E[\epsilon_{ij-}])}, \tag{4.31}$$

where the expectation is taken over the input distribution. For large F, the mean ϵ_{ij+} required for saturation may be quite low, confirming our intuitions, and suggesting that the convergent Hebbian rule may even fail in the presence of a large number of *mildly* erroneous feeders. This analysis may be seen as further support for some level of distributional bias to ensure a sparse feeder space, reducing the likelihood of saturation through such effects.

Correlation-based rules, while particularly useful in the acquisition of conjunctions of the input channels, nevertheless suffer from the absence of a direct linkage between an individual weight update and the global performance of the hypothesis. While we have compromised to the extent of introducing an error-linked threshold update, saturation problems such as those discussed in the preceding paragraphs may prevent correct classification performance. Some possible solutions to these problems are explored in the following section.

4.5 Robust Learning

Most large scale learning systems may experience difficulties in handling excessive input activation and weight saturation, particularly in the presence of partially correlated – but nevertheless irrelevant – attributes. Such effects usually delay learning of the target concept, or perhaps prevent convergence altogether. Indeed, Valiant (1996) argues that no learning rule should be considered as a candidate mechanism for knowledge acquisition within a natural learning system unless it is able to tolerate some level of noise and a abundance of irrelevant attributes. Such robust updates must heavily penalise contributions which result in incorrect firing at the concept unit whilst preserving or enhancing the weights from more desirable inputs. In this section, we shall examine two approaches to the problem, one based upon a rather brutal excision of unwanted inputs and the other a softer scaling approach with near-optimal performance for the task.

4.5.1 The Elimination Rule

One strategy for dealing with irrelevant attributes is suggested by Valiant's earlier work on the acquisition of Boolean disjunctions within the Probably Approximately Correct (PAC) framework (see for example Natarajan (1991)). In this approach, the system attempts to learn the disjunction of up to n input variables, and the expression

$$x_1 \vee x_2 \vee x_3 \vee \ldots \vee x_n$$

forms an initial hypothesis, which is subsequently tested against both positive and negative examples of the target. Learning is here a matter of elimination, the hypothesis being refined to exclude inputs which fire when the target does not[15].

Implementation is straightforward within the random network system, with weights between feeders and the concept node initially set to unity, and annihilated if the feeder represents a false positive with respect to the target on *any* pattern. The approach is predicated on the assumption that false negatives are of no consequence (as we are using only positive weights) and that the feeder set contains one or more feeder units whose disjunction is sufficient to cause the hypothesis to fire in agreement with the target – while none of this subset of units provides the false positive response which would lead to its elimination. If no such subset of the feeder set exists, then little may be said about the quality of the hypothesis.

[15] A similar algorithm exists for the acquisition of Boolean *conjunctions*, relying upon the elimination of inputs which fail to fire on a positive example. Note that this effect may be realised through the update of equation 4.20 as long as the problems identified in Section 4.4.4 do not intervene.

A number of variations are possible in the treatment of inputs which survive the elimination process and make some contribution to the network hypothesis – the choice of strategy being governed by the nature of the combination we wish to realise. Patently, if we consider only disjunctions then the surviving weights need not be altered, and any positive value of T_i will suffice for the unit to represent the concept. This form of the elimination rule provides spectacular convergence – within a single epoch – to the final hypothesis, regardless of its quality. This situation is reminiscent of the one-shot acquisition of concepts discussed in Chapter 2, and relies similarly on the emergence of uncomitted 'binder' units able to denote the concept through the combination of exisiting committed knowledge.

However the recruitment layer may be made more general if we use the update of equation 4.20 for the surviving weights. As noted in Section 4.4.1, this rule causes weights to converge to the ratio $E[t_i f_j]/E[t_i]$, ensuring that any input which fires in conjunction with the target ultimately receives a fixed non-zero weighting. If the unit threshold is initially small, and the update of equation 4.24 is employed with a small value of μ, the node may comfortably implement either a disjunction (favoured by the initially low threshold value) or a conjunction according to need, the in-degree being controlled by a combination of the pruning step and threshold adjustment.

Table 4.6. The contingency table for the elimination rule.

Target Value	Feeder $f_j(x)$	
$t_i(x)$	0	1
0	w_{ij} unchanged	$w_{ij} := 0$
1	$w_{ij} := (1-\zeta)w_{ij}$	$w_{ij} := (1+\zeta)w_{ij}$

The contingency table 4.6 presents the rule in this more general form, but the simpler disjunction rule may be recovered immediately by taking $\zeta = \mu = 0$.

Yet while the elimination rule described above is successful for a significant range of recruitment problems, it is unnecessarily severe in its treatment of 'mildly-erroneous' feeders which might otherwise be exploited to realise concepts with a finer balance between positive and negative examples. The perceptron rule of Rosenblatt (1961) provides a sensible starting point for discussion of these issues, with annihilation of weights being replaced by a subtler adjustment in response to errors. Over repeated pattern presentations, the offending weight is set to a value governed by the ratio of mistakes on positive examples to mistakes on negative examples during learning.

The classical perceptron rule is guaranteed to converge in a finite number of updates to a solution in which no mistakes are made provided that such a solution exists, and that all pattern vectors remain bounded. Yet while the method is ultimately robust, convergence may be impaired by the presence of attributes which are not present in the concept, and the number of mistakes (and thus updates) is thought, on the basis of an empirical study (Hampson and Volper, 1986), to grow linearly with the number of these irrelevant attributes. As large-scale random networks provide an extreme instance of this problem, another method must be found which preserves potentially helpful feeders, while pruning genuinely irrelevant inputs with a similar level of effectiveness as the elimination strategy. One suitable class of update, the Winnow rules of Littlestone, is considered in the following section.

4.5.2 Littlestone's Winnow Rules

Littlestone (1988), motivates the development of the Winnow rules through discussion of a pattern recognition problem in which a large class of feature detectors are available, but the system has little knowledge of their respective importance to the task. What is necessary is that the learning rule rapidly *winnow* the 'wheat from the chaff', thus constructing a solution free from interference from irrelevant features, with the goal of the learner being to make as few mistakes as possible.

The same approach is directly applicable to the more general problem of representing a class of Boolean knowledge:

> "In a special case of this situation, one may design a library of concepts specifically to ease learning of a certain class of complex functions. In this case, one chooses concepts for the library that allow representation of any function in the class as a simple function of the library concepts", (Littlestone, 1988).

Such a computational setting has obvious parallels with the present work, our library being that provided through the feeder units of the random black box, the knowledge base fostered through distributional bias toward a desired class of concepts. Littlestone illustrates this idea through the example library of 'conjunctions of no more than k literals', each k-DNF concept thus becoming a straightforward disjunction of the library terms.

The first Winnow rule, hereafter denoted *Winnow1*, was introduced in the context of learning a monotone disjunction, i.e. an expression of the form:

$$f(x_1, x_2, \ldots, x_n) = x_{i_1} \vee x_{i_2} \vee \ldots \vee x_{i_k}, \tag{4.32}$$

in which the input is defined over n attributes, but a subset of size k of these inputs – in which none of the literals is negated – is sufficient to determine the concept. The *Winnow1* update is similar to the elimination rule developed earlier, both owing much to Valiant's earlier PAC algorithm, but this time the update is governed by a lack of agreement between the hypothesis and target, and weight change occurs *only* when a mistake is made. As before, we present the rule in contingency table form (table 4.7), where $\zeta > 1$ is a multiplicative learning rate.

Table 4.7. The contingency table for the Winnow1 rule.

Hypothesis $h_i(x)$	Target $t_i(x)$	Update	Comment
1	0	$w_{ij} := 0, x_j = 1$ w_{ij} unchanged, $x_j = 0$	Elimination
0	1	$w_{ij} := \zeta w_{ij}, x_j = 1$ w_{ij} unchanged, $x_j = 0$	Promotion

As in the elimination rule, the penalty for disagreement is extremely high, the connection being annihilated with no prospect of recovery. The second Winnow rule, *Winnow2*, preserves the multiplicative update of the promotion step, but is less severe upon inputs which do not specify the task, providing a demotion step which is the formal inverse of the promotion update. In this way, inputs which contribute to a similar number of false positive and false negative errors may retain a serviceable weight value, while genuinely irrelevant inputs are assigned weights well under the promoted values (table 4.8).

Table 4.8. The contingency table for the Winnow2 rule.

Hypothesis $h(x)$	Target $t(x)$	Update	Comment
1	0	$w_i := w_i/\zeta, x_i = 1$ w_i unchanged, $x_i = 0$	Demotion
0	1	$w_i := \zeta w_i, x_i = 1$ w_i unchanged, $x_i = 0$	Promotion

4.5.3 Analysis of the Winnow Rules

As noted previously, one key objective of the Winnow rules is that the learner acquire the target while making as few mistakes as possible, notwithstanding the presence of a large number of irrelevant attributes. Indeed, Littlestone (1988) is able to prove that for a number of important classes $\mathcal{F}$ of linearly separable targets, the *Winnow2* rule may realise mistake bounds within a multiplicative constant of the theoretical optimum $opt(\mathcal{F})$[16]. More formally,

[16] It is straightforward to show that for any non-empty target class $\mathcal{F}$,

$$VC(\mathcal{F}) < opt(\mathcal{F}) \leq \log |\mathcal{F}|,$$

where $VC(\cdot)$ is the Vapnik Chervonenkis dimension of the class.

the *Winnow2* algorithm may be applied, through appropriate choices of $\zeta = \zeta(\delta)$ and $T_i = T_i(\delta)$, to *any* Boolean target class $\mathcal{F}$ satisfying the following δ-separation condition.

Definition 1. *δ-Separability*
Say a class of Boolean concepts $\mathcal{F} \subset B^n$ is δ-separable if $\forall f \in \mathcal{F}, \exists w_1, \ldots, w_n$ such that for each positive example x^+ of f, i.e. where $f(x^+) = 1$,

$$\sum w_i x_i \geq 1$$

while for each negative example x^-, for which $f(x^-) = 0$,

$$\sum w_i x_i < 1 - \delta.$$

Thus, the *Winnow2* rule may in principle operate successfully upon those Boolean target classes in which a clear (linear) separation may be seen between positive and negative examples of each target. Littlestone provides the caveat that the mistake bounds derived may be practicable "only for those [classes] for which δ is sufficiently large", a matter of limited concern in the present context as classes which are widely δ-separable include both conjunctions and disjunctions, the exceptional pattern in each case occupying a 'hyper-corner' of the pattern space.

It is worth remarking at this point on the wide applicability of the Winnow update in cognitive computation, and especially in the context of representations involving large populations of nodes and – potentially – disparate features rather than the more compact and stable hidden unit spaces of the backpropagation era. In particular, Gunay's work on continous time recruitment (see Part II) relies on a modified form of the Winnow update tailored to the neuroid context.

We consider this update for the case of temporal unsupervised memorisation (UMT). Weight change follows the promotion and demotion steps of table 4.8, but the learning rate is governed by the state of the neuroid, as reflected in the value of the *fitness criterion counter*, C_i, which is incremented whenever the presynaptic activity exceeds some calculated threshold. The neuroids pass from an uncommitted, *available for memorisation* state (AM) to an intermediate state (AM1) when input is elevated sufficiently to cause firing[17], remaining in this state until the pattern is seen to be stable. At this point, the neuroid transitions to the committed state (UM).

Establishing pattern stability is the role of C_i, which is incremented whenever the membrane potential exceeds some threshold D_i and is decremented otherwise. When the counter achieves some arbitrary value which Gunay labels α, the neuroid is held to have achieved memorisation. Weight adjustment is scaled by the relationship between C_i and α, and designed so that the

[17] Gunay follows Valiant in normalising the input activation so that unity is the firing threshold.

stable firing of a single input is sufficient to cause firing in the postsynaptic neuroid.

The rule then takes the form

$$w_i := w_i / C_i \alpha, \quad \textit{Demotion} \tag{4.33}$$

$$w_i := C_i \alpha w_i, \quad \textit{Promotion} \tag{4.34}$$

where the update is selected according to the coincidence of presynaptic and postsynaptic firings. If the pattern is stable, $w_i \to 1$ as $C_i \to \alpha$ as required; otherwise the weight decays rapidly. In the case of supervised inductive learning, the update is virtually identical to that of table 4.8, although weight change is permissible only while the neuroid exists in an error correction state. More details may be found in appendix C of (Gunay, 2003).

4.6 Summary of Rules

At this point, it is appropriate to summarise the updates which are employed in the simulations of the following sections and in those of Chapter 5. In what follows rules are presented in the form which allows the most concise statement, whether that form is a single equation or a succession of conditionals. For each rule, we provide

- a short, descriptive mnemonic;
- initial values of the threshold and weights;
- information on weight bounds, if applicable;
- a concise statement of the update; and
- possibly a short comment on its usage.

While we do not explicitly specify values of the learning rates ζ and μ, the choices being to some extent prescribed by the target and network realised, the Hebb-like rules produce far better convergence if a small value of ζ is used, and the value $\zeta = 0.0001$ is employed in many of the subsequent simulations. The multiplicative/divisive update of the *Winnow* rule requires a choice of $\zeta > 1$, and we shall normally take $\zeta = 2$. A choice of $\mu \in \{0.1, 0.01\}$ is adequate for most problems for which a threshold update is required. The rules are specified as follows:

Learning Rule 1. *Symmetric Supervised Hebb:* **SSHebb**.

Initialisation:

$$T_i \in (0, 1);$$
$$w_{ij} \in (-1, 1);$$

Weight Update: *If* $\tau_i = 2t_i - 1$ *and* $\xi_j = 2f_j - 1$, *then*

$$w_{ij} := w_{ij} + \zeta \tau_i \xi_j.$$

Weight Bounds: *Enforce* $|w_{ij}| \leq 1$.

Threshold Update:

$$T_i := T_i + \mu(h_i(x) - t_i(x)).$$

Comment: *Perhaps the least plausible of the quasi-Hebbian approaches listed, allowing the update to change the sign of the weights and employing weight bounding through truncation at* ± 1.

Learning Rule 2. *Convergent Supervised Hebb:* **CSHebb**.

Initialisation:

$$T_i \in (0, 1);$$
$$w_{ij} \in (0, 1);$$

Weight Update:

$$w_{ij} := w_{ij} + \zeta t_i(f_j - w_{ij}).$$

Weight Bounds: *Each* $w_{ij} \to w_{ij}^* \in (0, 1)$.

Threshold Update:

$$T_i := T_i + \mu(h_i(x) - t_i(x)).$$

Comment: *Neurally grounded local update.*

Learning Rule 3. *Elimination:* **Elim**.

Initialisation:

$$T_i := 0.1;$$
$$w_{ij} := 0.2$$

Weight Update:

$$\begin{aligned} &\textit{if } (t_i = 0) \wedge (f_j = 1) \textit{ then} \\ &\quad w_{ij} := 0 \\ &\textit{if } w_{ij} \neq 0 \textit{ then} \\ &\quad w_{ij} := w_{ij} + \zeta t_i(f_j - w_{ij}). \end{aligned}$$

Weight Bounds: *Weights either eliminated or convergent as in* CSHebb.

Threshold Update:

$$T_i := T_i + \mu(h_i(x) - t_i(x)).$$

Comment: *Threshold and weight values initially set to facilitate disjunctions of the feeders, allowing the threshold adjustment to select the appropriate degree if a conjunction is implemented instead. The pure disjunction approach* *is recovered if $\zeta = \mu = 0$, in which case both the threshold and weights may be set to unity as suggested in the text.*

Learning Rule 4. *Winnow2:* **Winnow.**

Initialisation:

$$T_i := 1;$$
$$w_{ij} := 1$$

Weight Update:

$$\begin{aligned}&\textit{if } (t_i = 0) \wedge (h_i = 1) \textit{ then}\\&\quad \textit{if } (f_j = 1) \textit{ then}\\&\quad\quad w_{ij} := w_{ij}/\zeta\\&\textit{if } (t_i = 1) \wedge (h_i = 0) \textit{ then}\\&\quad \textit{if } (f_j = 1) \textit{ then}\\&\quad\quad w_{ij} := \zeta w_{ij}\end{aligned}$$

Weight Bounds: *Not required.*

Threshold Update:

$$T_i := T_i + \mu(h_i(x) - t_i(x)).$$

Comment: *The initialisation given here is suitable for disjunction problems, and may be slow for other tasks. As noted previously,* Winnow2 *is more sensitive to the choice of threshold and learning rate than* Winnow1.

The application of these rules is considered in the following sections.

4.7 Simulation Results

4.7.1 Introduction

In the remaining sections of this chapter we explore the application of the sparse random network approach to a variety of applications, ranging from elementary Boolean concepts of moderate degree through to well known benchmarks from the UCI repository. As hinted in the introduction to this chapter, our goals are twofold:

- To explore the range of elementary computations realisable under consraints of sparsity and plausible learning rules; and
- To assess the performance of the random knowledge base approach when applied to standard benchmark problems.

In some sense these apparently disparate goals are unified by the converging constraints approach to cognitive computation, recognition that the computational sciences may make profound contributions to elucidating both neural function and the underlying mechanisms of information processing – almost independent of the contemporary state of neural models. The structure of the following sections reflects this ideal. Where possible, and particularly for elementary Boolean concepts, we examine related versions of the same problem, relying initially upon the more plausible *CSHebb* update, resorting to more powerful techniques such as the *Winnow* rules only as demanded by problem size. Similarly, while we always respect the sparse connectivity and low bandwidth constraints of Chapter 3, we may at times adopt differing threshold and weight distributions across layers in order to facilitate bindings of differing degree. For the more application-oriented tasks, we shall not scruple to relax some of the more demanding of these constraints, but even these experiments act to bound the range of computations realisable in neural systems. Each computational success obtained through a modest relaxation of neural constraints may demonstrate a surprising learnability, identifying those constraints of most significance, and fostering an exploration for equivalent circuits and updates which better respect the impositions of nature.

Not all of the learning rules of the previous section feature prominently in the sequel. While the elimination rule is novel – and competitive in many situations – its use may be justified only on computational grounds, and as its performance is inferior to that of the second Littlestone rule (*Winnow*), we shall employ *Elim* only for the demonstration task of $x_1x_2 + x_3x_4$ in Section 4.7.3. In a similar fashion, *SSHebb* is reserved solely for the negative weight architectures of the first half of Section 4.7.4.

Two indicative sets of network conditions from either extreme of the values reported in Chapter 3 are used to demonstrate the range of network conditions:

- ***C3:*** connectivity at 3%; $\alpha = 0.20$, $\xi = R/N = 0.13$;
- ***C16:*** connectivity at 16%; $\alpha = 0.47$, $\xi = 0.25$.

The approach to reporting simulation results differs here from that common in machine learning, principally to take account of the variations in network structure across realisations. For the elementary Boolean concepts and k-DNF problems we shall report the range of successes and failures, for a representative number of trials. More conventional criteria are adopted for the standard benchmark problems, and the precise details are provided in the relevant sections.

Initially, we shall demonstrate the power of our methods through the acquisition of highly unbalanced concepts, n_0-input conjunctions and disjunctions for which, respectively, negative examples greatly outweigh the positive examples and vice versa. For such concepts, the notion of generalisation is virtually meaningless, and the question is whether correlation-based updates may overcome the contradictory pressures of the threshold adjustment, which drives the hypothesis towards a constant output consistent with the majority.

These initial investigations are followed by low-dimensional disjunctive normal form or DNF problems, highlighting one approach to the specialisation of the knowledge base to particular problem domains. We subsequently consider the vexed concepts of *XOR* and low-dimensional *PARITY*, which were long the bane of the connectionist community and are especially problematic for the random network approach with increasing degree. The questions of generalisation and tolerance in the presence of noise are explored subsequently through the well-known *LED* and *Monk's* problems, drawn from the UCI machine learning benchmark database (Asuncion and Newman, 2007).

4.7.2 Conjunctions and Disjunctions

Our approach to multiple input conjunction and disjunction problems is hierarchical in nature, a view which is both philosophically consistent and pragmatic. As we have seen, direct recruitment of nodes of substantial in-degree is absurdly improbable, and we have thus made judicious use of threshold and weight distributional bias to ensure the existence of suitable *intermediate* level knowledge. Thus, it is especially important to provide, say, binary and quaternary conjunctions if we wish to recruit conjunctions of degree 6 or higher.

Due to the potential variation in basis functions, the selection of a fixed threshold for the output layer is inappropriate, and the update mechanism must be allowed to choose a value appropriate for a solution in terms of the knowledge provided. Boolean concept recruitment is further facilitated through the choice of the rounded sigmoid, $\sigma_R(y)$, as activation function for both the forward and output layers of the network.

We begin by examining conjunctions and disjunctions which may be realised with the simplest of random networks, those in which the knowledge base is derived through connections from the inputs to a single *forward* layer, with weight adjustment taking place on the linkages from this forward layer to the concept output. Subsequently, we shall incorporate one or more hidden layers in the network in order to investigate their effect on the representation of more complex knowledge.

The simulation networks of this section share the following properties:

- A centrally positioned input layer of size $n_0 = 4, 6, 8, \ldots$;
- A centrally positioned single-unit output or concept layer;
- A *forward* layer containing $n = 100$ or 200 units and approximately $n\xi\alpha$ feeder units[18];
- Zero or more hidden layers with similar size and properties to the forward layer;

[18] While some of the concepts considered may not require a forward layer of this size, a choice of at least $n = 100$ is sufficient to avoid end effects without unduly limiting performance.

- Connection parameters at the $C3$ or $C16$ levels – representing the extremes of our allowable range; and
- One or more of the following learning rules:
  The *CSHebb* update with learning rates of $\zeta = 0.0001$ for the weights, and $\mu \in \{0.01, 0.1\}$ for the thresholds;
  The *Winnow* update, with $\zeta = 2$ and $\mu \in \{0.0, 0.1\}$.

Note that all network layers other than the output or concept layer should be regarded as part of the random black box of the preceding chapter. In the tables of simulation results, we shall report the number of presentation epochs required for convergence of the weight adjustment between this black box and the concept layer. Usually, this convergence is rapid, and occasionally for simpler concepts the network initialisation is sufficient to realise the concept without weight adjustment. In this case we shall report the unusual figure of 0 epochs. At the other extreme, ∞ indicates a failure to converge.

4.7.2.1 Conjunctions of degree 4,6 and 8

The 4-input disjunction is a problem handled with little difficulty at the $C16$ level but less successfully treated at the $C3$ level (a constant output of 0 being obtained). The stopping criterion is here a bit error of zero, while unsuccessful training is terminated after 1000 epochs (∞). Successful convergence was extremely rapid in all cases.

Marginally inferior performance is observed for *6AND*, the intermediate knowledge remaining adequate for the task. A further increase in problem complexity as far as the 8-input conjunction suggests the use of parameters favouring intermediate knowledge of degree 6 or more, but there is little point to this strategy, such nodes being improbable even at the $C16$ level. While some success for the 8-input problem is achievable with the same network, (table 4.9), a (potentially) more extensible approach is provided by the *inter-binding* network considered shortly.

Table 4.9. Results for the 8-input conjunction problem for the *CSHebb* Learning Rule, employing the same networks as before.

Network Architecture:	Input $\rightarrow$ Fwd $\rightarrow$ Output
Forward Layer:	$n = 100$; $W = 1$, $T \in \{2, 4\}$; $\sigma_R(y)$;
Learning:	CSHebb; $\zeta = 0.0001$; $\mu = 0.1$;

Connectivity	Seed #	Bit Error	RMS Error	Epochs
$C16$	1	0	0.00	6
$C16$	2	1	$3.91E-3$	∞
$C16$	3	0	0.00	2
$C16$	4	0	0.00	5
$C16$	5	0	0.00	11

4.7.2.2 Disjunctions of Degree 4, 6 and 8

The recruitment of relatively low-degree disjunctions (again exemplified by the 4 and 6 input problems) is similarly straightforward, although in this case the control of intermediate level knowledge is entirely in the hands of the connectivity regime.

Tables 4.10 and 4.11 show, respectively, 'learning' results for the 4 and 6-input problems. While some success is possible at the *C3* level, *C16* networks are consistently initialised to states which implement the required disjunctions, although one may expect additional, potentially less desirable concepts to also be realised. This simple approach can also be applied to 8, 12, 16 and even 20 input problems – see table 4.12 – with *no* failures at the *C16* level, although no successes were obtained at the *C3* level for $n_0 > 8$.

Table 4.10. Results for the 4-input disjunction problem for the *CSHebb* Learning Rule.

Network Architecture:	Input → Fwd → Output			
Forward Layer:	$n = 100$; $W = 1$, $T = 1$; $\sigma_R(y)$;			
Learning:	CSHebb; $\zeta = 0.0001$; $\mu = 0.1$;			
Connectivity	Seed #	Bit Error	RMS Error	Epochs
$C3$	1	0	0.00	1
$C3$	2	1	$6.25E-2$	∞
$C3$	3	1	$6.25E-2$	∞
$C3$	4	0	0.00	0
$C3$	5	1	$6.25E-2$	∞
$C16$	1	0	0.00	0
$C16$	2	0	0.00	0
$C16$	3	0	0.00	0
$C16$	4	1	0.00	0
$C16$	5	1	0.00	0

While the successes of these sections are impressive on their own terms, one might expect that more complex binding problems over far larger domains may be assisted by the provision of further intermediate knowledge through one or more hidden layers. One approach – inspired by Feldman's random interconnection network – is considered in the next section.

4.7.2.3 The Random 'Inter-Binding' Network

As noted in Chapter 3, Feldman's random interconnection network (Feldman, 1982) was introduced as a mechanism for ensuring, with high confidence, the

Table 4.11. Results for the 6-input disjunction problem for the *CSHebb* Learning Rule.

Network Architecture:		Input $\rightarrow$ Fwd $\rightarrow$ Output		
Forward Layer:		$n = 100$; $W = 1$, $T = 1$; $\sigma_R(y)$;		
Learning:		CSHebb; $\zeta = 0.0001$; $\mu = 0.1$;		
Connectivity	Seed #	Bit Error	RMS Error	Epochs
$C3$	1	1	$1.56E-2$	∞
$C3$	2	0	0.00	0
$C3$	3	1	$1.56E-2$	∞
$C3$	4	1	$1.56E-2$	∞
$C3$	5	1	$1.56E-2$	∞
$C16$	1	0	0.00	0
$C16$	2	0	0.00	0
$C16$	3	0	0.00	0
$C16$	4	0	0.00	0
$C16$	5	0	0.00	0

Table 4.12. Results for the 8, 12 and 16-input disjunction problems with no hidden layers.

Network Architecture:		Input $\rightarrow$ Fwd $\rightarrow$ Output		
Forward Layer:		$n = 100$; $W = 1$, $T \in \{2, 4\}$; $\sigma_R(y)$;		
Problem:		8-OR		
Connectivity	Seed #	Bit Error	RMS Error	Epochs
$C3$	2	0	0.00	0
$C16$	$1-5$	0	0.00	0
Problem:		12-OR		
Connectivity	Seed #	Bit Error	RMS Error	Epochs
$C16$	$1-5$	0	0.00	0
Problem:		16-OR		
Connectivity	Seed #	Bit Error	RMS Error	Epochs
$C16$	$1-5$	0	0.00	0
Problem:		20-OR		
Connectivity	Seed #	Bit Error	RMS Error	Epochs
$C16$	$1-5$	0	0.00	0

existence of at least one connection path between selected nodes at either end of a locally sparse multi-layer network. Our present needs are similar, in that we wish to construct a representation of a concept of large in-degree in terms of intermediate nodes whose in-degree is in practice bounded by a low probability of connection. In principle, we may adopt this strategy for both higher-order conjunctions and disjunctions, although initially we shall examine the former class of problems.

Thus, for tasks such as the 8 input AND considered earlier, we must choose radius and threshold parameters to favour 2 (and perhaps 4) input conjunctions at each layer of the network, providing at most $\lceil \log n_0 \rceil$ layers if the concept has n_0 inputs.

The method is illustrated by the results of table 4.13 for the 8-input conjunction, with rapid convergence to the correct solution in the majority of cases. The training failures for this task indicate the fundamental tension between the default generalisation as prescribed through the data – that to a constant output of 0 – and the likely network configurations as determined through network distributional bias – that of a composition of conjunctions. Attempts to extend this approach to 12 and 16 input conjunctions (with significantly larger layer sizes of 200 and 500) were unsuccessful in the presence of the entire training set – the single positive example being insufficient to overcome the weight of negative examples. While 'generalisation' to the concept may occur if the number of negative examples is reduced, such performance depends trivially upon exposure to the single positive example and no systematic analysis has been undertaken.

Application of the approach to multiple input disjunctions is again more successful, with a happy coincidence between default generalisation and architectural bias. Table 4.14 summarises results for the 8, 12 and 16 input disjunction problems under the interbinding approach, for layer sizes of $n = 200$ and the $Winnow$ learning rule[19]. Note the rapid convergence to the solution in each case.

Table 4.13. Results for the 8-input conjunction problem for the *CSHebb* Learning Rule, employing the inter-binding approach.

Network Architecture:		Input $\rightarrow$ Hid $\rightarrow$ Fwd $\rightarrow$ Output		
Forward Layer:		$n = 100$; $W = 1$, $T \in \{2, 4\}$; $\sigma_R(y)$;		
Hidden Layer:		$n = 100$; $W = 1$, $T \in \{2, 4\}$; $\sigma(y)$;		
Learning:		CSHebb; $\zeta = 0.0001$; $\mu = 0.1$;		
Connectivity	Seed #	Bit Error	RMS Error	Epochs
$C16$	1	1	$3.91E-3$	∞
$C16$	2	0	0.00	5
$C16$	3	0	0.00	8
$C16$	4	0	0.00	4
$C16$	5	1	$3.91E-3$	∞

19 In the trials undertaken, a three layer network was sufficient for the higher degree problems.

Table 4.14. Results for the 8, 12 and 16-input disjunction problems under the interbinding approach.

Network Architecture:		Input → Hid → Fwd → Output		
Forward Layer:		$n = 200$; $W = 1$, $T = 1$; $\sigma_R(y)$;		
Hidden Layer:		$n = 200$; $W = 1$, $T = 1$; $\sigma_R(y)$;		
Learning:		Winnow; $\zeta = 2.0$; $\mu = 0.0$;		
Problem:		8-OR		
Connectivity	Seed #	Bit Error	RMS Error	Epochs
$C16$	$1-5$	0	0.00	7
Problem:		12-OR		
Connectivity	Seed #	Bit Error	RMS Error	Epochs
$C16$	$1-5$	0	0.00	7
Problem:		16-OR		
Connectivity	Seed #	Bit Error	RMS Error	Epochs
$C16$	$1-5$	0	0.00	7

While it is tempting to conclude that we may represent disjunctions of progressively higher degree through the introduction of extra hidden layers, there are substantial performance penalties involved in this process, with each additional hidden layer contributing as much as 90% of the single layer execution time. While the existence results cited in Section 4.3 suggest that additional layers offer no representational advantage over a larger single layer approach, there is some attraction in the notion that the recruitment task may be simplified if members of the feeder set realise high-order disjunctions through the use of additional hidden layers.

However, our experiments over both conjunctions and disjunctions suggest that more reliable recruitment may be obtained through the action of the learning rule over a feeder space whose members realise low-order functions over some subset of the inputs – especially given algorithms such as *Winnow*, which are provably robust in the presence of irrelevant attributes. This conclusion leads us naturally to identify a class of concepts for which our networks may perform remarkably well, the k-DNF expressions of the following section.

4.7.3 k-DNF Expressions for Small k

In this section we consider the specialisation of the network to accommodate k-DNF expressions, concepts which may be written as the disjunction of logical terms (conjunctions) of at most k of the input variables. Our initial intuitions favoured the construction of a network with one hidden layer (biased toward the representation of conjunctions) feeding to the forward layer (biased toward disjunctions) with selection across the forward layer accomplished as before.

Again, however, results with this multi-layer approach proved disappointing, and more reliable and rapid recruitment may be accomplished through the use of a conjunction-biased forward layer, coupled with output weights and threshold biased toward a disjunction over the feeder space. Recruitment then proceeds by *Winnowing* the relevant from irrelevant attributes, with convergence normally achieved *very* rapidly.

Here, we demonstrate the approach through the tasks $x_1x_2 + x_3$ and $x_1x_2 + x_3x_4$, both of which are achievable with our minimal 100 node forward layer. Input weights are set to unity, with forward layer thresholds drawn randomly from $\{2, 4\}$, in anticipation of the larger problems of subsequent sections. Table 4.15 shows learning results for the former problem at both the $C3$ and $C16$ connectivity levels, with convergence occurring in all cases within a handful of epochs. The fractionally more complex task of $x_1x_2 + x_3x_4$ (table 4.16), provides a greater challenge at the $C3$ level, but is again comfortably handled at the $C16$ level with the *Winnow* update.

This latter problem also provides a good illustration of the strengths and weaknesses of the *Elim* rule (table 4.17). While this update converges *extremely* rapidly – typically within a single epoch – it will seldom deliver perfect memorisation, with the excision step almost inevitably resulting in knowledge wastage. With the exception of seed #5, for which the network oscillates between 7 false negatives and 9 false positives, in each case the minimum error hypothesis is obtained within one epoch, and we report a small bit error due entirely to false negatives. Evidently, as may be seen from table 4.16, the *Winnow* update is more successful at utilising mildly erroneous feeder concepts, limiting their detrimental effect and avoiding false positives while retaining their services in representing positive network responses on other patterns.

The next section examines the *XOR* and *PARITY* problems, the larger *LED* and *Monk's* problems being deferred until later. While the *PARITY* problems are included primarily to illustrate the limitations of the random network approach, each task may also be expressed as a larger problem over its inputs and their respective complements, thus providing an intermediate level challenge for the system.

4.7.4 XOR *and* PARITY *Problems*

The *XOR* and *PARITY* problems pose significant representational difficulties, and are a foretaste of the *Monk II* problem later in the chapter. Moreover, the problems are of some historical interest.

Table 4.15. Results for the 3-input DNF problem $x_1x_2+x_3$, for both the $C3$ and $C16$ connectivity regimes.

Network Architecture:		Input → Fwd → Output		
Forward Layer:		$n = 100$; $W = 1$, $T \in \{2, 4\}$; $\sigma_R(y)$;		
Learning:		Winnow; $\zeta = 2.0$; $\mu = 0.0$;		
Connectivity	Seed #	Bit Error	RMS Error	Epochs
$C3$	1	0	0.00	2
$C3$	2	0	0.00	2
$C3$	3	0	0.00	2
$C3$	4	0	0.00	3
$C3$	5	0	0.00	3
$C16$	1	0	0.00	4
$C16$	2	0	0.00	3
$C16$	3	0	0.00	4
$C16$	4	0	0.00	4
$C16$	5	0	0.00	3

Table 4.16. Results for the 4-input DNF problem $x_1x_2 + x_3x_4$, for both the $C3$ and $C16$ connectivity regimes.

Network Architecture:		Input → Fwd → Output		
Forward Layer:		$n = 100$; $W = 1$, $T \in \{2, 4\}$; $\sigma_R(y)$;		
Learning:		Winnow; $\zeta = 2.0$; $\mu = 0.0$;		
Connectivity	Seed #	Bit Error	RMS Error	Epochs
$C3$	1	0	0.00	2
$C3$	2	0	0.00	2
$C3$	3	3	$1.08E-1$	∞
$C3$	4	3	$1.08E-1$	∞
$C3$	5	2	$8.84E-2$	∞
$C16$	1	0	0.00	3
$C16$	2	0	0.00	3
$C16$	3	0	0.00	3
$C16$	4	0	0.00	3
$C16$	5	0	0.00	3

Table 4.17. Results for the 4-input *DNF* problem $x_1x_2 + x_3x_4$ using the *Elim* update, at the $C16$ connectivity level.

Network Architecture:		Input $\rightarrow$ Fwd $\rightarrow$ Output		
Forward Layer:		$n = 100$; $W = 1$, $T \in \{2, 4\}$; $\sigma_R(y)$;		
Learning:		Elim; $\zeta = 0.0001$; $\mu = 0.01$;		
Connectivity	Seed #	Bit Error	RMS Error	Epochs
$C16$	1	1	$6.25E - 2$	1
$C16$	2	2	$8.84E - 2$	1
$C16$	3	3	$1.08E - 1$	1
$C16$	4	2	$8.84E - 2$	1
$C16$	5	$7 - 9$	$1.5 - 1.8E - 1$	∞

As remarked earlier, the *XOR* problem had a principal role in Minsky and Papert's devastation of the connectionist enterprise, and is thus a natural test problem for any new learning system, and such tests are commonly extended to include *PARITY* problems of higher dimension.

Two approaches were considered in the present work, exemplified respectively by the *HU1* (low number of hidden units; shortcut connections between input and output layers) and *HU2* (larger number of hidden units; strictly feedforward connections) architectures for the *XOR* problem introduced in Chapter 3. Learning performance was generally poor, requiring very substantial network resources and extensive training.

Reliable recruitment with the *HU1* architecture requires layer sizes of at least 1000, training for between 250 and 2500 epochs, and the use of a input assemblies of size 5 and 10. While similar experiments with the 4 and 6-input *PARITY* tasks produce occasional successes, these are achievable only with very large network layers ($n > 5000$), and with careful attention to the network connectivity.

Subsequent simulations inspired by the *HU2 XOR* architecture fostered 2-input conjunctions and disjunctions at the forward layer were similarly unsuccessful, even with layer sizes of up to 1000.

However, some modest success is achievable if the problem is transformed to one over the 8 inputs $\{(x_i, \bar{x}_i), i = 1, \ldots, 4\}$. A trial run of a similarly transformed *XOR* problem yields recruitment in 4 epochs for every trial at the $C16$ level, but the 4-*PARITY* task is limited to two successes, as may be seen in table 4.18.

However, extensions of this approach to the 6-input problem are not successful and the computational evidence confirms our earlier speculation that the random network approach is not suited to the representation of higher-order *PARITY*-like tasks.

The next sections consider problems for which we may have a greater expectation of success, the well-known *LED* and *Monk's* benchmarking tasks,

Table 4.18. Results for the transformed 4-*PARITY* problem at the $C16$ connectivity level.

Network Architecture:	Input → Fwd → Output			
Forward Layer:	$n = 100$; $W = 1$, $T \in \{2, 4\}$; $\sigma_R(y)$;			
Learning:	Winnow; $\zeta = 2.0$; $\mu = 0.01$;			
Connectivity	Seed #	Bit Error	RMS Error	Epochs
$C16$	1	7	$1.65E-1$	∞
$C16$	2	0	0.00	33
$C16$	3	0	0.00	10
$C16$	4	4	$1.25E-1$	∞
$C16$	5	7	$1.65E-1$	∞

although successes in the latter case are limited to the first and third problems – with the *Monk II* experiments (Section 4.7.7) providing further support for the conclusions of the previous paragraph.

4.7.5 Baseline Simulations for the LED *Problems*

The *LED* classification problems were introduced by Breiman et al (1984), with data generation programmes later made available by Aha (Asuncion and Newman, 2007). The task is based on the familiar light emitting diode (or liquid crystal) representation of the decimal digits through the presence or absence of a number of horizontal and vertical segments. Figure 4.4 shows the encoding of the digit '5', with the segment labels corresponding to the order of the attributes, giving '5' the vector representation 1101011. In our approach, the input domain is again expanded to include explicitly the complement of each variable, leading to 14 network inputs[20].

This task is usually made substantially more difficult through the introduction of noise in the examples, with segments associated with a digit being flipped randomly with some prescribed probability (5% and 10% are common choices). Initially, however, we shall work with uncorrupted data, our task – as in the succeeding treatment of the *Monk I* problem – being to identify network conditions such that the concept may be rapidly and reliably memorised. Such networks then provide a basis for the subsequent investigation of generalisation and fault tolerance (see Section 4.7.9).

Table 4.19 shows learning results at the $C16$ level for the 14-input encoded decimal digits, the network having 10 sparsely-coded outputs. In other respects, the network configuration is similar to that employed in the earlier *DNF* simulations, excepting that the forward layer now consists of 500 nodes. Learning is completely reliable and *extremely* fast in all cases, requiring as few as 10 epochs for convergence.

[20] The corresponding vector for '5' is then 10100110011010.

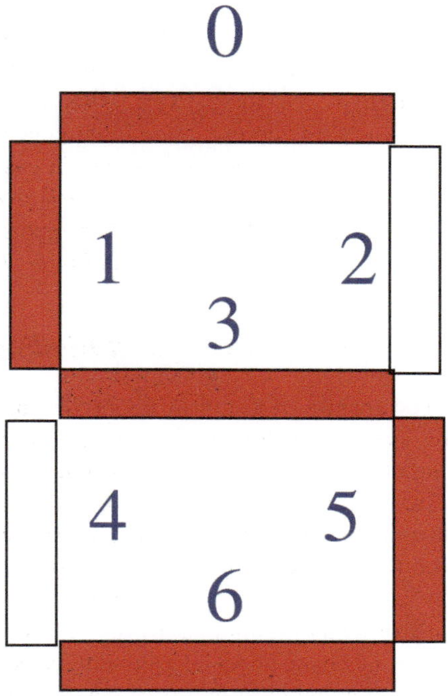

Fig. 4.4. The LED representation of '5'. The numbers indicate the ordering of the segment attributes.

Table 4.19. Results for the input+complement *LED* task at the $C16$ connectivity level.

Network Architecture:		Input $\rightarrow$ Fwd $\rightarrow$ Output		
Forward Layer:		$n = 500$; $W = 1$, $T \in \{2, 4\}$; $\sigma_R(y)$;		
Learning:		Winnow; $\zeta = 2.0$; $\mu = 0.01$;		
Connectivity	Seed #	Bit Error	RMS Error	Epochs
$C16$	1	0	0.00	11
$C16$	2	0	0.00	13
$C16$	3	0	0.00	14
$C16$	4	0	0.00	10
$C16$	5	0	0.00	20

4.7.6 *Baseline Simulations for* Monk I

As noted in Chapter 3.4, the *Monk's* problems were introduced by Thrun and his co-workers (Thrun, 1991) to facilitate comparison between various machine learning algorithms[21]. The first of these problems (*Monk I*) is defined through the expression `(head_shape=body_shape) or (jacket_color=red)`, and is easily represented as a *DNF* expression. Recruitment within the random network context is more difficult than one might expect, however, as the forward layer must support conjunctions between corresponding pairs of the six `head_shape` and `body_shape` nodes, whilst overcoming the potentially deleterious effects of additional inputs. For the DNF networks of this section, reliable performance requires a forward layer size of 5000, although irrelevant conjunctions are suppressed in as little as a single epoch.

Table 4.20 shows the results of simulations at the $C16$ level over the full 432 patterns (equally split between positive and negative examples). These results again demonstrate completely reliable and very rapid recruitment of the concept. While the use of a forward layer of size 5000 appears excessive, performance remains markedly superior to backpropagation based approaches – although modern learning systems such as the Support Vector Machine are more competitive.

Table 4.20. Results for the *Monk I* task at the $C16$ connectivity level.

Network Architecture:		Input $\rightarrow$ Fwd $\rightarrow$ Output		
Forward Layer:		$n = 5000$; $W = 1$, $T \in \{2, 4\}$; $\sigma_R(y)$;		
Learning:		Winnow; $\zeta = 2.0$; $\mu = 0.01$;		
Connectivity	Seed #	Bit Error	RMS Error	Epochs
$C16$	1	0	0.00	6
$C16$	2	0	0.00	1
$C16$	3	0	0.00	3
$C16$	4	0	0.00	9
$C16$	5	0	0.00	11

For the sake of comparison, a number of simulations were undertaken using backpropagation-based training for the $17-2-1$ architecture. Stagnation in a local minimum was frequently observed, with bit errors as high as 72 after $6-700$ epochs ($\zeta = 0.1$) and $1500-3000$ epochs ($\zeta = 0.001$). The best performed trial converged to 100% classification accuracy after 40 epochs, and we note that Thrun (1991) reported 390 epochs for the $17-3-1$ architecture over a smaller, but still balanced, training set. Our simulations demonstrate that the backpropagation algorithm consistently requires many more training

[21] A brief description of the attributes was given in Section 3.4.2.2.

epochs than the present approach, while frequently returning far from optimal solutions. Moreover, weight adjustment is dramatically slower than in the present approach.

However, any optimism that these successes will be repeated on the *Monk II* task is sadly misplaced, as we shall see in the following section.

4.7.7 The Monk II *Problem*

The second of the *Monk's* classification tasks is conceptually similar to the *PARITY*-like problems of Section 4.7.4, being defined through the proposition

"exactly two of the six attributes have their first value",

a description for which an equivalent *DNF* expression is difficult to determine. As in Section 4.7.4, our approach is to expand the input representation through augmentation with the complement of each variable, leading to 34-bit input vectors. Experimentation with this problem proved thoroughly unsuccessful – even with layer sizes of 10 – 20000 – with the network typically supporting an hypothesis of *ZERO*. Such failures are to be expected given the evidence of the earlier section, with the present task being of similar difficulty and of significantly higher dimension.

4.7.8 Baseline Simulations for Monk III

Like the *LED* problems, the third *Monk* problem is a comparatively straightforward task made more challenging through the addition of (mild) classification noise. Initially, however, our objective is to determine a network configuration suitable for the later experimentation, and for this purpose we have employed as our training set the 316 noise-free examples which will constitute the test set in Section 4.7.9.

The underlying task may be expressed through the disjunctive normal form expression `(jacket_color=green and holding_a_sword) or (jacket_color!=blue and body_shape!=octagon)`, and due to the negations in the second term we employ an additional complement node for each of the network inputs. While there are considerable similarities between the network used here and that of Section 4.7.6, early experiments at the $C16$ level proved troublesome, with frequent saturation – due primarily to extra input activation from the complement nodes – and a consistent failure to acquire the target. In contrast, performance at the $C3$ level is extremely good, with rapid acquisition of the concept for each of the seed values (table 4.21), and this latter configuration is used in the simulations of Section 4.7.9.

Table 4.21. Results for the noise-free *Monk III* task at the $C3$ connectivity level.

Network Architecture:		Input → Fwd → Output		
Forward Layer:		$n = 5000$; $W = 1$, $T \in \{2, 4\}$; $\sigma_R(y)$;		
Learning:		Winnow; $\zeta = 2.0$; $\mu = 0.01$;		
Connectivity	Seed #	Bit Error	RMS Error	Epochs
$C3$	1	0	0.00	15
$C3$	2	0	0.00	19
$C3$	3	0	0.00	21
$C3$	4	0	0.00	8
$C3$	5	0	0.00	17

4.7.9 Generalisation and Noise-Tolerance

In this section, we examine the performance of our now familiar random network architectures on the *Monk's* and *LED* problems, with respect to variations in training and test set size, and in the presence of both mild and significant noise levels.

Generalisation performance was examined initially through the *Monk I* problem, with the 432 patterns being randomly partitioned into training sets of size $50 - 400$ in increments of 50, the remaining patterns being used to form the test set in each case. Five data selection trials were performed for each set size, with the whole process being further replicated through the use of the 5 distinct connection and weight seed pairings. The stopping criterion was in all cases a bit error of zero over the training set – coupled with a maximum iteration guard of 50 epochs – which was achieved for all runs within 15 epochs and frequently within 7.

Figure 4.5 shows the resulting learning curves for the problem, exhibiting strong generalisation performance from training sets of size fraction 0.35 or greater. Each seed-specific curve shown is the mean of its 5 selection runs, with the additional (Mean) curve being the mean over these derived figures. Error bars of one standard deviation are shown for each data point of this latter series, with the others suppressed to avoid clutter.

Figure 4.6 provides slightly more insight into the problem, showing an extremely high rate of false positives for the smaller training sets. Such a phenomenon is readily explained by the strong weight and threshold bias toward a disjunction over the entire feeder space, and increased training set size provides the numerous additional negative examples needed to reduce the number of feeders supporting the concept. In contrast, the level of false negative errors remains low throughout the experiments, with a mild decrease as the training set fraction approaches 1.0.

Low level noise is then examined through the *Monk III* problem as presented in Thrun (1991), with the output classification flipped on 6 of the 122 training examples, and the correct version of these patterns added to the remaining 310 examples to form the test set. For four of the five seed selections, the network converged to a representation consistent with the training data (bit error of zero), while the run for seed #3 returned a bit error of 0.09 after 50 epochs. Table 4.22 shows classification performance over the test set, with a mean generalisation error of around 19%, significantly higher than that reported in the original study, albeit after extremely rapid training.

Table 4.22. Generalisation results for the *Monk III* task at the *C*3 connectivity level.

Network Architecture:		Input $\rightarrow$ Fwd $\rightarrow$ Output			
Forward Layer:		$n = 5000$; $W = 1$, $T \in \{2, 4\}$; $\sigma_R(y)$;			
Learning:		Winnow; $\zeta = 2.0$; $\mu = 0.00$;			
Connectivity	Seed #	Class. Error	RMS Error	False +ves	False -ves
$C3$	1	0.15	$2.19E-2$	0.08	0.07
$C3$	2	0.21	$2.57E-2$	0.14	0.07
$C3$	3	0.26	$2.88E-2$	0.13	0.14
$C3$	4	0.19	$2.47E-2$	0.09	0.10
$C3$	5	0.16	$2.24E-2$	0.09	0.07
Mean:		0.19	$2.47E-2$	0.11	0.09
S.D.:		0.04	$2.79E-3$	0.03	0.03

A more substantial test is provided by the *LED* tasks at a segment noise level of 10% i.e. each segment of a pattern may be flipped with probability of 0.1 – although around 9.5% noise was achieved over the 500 examples produced by Aha's data generation code. As in the baseline simulations of Section 4.7.5, we employ 14 network inputs, consisting of the 7 segment attributes paired with their respective complements.

In a similar fashion to the *Monk I* simulations, the data set was randomly partitioned into training sets of sizes 100 to 500 (with an increment of 100), with the remaining patterns forming the test set in each case. Five selection trials were performed for each run, with further replication provided by the five network seed pairings.

Due to the higher level of noise in the *LED* data set, no training run achieved a bit error of zero, with termination in each case at the bound of 100 epochs. Examination of the errors logged at the end of each training epoch strongly suggests that unsuccessful training in the present framework leads to a (bounded) random walk for the error, with the distribution established rapidly and exhibiting low variance. Moreover, analysis of the error

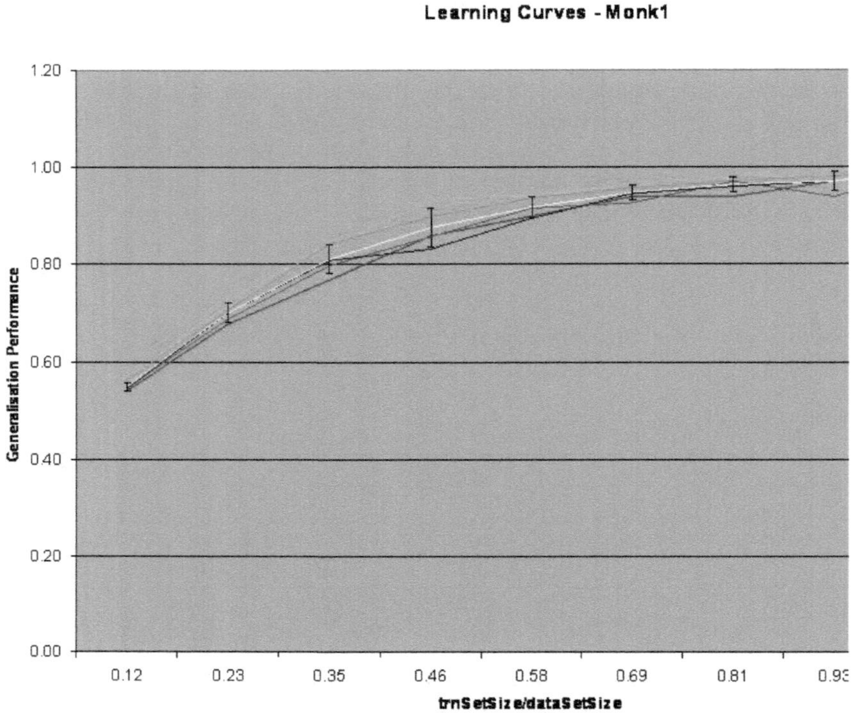

Fig. 4.5. Learning curves for the *Monk I* problem.

series show that the 100 epoch termination error was in all cases within two standard deviations of the mean, and in around half of the trials within one standard deviation[22].

Figure 4.7 shows variations in mean bit error with increasing training set size, normalised with respect to the number of output units (in this case 10). Each curve represents the mean of the 5 seed-specific series, which in turn are derived from 5 random partitions of the data at each training set size. As before, the error bars indicate one standard deviation either side of this mean. Training results indicate a comparatively low error for the smaller set sizes, with the gap between model capacity and the data becoming more pronounced as the size fraction increases. The test set error is particularly high over the smaller sets, and some explanation of this is again provided by a high rate of false positives (Figure 4.8), which decreases markedly as the

[22] The mean in each case was taken over the 100 values logged during training i.e. we exclude the error for the initial network state (epoch 0).

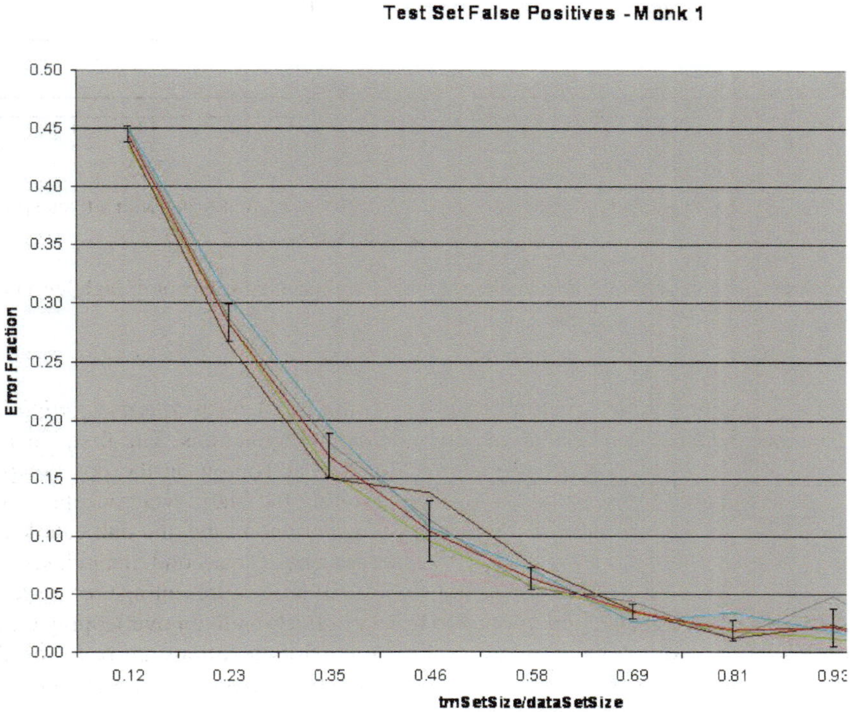

Fig. 4.6. Variations in false positive errors for the *Monk I* problem test sets.

size fraction approaches 1.0, although the overall error remains disturbingly high.

The situation is complicated by the frequent appearance of extra positive classifications on patterns which nevertheless report high activation for the target output. Such patterns show up as false positives, consequently increasing the overall bit error. Closer scrutiny suggests that our *LED* networks fail to report a correct classification for between 30 and 40% of the test examples, with all except the figure for a size fraction of 0.6 lying within a narrow corridor of 35%, a figure 5% – 10% higher than the best reported by Aha (Asuncion and Newman, 2007)[23].

Evidently, however, our approach suffers from a substantial and consistent over-generalisation, as seen in table 4.23. False positive reports in addition to a correct classification occur in 16.7% of the examples, with on the average 1 to 2 erroneous selections accompanying the appropriate bit. While such

[23] Aha summarises results from authors using a variety of approaches, with classification rates between 71 and 73.3%. The optimal Bayes rate for the problem is given as 74%.

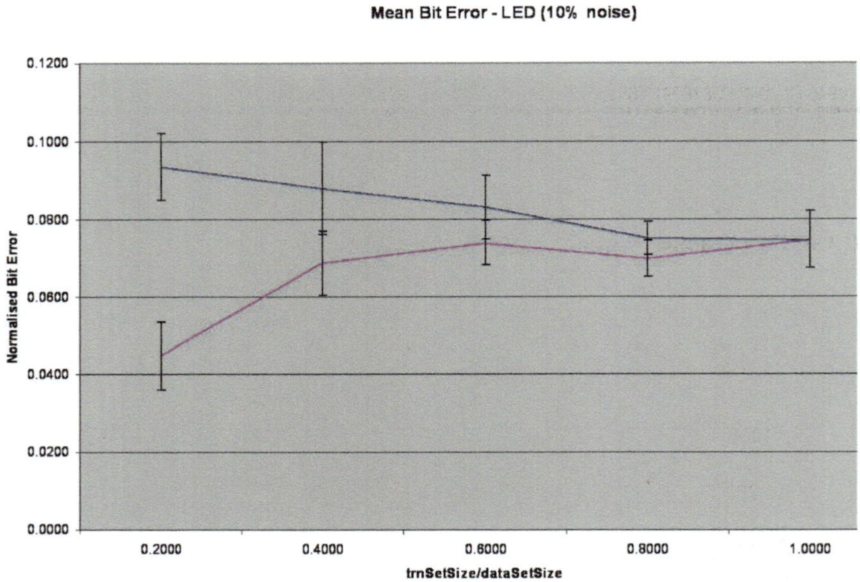

Fig. 4.7. Variations in bit error for the *LED* problem training and test sets.

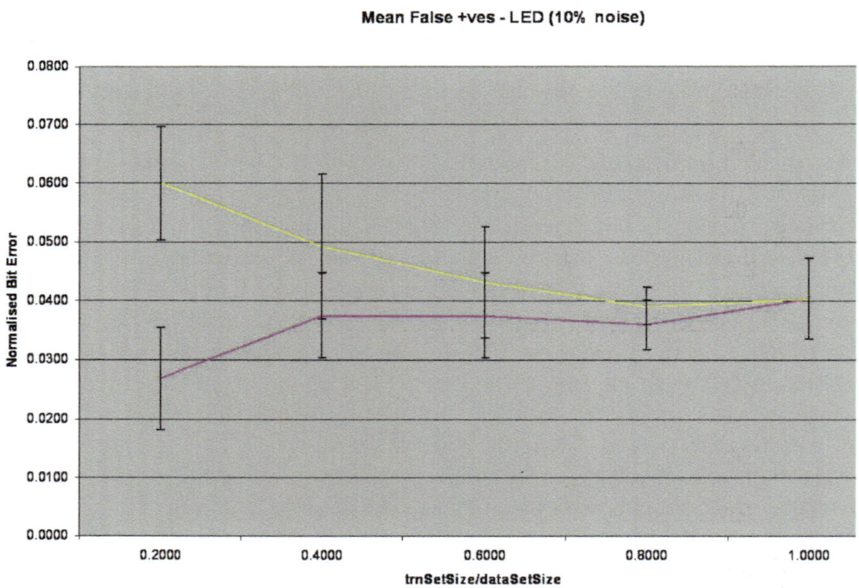

Fig. 4.8. Variations in false positive errors for the *LED* problem training and test sets.

Table 4.23. Generalisation results for the *LED* task at the 10% noise level, showing the rate of erroneous positive classifications.

Size Fraction	Correct Class.	Additional +ve. Class.	Mean False Pos Bits
0.2	0.698	0.168	0.215
0.4	0.630	0.186	0.158
0.6	0.579	0.169	0.124
0.8	0.666	0.144	0.138
Mean :	0.643	0.167	0.159

statistics rule out the trivial over-generalisation of a constant firing response, it is nevertheless a weakness of our framework, and one which is difficult to address while retaining the power of the initial feeder disjunction.

4.8 Adult Neurogenesis

As noted in Section 3.4.1, there is limited cause for optimism that adult neurogenesis – even confirmation of *cortical* neurogenesis – would provide a significant boost to the models introduced here. However, while any discussion of these issues must necessarily remain speculative, a very interesting set of computational models of the effect of hippocampal neurogenesis has emerged since the work of Becker (2005). Notwithstanding the apparent advantages of experimental confirmation, the field is sufficiently immature that even here some caution is warranted, and this is reflected in the nature of the studies and in the careful discussion of their findings.

Central to this work is the hypothesis that pools of immature neurons act in some way to extend the capacity of the hippocampus to encode new memories while protecting against the disruption of the old. Both Becker (2005) and Aimone and Wiskott (2008) explore this effect as an increase in capacity through a corresponding increase in the dimensionality of an associative memory representation. In this light, the effect of new neurons is critical, with their absence predicting some decay or washout of existing kowledge. However, as noted by Aimone et al (2009), and discussed in the concludimg chapter, the experimental evidence from knockdown studies provides mixed evidence for adult neurogenesis as an essential pre-requisite for learning and memory. The picture painted in this recent study of a dentate gyrus network is more complex, suggesting an important role for 'adult born' neurons in memory formation through encoding of similarity of memories acquired at the same time, while supporting a distinction between memories in familiar and novel contexts.

Of particular relevance for recruitment is the variation in neuronal and synaptic characteristics during the maturation process, with observations including a lower threshold and higher amplitude LTP (Ge et al (2007), Schmidt-Hieber et al (2004), reviewed in Aimone et al (2009)). The simulations of this chapter are necessarily a connectionist abstraction of neural processes, with a particular focus on a single *goal* or concept unit as the locus of plasticity. In this context, additional neurons plainly allow representation and learning of additional concepts, but in the present architecture, this is equivalent only to maintaining n learning systems instead of one. Of far more interest is the provision of one or more new neurons to an established block of units engaged in competition, with a range of units victorious in response to associated sets of input patterns. In this sense, the units act as a set of discriminatory pattern detectors, with task-specific weight adaptation emerging over time.

The introduction of a new, highly plastic and excitable neuron to the circuit is thus an attractive mechanism for the realisation of additional concepts without the destruction of the more settled competitive layer. Above all, the greater plasticity and additional excitation provided in the early phases of neural life may indeed be expected to ensure that the new unit is *competitive* as a new pattern is established. Given the srength of this initial plasticity, its decline with time to more mature levels, and the complex interplay between the set of units as the dynamics evolve, these ideas offer an intriguing avenue for computational simulation – even if, as for all other explorations of neurogenesis, we must remain a little cautious.

We shall revisit the question of adult neurogenesis in recruitment in the concluding discussion of Chapter 9.

4.9 Discussion

In Section 4.3.1, we examined the potential for over-generalisation through collateral recruitment of feeder concepts f_j whose firing is highly correlated with the target over the training examples, but which exhibit a substantially higher firing rate than the target over the generalisation set. In particular, we noted the the risk posed by such feeders whose training set firing is correlated with some constituent conjunction of the target, and the likelihood that such nodes would be captured through learning rules which naturally favour acquisition of the constituent.

While we do employ the *CSHebb* update in simulations involving the benchmark problems, such a situation is a good illustration of the failings of this rule identified in the analysis of Section 4.4.4. The primary weakness of the *CSHebb* update lies in the extent to which weight change is governed by the target firings, with no penalty exacted for the appearance of a false positive with respect to the training set. As the relevant weight will converge to

$$\frac{E[f_j|t_i = 1]}{E[t_i = 1]},$$

over-generalisation is virtually assured. In contrast, under the *Winnow* update, a preponderance of false positives i.e.

$$E[f_j|t_i = 0] \gg E[f_j|t_i = 1],$$

will ensure that the node is discarded from the hypothesis.

Notwithstanding its superiority over the correlation update, *Winnow* is not immune to the problems identified above. Only careful management of the conjunction layer – in particular ensuring that 'desirable' conjunctions exist at an entropic maximum within the Hamming space defined over their positive examples – will allow this risk to be minimised.

Such control is relatively straightforward for Boolean concepts over the entire domain $\{0, 1\}^n$, as each distinct feeder conjunction f must differ from its competitors in at least one input, resulting in a normalised Hamming distance of at least 0.5, if the calculation is taken over the complete set of positive examples of f. However, this comfortable result is destroyed if we expand the inputs to include their complements, and this risk of over-generalisation appears to be the unavoidable price for the additional representational capability.

The situation is less easily determined if the input is sparse coded, although there is little evidence that the issue is troublesome for the *Monk I* problem, with strong generalisation exhibited once sufficient negative examples appear in the training set to refine the initial hypothesis. Application to the *Monk III* and *LED* problems – in which the expanded input layer is coupled with, respectively, low and high levels of pattern noise – leads to a different outcome, the dominance of false positive errors of the *Monk I* simulations being replaced with a regime in which comparable levels of false positive and false negative errors are observed, probably reflecting high activations associated with the complemented inputs.

In the present chapter we have seen the successful application of our approach to the important domain of DNF-based Boolean knowledge, while its limitations as a general purpose classifier have been identified through higher degree*PARITY* and the related problem of *Monk II*. Within a purely cognitive domain, however, our approach provides a very useful framework for investigation of learning and knowledge representation – especially for the qualitative (and perhaps ultimately quantitative) simulation of developmental processes such as early language acquisition. The application of our network models to such a domain is the subject of the final chapter of Part I of this volume.

Chapter 5
Cognitive Applications

James M. Hogan

5.1 Introduction

In the final chapter of Part I of this book, we examine the application of connectionist models of recruitment to problems at the core of cognitive science, the integration of visual and linguistic information and the acquisition of an elementary semantics of space and facial expressions. Most significantly, the tasks of this chapter transcend barriers of race and culture, forcing computational systems to confront deeper issues of representation and generalisation. In this respect they lie firmly within the framework of the L_0 project of Feldman et al (1990), the *"touchstone task for cognitive science"* devised in response to the excessively domain specific computational models of cognition prevalent at that time.

The connectionist models of the 1980s provided some important insights into human cognitive behaviour, and some impressively natural generalisation within restricted domains. Yet there remained an awkward suspicion that some of the more difficult aspects of representation and concept acquisition had been abstracted away, and with them the essence of the cognitive phenomena being modelled. L_0, a grounded early language acquisition task, demanded computational mechanisms which could operate independently of linguistic convention, even if some specialisation might be required for a particular language.

In nature, the semantics of spatial relations is grounded in the perception of the visual environment, and computational learning systems gain in plausibility by respecting neural mechanisms of visual processing. In subsequent sections, we shall outline a structured connectionist model assembled from abstract versions of these components, each emerging from random networks of the class described in the preceding chapters. Random prior knowledge – especially at the level of structure – underpins the binding of disparate inputs, the development of receptive field representations of 'retinal' images, and encoding of sequences of attended objects within the various input scenes. In

J. Diederich, C. Günay, and J.M. Hogan: Recruitment Learning, SCI 303, pp. 137–179.
springerlink.com

contrast to the test problems previously encountered, much of the information processing is here devoted to the *binding* of transient representations – in essence a large-scale conjunction – in order to facilitate acquisition of an encoded semantics. This representation allows us to extend the earlier feed-forward network structures to encompass the simple recurrent network architecture of Elman (1990), and to investigate the existence of random *context* able to support the representation of regular languages. As outlined below, acquisition of the semantics of each relation is here equivalent to inducing a restricted visual grammar.

Spatial relations, such as the English *above*, are encoded through the relative position of a (possibly) mobile object called the *trajector* (TR) and a static reference object called the *landmark* (LM) (Langacker, 1987)[1]. The L_0 task, in its original form, was to construct a computer system to perform Miniature Language Acquisition, without reliance upon "forthcoming results in related domains" to resuscitate an otherwise inadequate model:

> *The system is given examples of pictures paired with true statements about those pictures in an arbitrary natural language.*
> *The system is to learn the relevant portion of the language well enough so that given a novel sentence of that language, it can determine whether or not the sentence is true of the accompanying picture.*

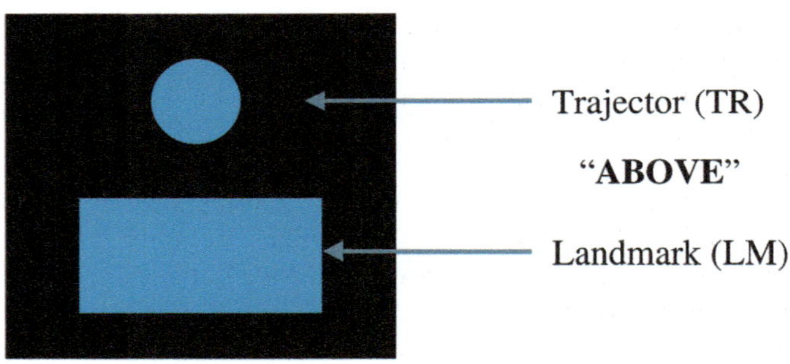

Fig. 5.1. Example image for the L_0 task, with which one might associate the English fragment "circle above rectangle".

The L_0 spatial relations task – first considered by Regier (1992) – requires that the model system learn to associate pictures such as Figure 5.1 with the appropriate linguistic term, such as the English lexeme 'above'. In essence, spatial semantics defines a partitioning of a set of object pictures of this type into classes prescribed by the underlying natural language. The model system

[1] In line with the L_0 conventions, we here adopt the terminology of the eminent cognitive linguist Ronald Langacker. The alternative terminology of *figure* and *ground* is often employed.

is then to learn this classification from positive examples (Chomsky, 1965) of these categories, which in English might include the static concepts: *above, below, left, right, in, out, on*, and *off;* and their dynamic equivalents: *above, below, left, right, around, in(to), onto, out of, through* and *over*. However, the situation is more complex, as some gradation of membership is desirable, particularly as some scenes are more *prototypical* exemplars of a given relation than others (Labov, 1973).

In the models which follow, the binding of feature and location is accomplished through time by the mechanisms of selective visual attention (see for example Niebur and Koch (1998)), the process which allows the human visual system to focus on the more *salient* aspects of the 10^8-10^9 bits per second (Olshausen and Koch, 1995) which are made available by the optic nerve. In our model, transitions in attentional focus effectively partition the scene into a series of image frames, each of which may be categorised according to the focus and the objects which exist 'in shot'. In essence, the binding realised encodes a symbol associated with the frame. The semantics of the spatial relation is then captured through a *frame grammar*, with positive examples of the concept corresponding to an admissible 'string' of symbols of the appropriate types. Prototypical examples allow a cleaner decision for acceptance, and so the model is able to capture some of the uncertainty inherent in the relation. Significantly, operating through attentional transitions also allows a unified treatment of the range of concepts potentially supported by the same static scene. Our prototypical example of Figure 5.1, for example, may be regarded as a prototypical example of both *above* ("Circle above Rectangle"; *TR=Circle; LM=Rectangle*) and *below* ("Rectangle below Circle"; *TR=Rectangle; LM=Circle*). The 'correct' selection is seen in the transitions of Figure 5.2, with the sequence of frames defining the simplest string accepted by the *above* grammar.

Cartoon inputs of this type significantly simplify the visual processing task, and the system need not consider feature-laden scene backgrounds and object interiors, confusion due to distractors, nor the role of luminance and colour. Nevertheless, the task remains far from trivial, particularly as we proceed from static relations such as *above* to more complex dynamic concepts such as *over*. Substantial variations in learnability are observable between relations, and the binding of representations of form and location remains complex.

It is well-accepted that perceptual representations in primates rely upon independent encodings of object features and properties in distinct anatomical areas, and that some mechanism is then required to associate or *bind* the representations together to facilitate processing of a particular object instantiation (Treisman, 1996). This observation is best illustrated by the separation of the object recognition (variously the 'what'/'object'/'ventral'/'occipito-temporal') and location ('where'/'spatial'/'dorsal'/'occipto-parietal') pathways of the visual system. Some linkage must thus be provided between object identification and location if spatially based semantics are to be encoded and processed.

Treisman's Feature Integration Theory (Treisman and Gelade, 1980) was introduced to explain the appearance of illusory conjunctions (i.e. cross-talk) in simple perceptual tasks when experimental conditions were chosen so as to limit the efficacy of selective attention. The mechanism proposed utilises a global location map as a common reference point to link disparate representations – activations corresponding to the attended object remain undiminished, whereas those away from this region are suppressed. It is our contention that transient enhancement corresponding to a selected object (and simultaneous inhibition of the remainder of the map) is sufficient to encode the semantics of many of the spatial relations considered, each lexeme corresponding to a particular *sequence* of dominating subregions of the visual field. Yet the selection or labelling of the appropriate object as LM or TR cannot be accomplished solely through the kind of saliency associated with a rapidly moving object in front of the eyes, or with the appearance of a stop light. Reliance only upon emergent or bottom-up selective attention of this nature is not sufficient, and some top-down mechanism akin to directed search must be employed, causing some enhancement of activation as a result of some *internal conception* of the stimulus.

Computational models of selective attention have long employed a matrix of units tuned to receptive fields covering the input scene. The *saliency map*, introduced by Koch and Ullman (1985), and similar in principle to the *master feature map* of Feature Integration Theory (Treisman and Gelade, 1980), is constructed from eight subsidiary input maps based upon orientation, intensity, chromatic components and temporal change, along with provision for "external" (i.e. top-down) inputs such as those arising from object assemblies. The most salient feature in the input field is then computed by means of a winner-take-all network over the map, selecting the unit with the highest activation and suppressing output from the remainder through inhibitory recurrent connections. A modulating signal from the map is re-propagated via recurrent connections to the unified pathway of primary visual cortex (V1). The winning location – and its associated object – thus enjoy substantial advantages in any competitive selection processes encountered, and this enhanced activation provides a plausible mechanism for the selection of feature binding assemblies (Treisman, 1996)[2].

Having briefly surveyed the neural basis of visual processing, and the established computational models of visual attention, we proceed in the next section to a high-level overview of the model, identifying where appropriate the linkages to the biological systems. Computations realised within each component are here described at a symbolic level. Subsequently, in Section 5.3, we examine the principal components in more detail, with a particular emphasis on the receptive field map (Section 5.3.1) – and its critical role in binding representations – and on the output network (Section 5.3.2), based on the simple recurrent network architecture of Elman (1990).

[2] These issues, and the plausibility of the model, are considered in far more detail in (Hogan, 2000).

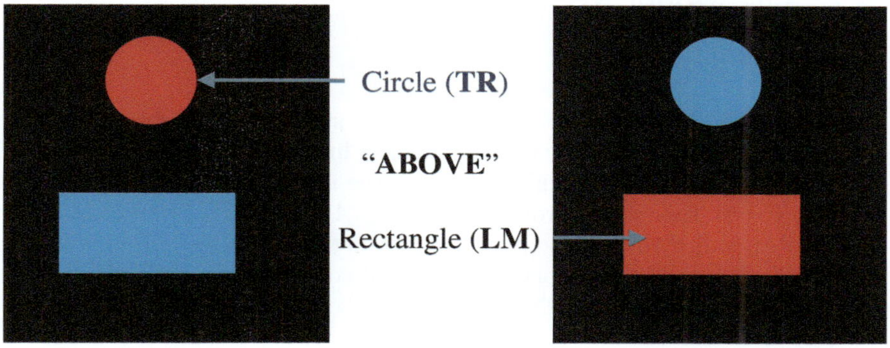

Fig. 5.2. The role of selective visual attention in parsing the scene into TR and LM tokens, as shown for the example of the English fragment "circle above rectangle".

After tracing the progress of example inputs through the system (Section refsect:l0Simulations), we proceed to describe in detail the acquisition of spatial relations (Section 5.5), exploring successively the nature of the input, prototypical and non-prototypical examples and the learning process itself. Section 5.6 describes an extension of the L_0 task and the receptive field machinery of Section 5.3.1 to the semantics of facial expressions. We conclude Part I of this work with the discussion of Section 5.7.

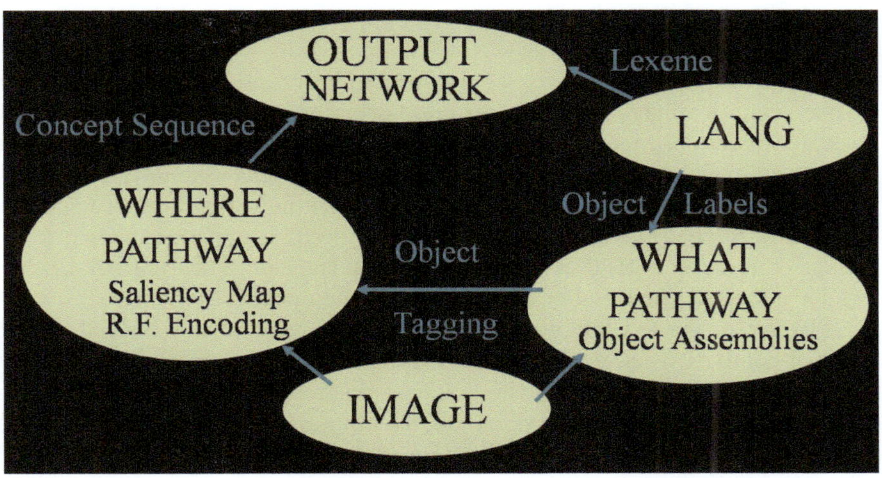

Fig. 5.3. The overall system architecture.

5.2 The Model Architecture

5.2.1 Overview

Figure 5.3 shows a schematic version of the system architecture. Learning here requires a robust association between a limited language fragment and a sequence of attentional highlights within the image. Thus, information necessarily enters the system from both the *IMAGE* (shown at the bottom of the figure), and through a language processing module (top right), a simple abstraction of working memory, a component common to models of top-down visual attention (Usher and Niebur, 1996).

Only the *WHERE* module receives genuine bitmap input, the *WHAT* assemblies receiving activation according to the range of objects present within the image. The *WHERE* component has responsibility for location coding and determination of saliency, and computationally we shall make little distinction between these roles, abstracting somewhat from the repropagation of activation described earlier. Object-based or top-down attention is modelled through the horizontal linkage between the *WHAT* and *WHERE* pathways, the system effectively regarding object tagging as another feature map input to the saliency mechanism. The *WHERE* pathway also includes dimension reduction through receptive fields 'above' the saliency map, providing a manageable sequence input to the recurrent *OUTPUT* network, where the association with the language fragment is ultimately acquired.

We now consider each of the system components in more detail.

5.2.2 The Recognition or WHAT *Pathway*

As noted in the introduction, we do not explicitly model extraction of each object from the image bitmap, a process normally associated with early vision (see, for example, Niebur and Koch (1998)). Object input is instead provided through a bank of unit 'assemblies', with the simplifying assumption that only a limited range of objects are permissible. Individual assemblies are clamped according to the presence of a particular object. This effect is shown for an image containing a `CIRCLE` and `RECTANGLE` in the centre of Figure 5.4, where we have limited the number of allowable iconic primitives to three.

The left of the same figure shows the primary role of the *WHAT* system, that of competitive selection of the object to be tagged for the current time step. Activation arising from the language fragment is modelled through assemblies similar to those used for object input, apprehension of the object description (for example a simple noun such as *circle*) being assumed sufficient to retrieve some iconic representation encoded as a result of previous exposure to the image. In computational terms, the visual object has been

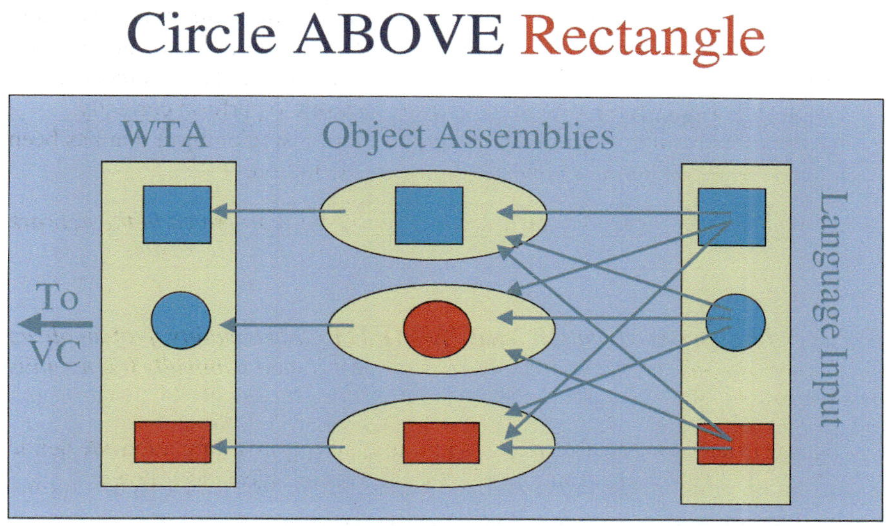

Fig. 5.4. The binding or WHAT module of the system.

tagged as a `CIRCLE` token, and the iconic `CIRCLE` representation activated, although the reality is less neatly partitioned.

The arrows to the right of Figure 5.4 indicate random projections from the language input, providing enhancement of activation to *each* of the available object assemblies. Tagging, a conjunction between the language and object units, requires the selection of a *winning* conjunction through the Winner-Take-All (WTA) network shown at the left (Feldman and Ballard, 1982), and unwanted, weaker conjunctions are discarded. Such selection and suppression mechanisms allow generalisation of the tagging system to more cluttered scenes or sophisticated linguistic phenomena – particularly as tagging is performed over time, which helps to avoid the potentially fatal problem of cross-talk.

The robustness of the cell assembly representation is here captured through multiple random projections from each unit to the binding network, ensuring with high probability that *at least one* connection with a particular binding unit is realised. An analysis of the associated Bernoulli trials was provided in Chapter 3..

The function of the binding subsystem is illustrated by the example of Figure 5.1 and the language fragment "circle above rectangle". For clarity, the number of scene frames in the example is limited to four, with change in the language input occurring after the second frame (table 5.1).

Table 5.1. Identification of trajector and landmark tokens.

Frame	Lang Circle	Lang Rectangle	Object Circle	Object Rectangle	Binding
0	1	0	1	1	$TR < Circle >$
1	1	0	1	1	$TR < Circle >$
2	0	1	1	1	$LM < Rectangle >$
3	0	1	1	1	$LM < Rectangle >$

5.2.3 Integration with the WHERE *Pathway*

Realisation of object tagging requires some integration of the feature and location based models of selective attention considered in Section 5.1. 'What'-to-'where' transfer is abstracted so that object input is effectively another feature map, eliminating the need for direct input from the object assemblies to the decision network, as binding has already been accomplished.

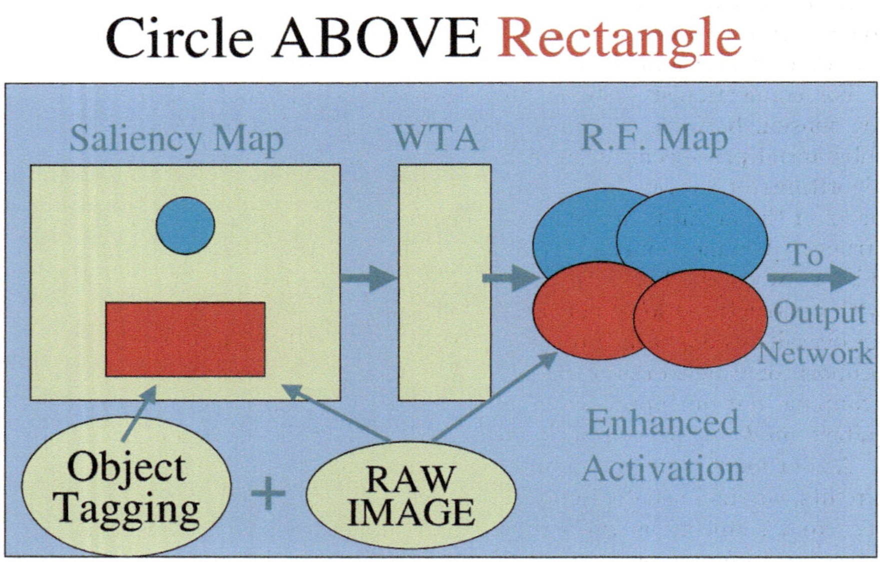

Fig. 5.5. The relationship between the binding subnetwork and the global saliency map.

Figure 5.5 shows the relationship between the binding subsystem of Section 5.2.2, here represented by the input *object tagging*, and the saliency map, the latter based heavily upon that described in (Niebur and Koch, 1996), but with fewer input feature maps (see Hogan (2000) for more details). As before, we may characterise this interaction by examining the bindings realised. The input sequence is unchanged from the previous example, and location input restricted to representative vectors x_1 (the rectangle) and x_2 (the circle)[3].

Table 5.2. Location binding for the trajector and landmark.

	Binding Network	Object Location Input	Implicit SM Binding
0	$TR < Circle >$	(x_1, x_2)	$TR(x_1)$
1	$TR < Circle >$	(x_1, x_2)	$TR(x_1)$
2	$LM < Rectangle >$	(x_1, x_2)	$LM(x_2)$
3	$LM < Rectangle >$	(x_1, x_2)	$LM(x_2)$

5.2.4 The Saliency Map and Receptive Field Encoding

As envisaged by Koch and successive co-authors, the saliency map must determine the most salient input region, and gate visual input so as to highlight this attended region for more detailed processing. Selective attention is thus sited conceptually amongst detection of elementary features, and decoupled from more sophisticated representations computed further along the visual pathway. Yet while the selection mechanism typically isolates the salient region for high-resolution representation and processing, modulation may also be reflected in a relatively low-resolution representation of a larger area; the region encompassing the attended object must still be highlighted at the expense of less salient parts of the field. As the modulating signal is thought to be directed back to primary visual cortex, such *reduced maps* may be computed at a number of points along the visual processing hierarchy, and combined with the necessary feature detectors according to the requirements of the relation to be encoded.

Our approach to spatial coding in the *WHERE* pathway is then as shown (albeit schematically) to the right of Figure 5.5. Conceptually, the saliency mechanism – the map and an associated winner-take-all structure – is used to identify the most salient location, with consequent enhancement of a small spotlight surrounding the corresponding position in various cortical retinotopic maps of similar resolution[4]. We shall refer to this representation as the *enhanced contrast map*. Reduction to a resolution more appropriate for encoding of gross spatial relations is then provided through receptive fields *over* this map, with consequent enhancement of activation of the fields encompassing the attended region. This is reflected in the schematic through

[3] In principle these might be viewed as a kind of centre of mass vector.

[4] No map of this kind is shown in the figure.

the colouring of the lower ellipses of the receptive field map, the activation heightened as a result of the attended rectangle.

Thus gated by the saliency mechanism, the receptive field map effectively forms the input layer for the *OUTPUT* system module.

5.2.5 *Lexeme Binding and the* OUTPUT *Module*

Categorisation of spatial relations takes place in a randomly connected simple recurrent network. In its purest form, the model exerts very tight control over the information which is passed to this decision network; as discussed above, object and location information is effectively gated by the saliency map. This decoupling of the components of the problem both simplifies and complicates the issue: binding at the output network requires a lower degree conjunction, but the lexeme is now in principle a complex temporal conjunction – requiring the use of a recurrent output network.

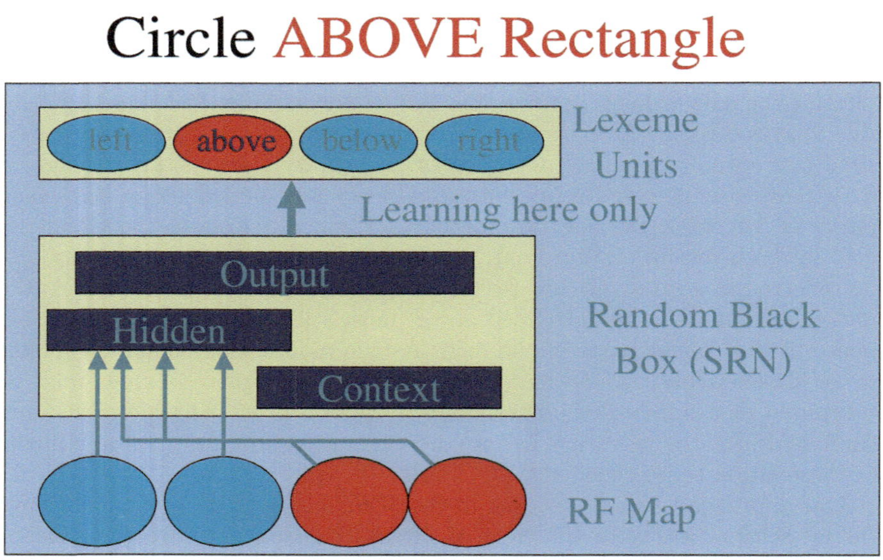

Fig. 5.6. The system output component.

Figure 5.6 shows a schematic of the output component. Here, input is provided from both the low-dimensional receptive field map and the language fragment, the latter providing the network targets during training. The module must now encode a binding of the form:

$$above < TR\_(x_1), LM\_(x_2) > .$$

Successful acquisition of such bindings is thus heavily dependent upon the structure of the saliency map. For a simple $n \times n$ feature map, the initial location may correspond to the TR (say), but subsequent transitions (to the most salient region of the LM, for example) are virtually unconstrained, legal transitions being available to any of n^2 locations. In essence, each TR position constitutes a valid initial symbol within a complex location-based grammar for the given concept, and while some rationalisation is possible, a model based upon this representation may lead to a vast number of network states. Some alternative representation of markedly lower dimensionality is therefore necessary.

Our solution is to propose an encoding in terms of a low-dimensional receptive field map, set formally over the contrast enhanced retinotopic map computed once saliency is determined. Thus, the attentional highlight has the effect of providing strong activation in one or more of the receptive fields, effectively an input symbol (drawn from our location grammar) to be processed by the output component.

Let us consider an example 2×2 receptive field map as shown as part of Figure 5.5. If the fields are labelled according to the matrix:

$$\begin{bmatrix} a & b \\ c & d \end{bmatrix},$$

then the static concepts may be encoded through a limited number of 'syntactically valid' strings or transitions within the location grammar, the symbols being empowered to determine the concept through their associated LM or TR binding.

Training is restricted to *prototypical* examples of the concept, and it is our expectation that the system will give some partial response (essentially with weaker confidence) for test examples which, while justifiably members of the class, do not satisfy the more stringent requirements of prototypicality. Further, some of the concepts to be handled by the system involve partial matches across more than one symbol, and it is useful to reference these combinations as a single concatenation of the underlying labels.

It is an assumption of the model that the positions of (at least the prototypical) examples are reasonably central due to the previous foveal alignment. However, for each concept, the training data created allows some (bounded) random variation from what might be regarded as 'true' prototypicality. For example, if the picture in Figure 5.1 is regarded as a perfect example of *above*, we may also accept the additional trajector positions shown in Figure 5.7.

Moreover, similar encodings may be proposed for the remaining tasks, but their effectiveness necessarily relies upon the extent to which the receptive fields may be localised. Our task is to ensure that high activation in the unit is indicative only of the presence of attended objects within some neighbourhood of the centre of the field. These issues are considered in the following

Fig. 5.7. Variation in trajector positions for prototypical examples of the concept *above*.

section, which explores the implementation of the two critical components of the system under random network assumptions.

5.3 Random Network Components

This section presents a more detailed look at the connectivity models used in the development of two critical components in the semantics acquisition model, the receptive field map and the output component. The latter structure follows a simple recurrent network architecture, but relies on random connectivity to realise the necessary information processing. While other components in the model exhibit a similar reliance upon sparse random connectivity, these may be seen as direct applications of the networks of Chapter 3. The components considered here have a familial relationship to the earlier connectivity models, but are specialised to the roles required.

5.3.1 The Receptive Field Map

In the preceding sections we have noted the complexity of a location grammar based directly on the saliency mechanism, and here consider a more compact representation. However, we must also ensure that any receptive field units together provide a localised and graded response to changes in object size and

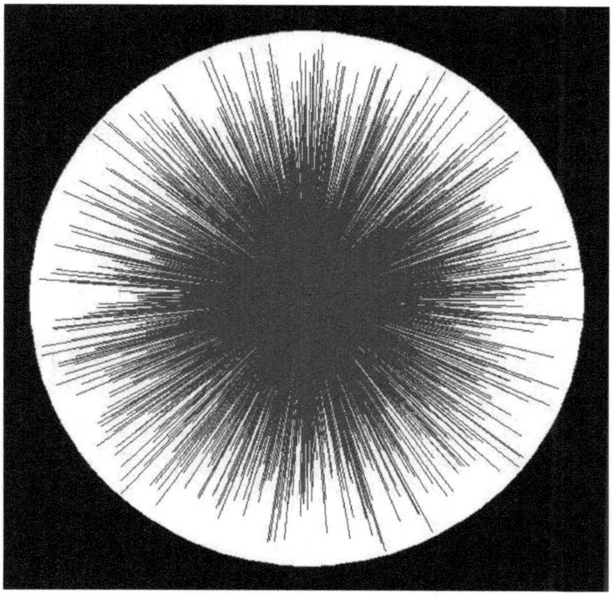

Fig. 5.8. The "canopy" receptive field.

position within the scene. Ideally, we require a response function constrained to respond strongly only within some tightly bounded region of the input space, while providing some limited indication of the presence of more distant objects in proportion to their size and intensity – what we might summarise as their *influence*.

Such a response is readily provided by receptive field units which receive random projections from a set of source nodes, under the familiar assumptions of decay in connectivity with increasing separation. The approach is a generalisation of the layered random network structures of the earlier chapters, in which the target unit exists at the centre, c, of a logical canopy over a large, but finite group of randomly positioned source nodes, as shown in Figure 5.8. Suppose that each node projects a connection of unit weight to the target with probability $\alpha(r)$, a nonincreasing function of the (normalised) radial separation between the source and target nodes[5].

If we define a connectivity function $C(r)$ to be the cumulative fraction of connections to the target within radius r of the centre, the fraction contained within an element of area may be expressed through the equation

$$\{C(r + dr) - C(r)\}d\theta = r\alpha(r)drd\theta. \tag{5.1}$$

[5] While there may be some angular variation in the image intensity, it is suppressed here for clarity, and we write $I = I(r)$.

The corresponding connection density $c(r)$ is then

$$c(r) = \frac{dC}{dr} = r\alpha(r), \tag{5.2}$$

where α must be chosen so that

$$\int_0^\infty c(r)dr = 1. \tag{5.3}$$

If we make the mathematically convenient selection of

$$\alpha(r) = e^{-\frac{r^2}{2}}, \tag{5.4}$$

then the connection density may be regarded as a weighting function for the map, with the effective influence of each object upon the receptive field critically dependent upon its separation from the unit centre. The field response is then formally the expected value of the intensity, weighted with respect to the connection density across the domain

$$v(c) = \int_0^\infty I(r)c(r)dr. \tag{5.5}$$

While the Gaussian response function of equation 5.4 is similar to those found in Radial Basis Function (RBF) networks, RBF units are normally positioned within the vector space of input patterns, providing a weighted response to the current input according to its separation from the unit centre. Such units are thus limited in the extent to which they may account for the influence of entire objects, and *domain responses* akin to equation 5.5 offer significant advantages in the present context.

However, RBF units commonly allow a more general footprint in the pattern space, with the single radius parameter, R, of the preceding analysis being replaced with an $n \times n$ transformation matrix. As this more general response is potentially useful, we shall develop these ideas more fully in Section 5.6.

5.3.2 The Output Network

Ultimately, semantic categories are represented within an unstructured, general-purpose output component, without the benefit of category-specific prior knowledge. Extraction of features – perhaps the extraction of *semantic primitives* – is farmed out to task-specific modules, limiting the role of the output network to acceptance or rejection of a particular sequence of frames. As our system relies upon temporal representations even for static categories, with context arising as a result of attentional parsing of the scene, it is essential that our chosen output component be capable of representing finite state automata of some complexity.

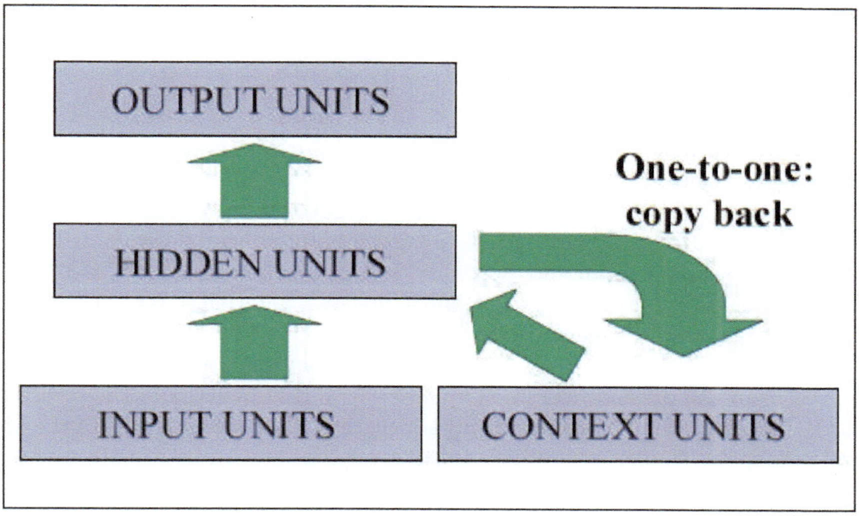

Fig. 5.9. The Simple Recurrent Network architecture of Elman.

However, the grammars realised and the strings accepted are necessarily limited by the human cognitive apparatus, particularly in respect of the length of the 'movie' presented. We may argue, for example, that the representation of a binary relation such as *above* requires at least two distinct frames at the level of the enhanced contrast map – one with attentional focus upon the trajector and the other with attentional focus upon the landmark. At the opposite extreme, while the use of high speed cinematography may reveal a sequence of frames which might be validly associated with the fragment `BULLET` *over* `HEAD` – indeed it may even be regarded as prototypical – we need not accept that such an example may be captured and processed by the human visual system. In consequence, all of the sequences considered in the experiments of Section 5.4 will have length bounded within the set $\{2, \ldots, L_{max}\}$, where L_{max} will remain unspecified, but is potentially large.

5.3.2.1 The Simple Recurrent Network

The Simple Recurrent Network or SRN architecture was introduced by Elman (1990) as a means of investigating time dependencies in language, whilst avoiding the more complex dynamics of the earlier Fully Recurrent Network (FRN). The SRN simplifies the earlier models in its treatment of state information, explicitly preserving the network hidden layer through fixed copy-back connections from each hidden unit to a corresponding context unit. By allowing complete connectivity in the reverse direction, i.e. between context and hidden layers, the architecture provides for the current state to be a function of both the previous state and the current network input:

$$x_i^{(h)}(t) = f(\sum_j w_{ij}^{(ih)} x_j^{(i)}(t) + \sum_j c_{ij} x_j^{(h)}(t-1) - T_i^{(h)}), \tag{5.6}$$

where the superscripts (h) and (i) indicate hidden and input layers respectively. Figure 5.9 shows a schematic of the architecture. As the copy-back connections are fixed and of unit weight, the SRN has the tremendous computational advantage of being trainable through standard back-propagation, there being no necessity for the unfolding of context which complicates the backpropagation through time algorithm for training the FRN (Williams and Zipser, 1989).

Elman's studies showed that some sort of elementary grammar could be acquired from the presentation of simple sentence fragments, and the structure of the representation revealed through hierarchical cluster analysis of the hidden unit space, enabling identification of groupings according to both category (type) and individual value (token)[6]. Examination of the state space thus allowed the conclusion that the network had implicitly acquired knowledge of word and sentence structure in the absence of constituent markers, offering apparently substantial computational support for linguistic empiricism.

Following Elman's results, researchers examined the effectiveness of the SRN architecture in representing finite state automata of various levels of complexity, both empirically and subsequently through programmatic insertion of state representations. Cleeremans, Servan-Schreiber, and McClelland (1989) established that the SRN could learn to represent small probabilistic FSAs, and Servan-Schreiber, Cleeremans, and McClelland (1991) extended this work to include grammars of greater complexity, showing that the network state representations obtained often exhibit multiple hidden state vectors corresponding to each state of the automaton, this apparent redundancy enabling the network to distinguish between different pathways into the state. This work paved the way for the later programmatic approaches, and highlighted the likely degradation of generalisation performance if fewer hidden units are employed – the reduced hidden layer supporting insufficient states to represent the 'extra' pathway information.

Kremer (1995), used a state splitting argument to confirm that the SRN architecture with threshold units can represent the state transition and output functions of *any* Deterministic Finite State Automaton (DFA), the class of automata in which there is exactly one transition from a given state associated with each input symbol (Hopcroft and Ullman, 1979). This result was subsequently generalised to the case of sigmoid units by Alquézar and Sanfeliu (1995). As every Non-Deterministic Finite Automaton

[6] Mapping of the network state transitions and extraction of the equivalent automaton is now routinely accomplished using principal component analysis (see for example Plunkett and Elman (1997)), but this method was not used in Elman's original study.

(NFA)[7] may be represented by an equivalent DFA, Kremer's result bounds the representation capacity of the SRN to include at least those languages accepted by finite automata, the *regular expressions* (Hopcroft and Ullman, 1979), the class of languages capable of representing the frame grammars of our model. Our network encoding is a randomised version of a similar scheme developed by Carrasco et al (2000), and this is considered in the following section.

5.3.2.2 Encoding Automata in Recurrent Networks

The representational equivalence between automata and neural networks has been of significant interest since the original papers of McCulloch and Pitts, and Minsky established as early as 1967 that any Finite State Machine was "...equivalent to, and can be 'simulated' by, some neural net" composed of McCulloch and Pitts neurons (Minsky, 1967). Direct *programming* through weight insertion first appeared through the work of Kremer (1995), but later encoding schemes and the present work benefitted significantly from the dynamical systems analysis of Omlin and Giles (1996a,b), which found that the location and nature of fixed points of the mapping are strongly dependent upon the choice of bias and self-recurrent weight. In particlar, the existence of two distinct stable state intervals – L and H – depends upon a recurrent weight of relatively large magnitude, with the location of the fixed point within these stable intervals determined by the threshold. The existence of a stable fixed point in both L and H suggests that each unit may encode one *bit* of information, and by extension, that a layer of n hidden units may encode up to 2^n distinct states. In practice, however, the ability of each unit to remain in a given position is governed by the extent to which it may be insulated from unwanted 'shocks' from the network context and input units, and the ideal 'binary' encoding is seldom if ever achieved through inductive learning. Similarly, such parsimony leads to substantial and perhaps insurmountable difficulties for the theoretician wishing to propose a robust encoding scheme, and it is common to use only *one-hot* or sparse representations for knowledge insertion.

The present work is at least in part driven by the requirements of the random network implementation, and the structural simplicity provided by one-hot encoding provides significant advantages over more distributed representations. Our model requires that each network output ultimately act as an acceptor for the location grammar associated with a particular category, and thus that the randomly initialised output network must support a number of potentially complex DFAs. As before, our networks must rely to a large degree upon the initial knowledge base provided at 'compile-time' and our task is again one of influencing the network through distributional bias.

[7] An automaton in which there may be zero, one, or more transitions from a state on the same input symbol (Hopcroft and Ullman, 1979).

Hogan (2000) presents an extensive discussion of these issues, and detailed coverage of the Kremer and Carrasco et. al. encodings. Here we focus on the nature of the state representation, and on its usefulness in the random network model. Successful representation of DFAs within recurrent networks of this type require that the state space be organised into mutually disjoint closed sets, with each state of the automaton mapped to the interior of exactly one of them (Casey, 1996).

Each network state is then best specified with reference to a particular 'hyper-corner' of the network state space, but this precludes the representation of automaton states of large in-degree. This limitation may be overcome by transforming the original DFA, M, to an equivalent, but less compact representation M' through the process of state-splitting (Kremer (1995), Carrasco et al (2000)). In this approach a distinct state is created in M' corresponding to each transition into a given state of M, the new states being distinguished according to the symbol accepted.

Simple recurrent network implementations of M' necessarily require far larger hidden layers than those of the second order networks, yet this aligns well with the random network approach, with sparse connections of fixed positive weight (here denoted ξ) and subtractive thresholds, in contrast to the dense connectivity and large fraction of inhibitory connections required by denser state representations. When coupled with a bounded in-degree, this approach implies far higher existence probabilities for the state units and a lower probability of contamination from unwanted inputs. Carrasco et al. provide a detailed analysis of the output and next-state functions, noting that the minimal weight value ξ necessary to guarantee correctness is an increasing function of the node in-degree allowable within the state and forward layers. While the resulting conditions differ in some minor respects, considerable simplification is possible through the authors' approach of enforcing the most restrictive of them.

Thus, sufficient conditions for the correctness of both next-state and output functions are provided by:

$$f(\xi(\chi\epsilon_l - 1/2)) \leq \epsilon_l, \tag{5.7}$$

and

$$f(\xi(\epsilon_h - 1/2)) \geq \epsilon_h. \tag{5.8}$$

where χ is an upper bound on the in-degree of the state and output nodes, and ϵ_l and ϵ_h define, respectively, the state intervals $[0, \epsilon_l]$ and $[\epsilon_h, 1]$. The minimum value of ξ for a given χ may then be found by writing

$$\xi = \frac{f^{-1}(\epsilon_l)}{\chi\epsilon_l - 1/2}, \tag{5.9}$$

and subsequently calculating, through functional iteration, the value ϵ_l^* for which $\partial\xi/\partial\epsilon_l = 0$. Substitution in equation 5.8 yields the corresponding value of ϵ_h.

Table 5.3, adapted from (Carrasco et al, 2000) shows values of ξ_{min} for a range of values of χ. Note the rapid trend of $\epsilon_l \rightarrow 0$ and $\epsilon_h \rightarrow 1$ with increasing χ. In the simulations which follow we shall choose non-zero weight values

$$\xi = \lceil \xi_{min}(\chi) \rceil,$$

with χ for all practical purposes bounded at around 6 by the diminishing existence probabilities derived in Chapter 3.

While no direct revision of these ideas is necessary for implementation within the random network framework, issues of reliability and management of extraneous connections must arise. However, these matters are best deferred until after consideration of the grammars appropriate to each category, and this issue is addressed in the next section.

Table 5.3. Minimal values of ξ and the corresponding choices of ϵ_l and ϵ_h for varying χ, as adapted from Carrasco et al. Note that only a single stable fixed point is realised for unit weights when $\xi \leq 4$.

χ	ξ_{min}	ϵ_l	ϵ_h
1	4^+	0.5^-	0.5^+
2	7.17799	0.0753324	0.965920
3	8.36131	0.0415964	0.982628
4	9.14211	0.0281377	0.988652
7	10.5829	0.0136861	0.994704
10	11.4664	0.00879853	0.996648

5.4 Presentation and Learning

This section concludes our discussion of the spatial relations tasks and their acquisition by recruitment learning. Here we make more explicit the descriptions of earlier sections, notably in respect of connectivity, pattern selection and structure. We begin by following the progress through the system of a prototypical example of *above* through both *WHAT* and *WHERE* processing, closely examining the effect of the object-TR and object-LM associations on representations at the receptive field map level (Section 5.4.2). Dynamic relations, as exemplified by *over*, are not introduced until this section, the distinction between the static and the dynamic being of limited significance in *WHAT* processing.

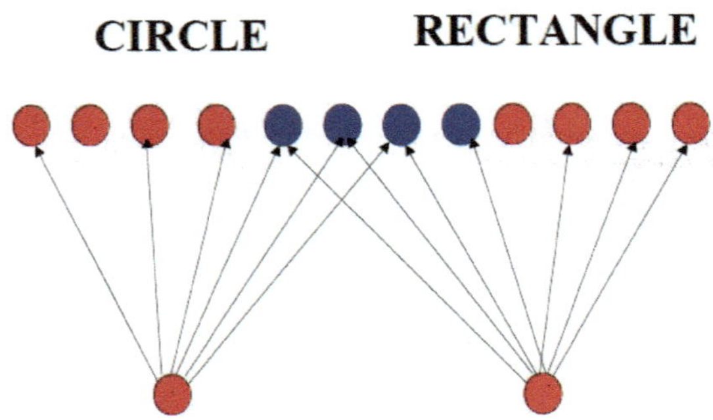

Fig. 5.10. The possibility of cross-talk between object representation assemblies `CIRCLE` and `RECTANGLE`, as illustrated by a boundary overlap of two units for each grouping.

5.4.1 The 'What' Pathway

In Section 5.2.2 we considered the parsing of the scene through top-down selection into trajector and landmark tokens (Figure 5.4), with the result formally shown in table 5.1. Here we examine more carefully the realisation of these associations through random network components. Object representation remains centred in the cell assembly – modelled by a layer of 10 units or more – but these are not individually clamped, relying instead upon random projections (of unit weight) from a single fixed input. Similar activity is propagated to the assembly from the corresponding language input.

In addition to these 'appropriate' connections, each assembly is assumed to receive projections of progressively decaying probability from 'neighbouring' inputs, introducing the possibility of cross-talk near assembly boundaries. Figure 5.10 shows the situation for the `CIRCLE` assembly, which shares a border with that of the `RECTANGLE`. Both of these assemblies will be activated through the presence of objects of their type in the input scene. However, we expect that some subset of the `CIRCLE` assembly will win any WTA competition as a result of enhancement of its nodal activation from the language input. Reliability of such representations may be analysed through the arguments of Chapter 3, the context being similar to the random chunking networks of Feldman (1982) and the frontier properties of Valiant (1994). However, in the present work we do not attempt to model a reinforcing coalition in the style envisaged by Feldman, and the combination of inputs – along with a WTA mechanism – must suffice to enable correct selection.

In the case of the `CIRCLE` assembly of Figure 5.10, our primary concern is that one of the `RECTANGLE` nodes may have its single source activation reinforced by stray active connections from one or more of the `CIRCLE` object and language inputs, thereby producing an activation comparable to or exceeding that of the enhanced `CIRCLE` assembly nodes. Any strategy for minimising the risk of cross-talk must depend upon the connectivity, local radius and assembly size assumed, and the possibility may be almost eliminated if we employ large assemblies (say $N \geq 100$) in conjunction with strong local connectivity and decay with radial separation. Activity within a central region of the assembly is thus well represented, with little chance of connection to the boundary nodes of a neighbouring region.

If the local radius R is adjusted so that $\xi = R/N$ approaches 1.0, the likelihood of cross-talk at the boundary nodes is substantially increased, albeit from a low base. For cross-talk to appear, at least one node near the boundary of a neighbouring region must receive *active* connections from at least 2 sources, with a strong likelihood of problems if 3 or more links exist[8]. Even assuming a generous $\xi = 1.0$, we note that such an occurrence has success probability of $O(\alpha^3)$ for degree 2 and $O(\alpha^5)$ for degree 3. For small α – perhaps under the C3 regime of Section 4.7 – there is only a slim chance (probability of $O(10^{-3})$ and $O(10^{-4})$, respectively) that such a node may exist. Choosing a large assembly size – in effect introducing additional Bernoulli trials – may significantly increase the probability that an appropriate degree-2 `CIRCLE` node will exist, and thus win the subsequent WTA competition. In the present work a choice of $N = 10$ has proven sufficient to ensure correct selection, a result consistent across multiple trials.

5.4.2 The 'Where' Pathway

In this section we consider in more detail encoding of position within the 'where' component of the model, examining performance at both the gross scale of the enhanced contrast map[9] (ECM), and at the reduced dimension of the receptive field map (RFM). As previously noted, the acquisition of the semantics of spatial relations terms depends critically upon such encodings, with the ECM supporting isolation of the attended object, and the RFM providing the dimension reduction necessary to make category representation a feasible learning task for the output network.

The enhanced contrast map differs from the saliency map in that enhancement and inhibition will not necessarily lead to the extremes of saturation and silence. There is no winner-take-all process in operation at this level, only localised addition of enhancing and inhibitory signals, and their effect must provide sufficient contrast to ensure a clear distinction between the attended object and its surrounds. This effect is naturally reflected in the receptive field

[8] Note that we may ensure that distinctions of such fine granularity are preserved by using node thresholds of around $T = 2$, thereby mapping the activation to the linear region of the sigmoid transfer function.

[9] As in Section 5.2.4, we remark that this representation shares computational resources with the saliency map.

representation overlaying the map, and the corresponding RF activations for `CIRCLE` *above* `SQUARE` and `CIRCLE` *over* `SQUARE` are shown respectively in tables 5.4 and 5.5[10].

Table 5.4. Receptive field map activations corresponding to the ECM representation of *above* for sigmoidal RF units. Note that the 0.12 responses indicate an input activation of 0.0.

0.12	0.82
0.12	0.12

Table 5.5. Receptive field map activations corresponding to the ECM representation of *over* for sigmoidal RF units. Note that the 0.12 responses indicate an input activation of 0.0.

0.73	0.12
0.56	0.12

As before, the units of the receptive field map are constrained to respond strongly only within some bounded region of the input space, while providing some limited indication of the presence of more distant objects in proportion to their influence. Following some prototyping, we assume a 2×2 grid of sigmoid units, each receiving random projections from units within the surrounding map region – itself defined as a unit square – as depicted in Figure 5.8. In contrast to the earlier models, the local radius is here calculated with reference to an equally spaced $n \times n$ map, each unit being separated from its nearest neighbour in each axis direction by a distance of $1/(n+1)$. As before, α decays strongly outside some local region of strong connectivity, here indicated by separation $\xi > \Xi$, according to the relation

$$\alpha(\xi) = \alpha_0 \exp(-\frac{1}{2}\xi^2 u(\xi - \Xi)), \tag{5.10}$$

where $\xi, \Xi \in [0, 1]$ and $u(\cdot)$ is the step function of equation 4.1. In the simulations which follow we have used $\Xi = 0.25$ throughout, in accordance with the specified 2×2 receptive field grid.

As noted in Section 5.3.1, the effect of such a regime is to weight the intensity of each object in response to its distance from the nominal center of the field, with normalisation of the activation accomplished through the usual sigmoid transfer function. While the representation may vary with the choice of connection seeds, this variation did not prove significant.

[10] We shall return to this issue in Section 5.5.4.

Perhaps more surprisingly, network performance is only mildly sensitive to variations in connection probability, with the regularity and homogeneity of the objects in the scene undoubtedly a contributing factor. Having established robust map representations, we now turn our attention to learnability within the output component.

5.5 Concept Acquisition

While concept learning is realised in this chapter in much the same fashion as before – through combination of certain units of a feeder basis – such hypothesis units are no longer limited to static concepts over the input space. Formally, each unit of the forward layer is itself an acceptor for some randomly generated visual grammar, and each concept unit must in turn act as a recogniser for the union of one or more of these grammars. In essence, our task is to realise a disjunction over the constituent feeders which is both spatial and temporal in nature.

The spatial disjunction arises primarily as a result of the choice of learning rules, it being likely that the *CSHebb* and *Winnow* rules of Chapter 4 will enhance connections from *all* feeder units activated upon presentation of the string termination symbol. However, the concept unit must also accept valid strings of markedly varying duration, a requirement likely to involve the combination of a number of the recogniser units from the forward layer, with each accepting strings of one or more distinct lengths. Regardless of the number of frames in the movie, acceptance must take place if the primitive conditions underlying the semantics of the category are presented during the sequence.

Following Regier (1992), we hold that the activation of the concept unit has meaning *only* upon termination of the current frame sequence, so that training may be undertaken only in the immediate aftermath of the last frame. Unlike Regier, however, we do not attempt to propagate the resulting concept error backward through time in order to facilitate adjustment of the network weights. In our approach, semantic knowledge develops through the exploitation of nascent structure. If learning is then a matter of selection from amongst existing alternatives, we must give due attention to the match between the distribution of valid input sequences and the distribution of elemental recognisers within the network.

5.5.1 Sequence Input

We have earlier considered natural limitations on the range of examples to be encountered by the model, asserting that all such movies must consist of $L \in \{deg, \ldots, L_{max}\}$ frames, where deg is the degree of the category relation – in our case at least two – and L_{max} is some potentially large, but as-yet-unspecified maximum. Semantic knowledge is held to be encoded in the transitions between attentional states, and provided each frame sequence embodies an appropriate shift or series of shifts, it should in principle be seen as a positive example of the relation. Computationally, however, this is unsatisfactory, making the unreasonable demand that the network attend only to the critical frames of an extended, even endless movie presentation. We thus provide explicit indication of the end of the current sequence, writing $\sigma_0 = \$$. In the case of static relations, little further consideration is necessary, and each valid sequence must have the form

$$f_0^{tr}, f_1^{tr}, \ldots, f_{m_{tr}}^{tr}, f_0^{lm}, f_1^{lm}, \ldots, f_{m_{lm}}^{lm}\$,$$

where the symbols f^{tr} and f^{lm} indicate, respectively, frames based on an attended trajector and an attended landmark, and $m_{tr} \approx m_{lm}$. The case of dynamic relations is more subtle, and some choice must be made as to whether the landmark should be represented initially in the absence of the trajector. Again some guidance may be obtained by appealing to selective attention. Noting the great weighting given to motion within the field in the determination of bottom-up saliency, we begin each of the dynamic sequences with the first appearance of the trajector, with saliency only passing to the landmark upon completion of trajector motion. Valid sequences associated with dynamic relations will thus have the form

$$f_0^{tr}, f_1^{tr}, \ldots, f_{m_{tr}}^{tr}, f_0^{lm}, f_1^{lm}, \ldots, f_{m_{lm}}^{lm}\$,$$

where m_{tr} may be significantly greater than m_{lm}.

5.5.2 Prototypicality and Framing

Discrete, visual-grammar based descriptions of both static and dynamic relations are facilitated through the lower case letter labels shown in table 5.6 below. As before, the use of a particular symbol serves to indicate a high activation state in the corresponding unit, and combinations of symbols may be used to indicate the presence of (especially attended) objects across more than one region of the map.

Table 5.6. Reference labelling of the visual sub-regions of the RF map.

a	*b*
c	*d*

In this section we provide a short summary of the grammar associated with the prototypical examples of each relation, with the understanding that dynamic relations may in turn be represented through grammars in which atomic elements are valid static relation strings. As we have chosen the minimal receptive field map resolution consistent with the domain, prototypical grammars for many of these relations may require the use of the combination symbols described above. Table 5.7 shows prototypical sequences for the relations *above, below, left of, right of, over* and *under*[11]. Weaker versions of a relation may also be specified, such as $a \rightarrow (cd)$ as a weaker version of *above*.

Table 5.7. A near-exhaustive list of prototypical visual strings for a number of the Regier tasks.

Relation:	Transition:
above	`TR_(ab)` → `LM_(cd)`
below	`TR_(cd)` → `LM_(ab)`
left of	`TR_(ac)` → `LM_(bd)`
left of	`TR_(a)` → `LM_(b)`
left of	`TR_(c)` → `LM_(d)`
right of	`TR_(bd)` → `LM_(ac)`
right of	`TR_(b)` → `LM_(a)`
right of	`TR_(d)` → `LM_(c)`
over	`TR_(c,a,(ab),b,d)` → `LM_(cd)`
over	`TR_(a,(ab),b)` → `LM_(cd)`
under	`TR_(c,(cd),d)` → `LM_(ab)`
under	`TR_(a,c,(cd),d,b)` → `LM_(ab)`

Note that while the dynamic relations are given above in *left-to-right* format, their *right-to-left* inverses[12] are equally valid exemplars, although the difference is of little consequence computationally.

Having examined the range of strings which may be regarded as prototypical examples of a particular spatial relation, we now turn our attention to their use as part of the network training regime, and this is considered in the following section.

5.5.3 The Training Regime

Training of the network on spatial relations tasks closely follows the approach chosen by Regier, with the additional requirement of repeated runs with modified random number generator seeds to allow for variations between network realisations. In principle, the procedure for each relation is to:

[11] Relations such as *in,out* and *through* must accept a wider range of transitions and for the sake of brevity are not presented.

[12] For example, each valid *over* sequence of the form $left\,of \rightarrow above \rightarrow right\,of$, has a corresponding *right-to-left* sequence $right\,of \rightarrow above \rightarrow left\,of$.

- Train the network solely on one or more prototypical examples of the desired relation.
- Test the network on the test set for that relation, including a wide range of positive, but not necessarily prototypical examples, as well as clearly negative instances as drawn from example sets for the other relations.

In this way the network must acquire some representation of a particular relation solely on the basis of positive examples of that relation, and one might anticipate some over-generalisation during the subsequent testing phase.

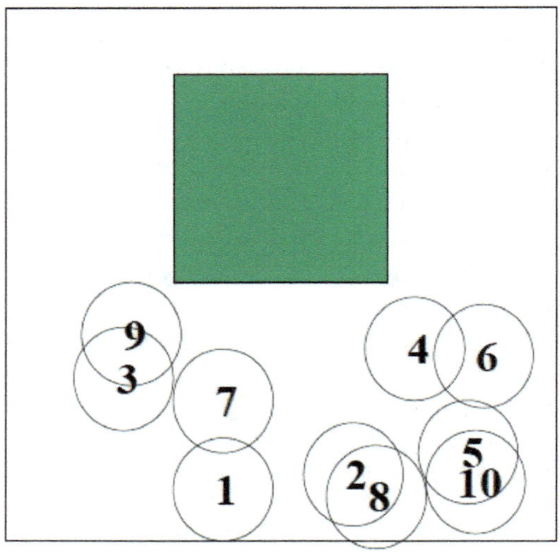

Fig. 5.11. The test set trajector positions for *below*, shown here with respect to a common landmark position.

Generalisation test sets were created by selecting approximately 60 random examples for each static relation, of which a majority were subsequently culled on the basis of similarity (as determined by eye) to other examples, leaving only a dozen or so tests for each relation. The approach is illustrated in Figure 5.11, which shows the test set for *below* in schematic form. The receptive field map activations for a two-frame sequence involving the off-centre trajector in position 3 are shown in table 5.8 below.

Training and test examples for the dynamic cases are far more difficult to craft than in the static case, due to the need to accurately calculate movement of the TR centre from frame to frame so that the motion is not jerky, and fewer movies have been created. An example is provided by circle *over* square as shown in Figure 5.12, in which the 6 frames of the movie have been superimposed onto the one figure, numbers indicating the appropriate trajector

Table 5.8. Receptive field map activations for *below* using the trajector in position 3 of Figure 5.11. In the left-hand frame the TR is attended, while in the right-hand frame the spotlight is on the LM.

0.12	0.12	0.97	1.00
0.68	0.12	0.12	0.12

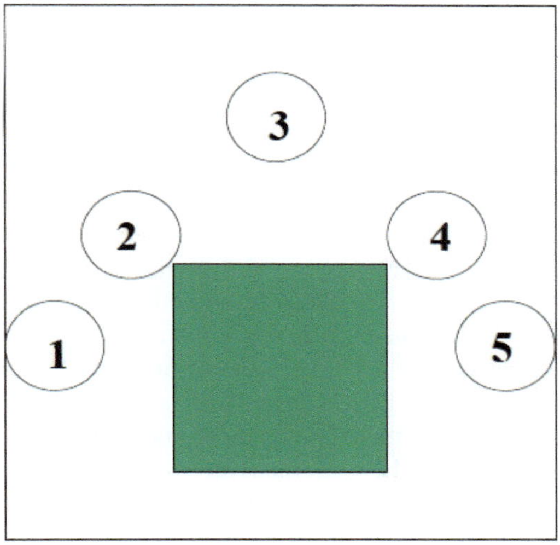

Fig. 5.12. The six-frame movie for CIRCLE *over* SQUARE, super-imposed onto the one image, with the final frame (the landmark only) unlabelled.

Table 5.9. Receptive field map activations for the six-frame CIRCLE *over* SQUARE movie. Frames 1-3 are in the first row, with 4-6 in the second.

0.12	0.12	0.73	0.12	0.94	0.44
0.56	0.12	0.56	0.12	0.12	0.12
0.44	0.96	0.12	0.82	0.12	0.12
0.12	0.12	0.12	0.50	1.00	1.00

position[13]. The corresponding RF map sequence is given in table 5.9, with the position and direction of movement of the TR easily discernible from successive frames.

[13] The sixth frame consists only of the landmark, and is not explicitly labelled.

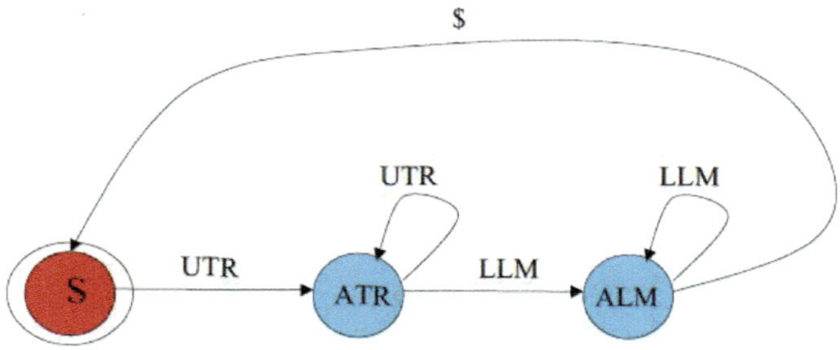

Fig. 5.13. The elemental visual grammar for *above*.

5.5.4 Semantic Relations

We begin this section by examining a simplified *above* grammar (shown in Figure 5.13), and its representation within the Elman architecture (Figure 5.14). Here, variations in position and size are subsumed within the catch-all tokens `UTR` (*upper trajector*) and `LLM` (*lower landmark*). The machine enters the *Active Trajector* or `ATR` state upon receipt of the first `UTR`, remaining in that state upon subsequent `UTR`s until a transition is forced by the appearance of the first `LLM`. In reality the task is markedly more complex, as we must allow for transitions between the various symbol-state pairings which together constitute the agglomerated *active trajector* (`ATR`) and *active landmark* (`ALM`) states.

State transitions within the network of Figure 5.14 are achievable only through simultaneous activation of the previous state and current input units, the weights and thresholds assigned according to the prescriptions of Section 5.3.2.2. Note in particular that a visual string may be accepted only if the network exists in state `ALM` when the termination symbol is presented.

Evidently, extending this architecture to encode the transitions of table 5.7 requires that we introduce a number of additional hidden and context units, and we may use the enumeration provided in the table as a guide for selection of connection parameters for the random network model, in particular to provide some bound upon the in-degree χ of the hidden units. Realisation of such structures within our framework exposes a fundamental conflict between the desire for representational purity – in the sense of limiting corruption of state information through extraneous input – and the necessity of generating state nodes of in-degree sufficient to support the relations, and in sufficient number to allow some robustness of the representation.

Such a conflict is dramatically heightened by the transient nature of the network state information, encoded through activations which must be re-established at every time-step through the co-operative firing of at least one input node and one context unit. To some extent we may alleviate this state decay by accurately following the encoding scheme, and ensuring saturated

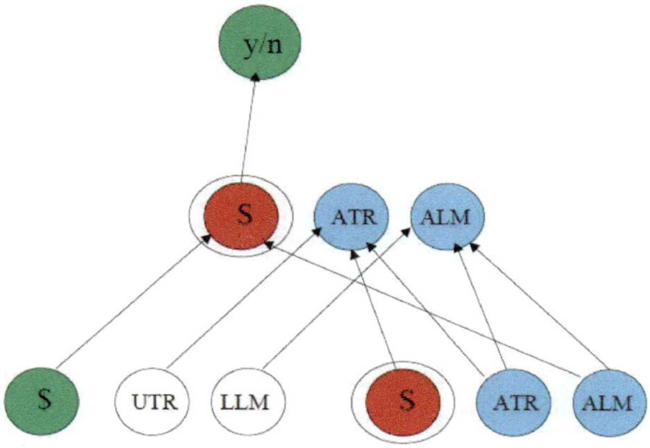

Fig. 5.14. A simple recurrent network which represents the visual grammar of Figure 5.13.

inputs through the use of rounded sigmoid units within the receptive field map. Under this approach, however, the possibility of a distributed symbol[14] leads to the further concern of fictitious state transitions, arising as a result of the conjunction of two input unit firings rather than as a result of the conjunction of an input symbol with an existing state[15].

It is at this point that the somewhat arbitrary inhibition and enhancement of enhanced contrast map unit activations, as foreshadowed in Section 5.2.4, and as reflected in the activations of the receptive field map, become genuinely significant. In particular, suppression of the inputs corresponding to the non-attended object reliably provides strong contrast between RF map frames, supporting the representation of spatial transitions and eliminating one potential source of erroneous state information.

Yet while discrimination at this level is fundamental to the performance of our model, it is not of itself sufficient to resolve the conflict identified in the preceding paragraphs, which turns upon the connectivity level supported by the model. Broadly speaking, sparse connectivity will tend to favour purity of representation at the expense of robustness, with the converse occurring for denser linkages. Striking a balance between the extremes is a non-trivial exercise, and may result only in a regime which provides unreliable hosting of corrupt representations. Moreover, sparse connectivity between context and state units greatly contributes to the decay of state information,

[14] For example, an attended landmark may be represented through two saturated inputs and it is difficult to justify a selection between them.

[15] It may appear tempting to limit this problem through additional resolution at the receptive field map, but such an approach only defers the inevitable, and one must expect to be faced with objects which are not neatly represented by a single, strongly activated unit.

imposing clear limitations on the length of sequences which may be successfully encoded.

As our early experimentation with the model showed some problems with over-generalisation, we shall here favour sparser connectivity, but we shall attempt to limit the resulting state decay by imposing some local structure within the output network. To this end, each receptive field map unit provides random connections to the output network state layer in the usual fashion, with radii chosen such that a given map unit is the dominant input to one sub-region of the layer. In this way, we provide a partial solution to the problem of state decay by concentrating state information associated with a particular input symbol within a single 'assembly' of units. This approach has the side benefit of greatly reducing the probability that fictitious state information will be formed as a result of a distributed input symbol.

The second criterion of purity of state representation is then addressed through extremely sparse connectivity between context and hidden layers, with probabilities chosen such that we expect only 1-2 inputs to each hidden node from the context layer ($\alpha \approx 0.01$). As will be seen in the subsequent simulation results, such a selection is sufficient to support appropriate transition encodings in a substantial majority of cases, but we cannot expect to address sequences of unbounded length.

5.5.5 Static Relation Simulations

Even with these constraints upon the network connectivity, simulations for static concepts using only prototypical training examples seldom require any weight adjustment to represent the concept, the concept layer bias toward disjunctions over the forward layer being sufficient to recruit an abundance of activated nodes. As one might expect, however, such a representation is a dramatic over-generalisation of the concept, and it is necessary in each case to modify the regime of Section 5.5.3 to provide training examples which negate the target concept. Here we use strong implicit negative evidence from directly antonymic relations. For example, we train our network on prototypical examples of *above* – for which we use a target response of 1.0 – and equally prototypical examples of *below* – for which we use a target response of 0.0 – but we do not employ additional relations with intermediate target responses. A schematic representation of the training images is provided in Figure 5.15.

The question of sequence length is addressed in this section through explicit training upon relation sequences of 3,5 and 7 frames, the bounds being provided below by *deg* – the smallest number of frames consistent with the relation – and above by the longest sequence which may be reliably captured for the present network parameter settings[16]. In essence, the role of

[16] Note that the sequence lengths include an additional frame due to the presentation of the agreed string termination symbol, which has the role of gating network output.

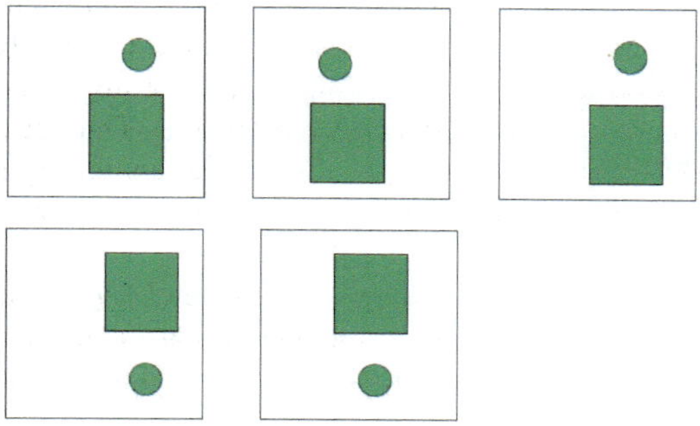

Fig. 5.15. Training examples used for the acquisition of the relation *above*.

our training sets is to refine the set of state paths admissible for the given relation, eliminating those which are found to be associated with directly contradictory relations.

In tables 5.10 through to 5.14 we present results demonstrating the learning and generalisation performance of our networks over the static concepts *above, below, left of* and *right of.* As expected, there is some similarity in the performance levels of antonymic concepts. As in Chapter 4, the results should be regarded as a representative sample of random network conditions rather than an exhaustive study.

Table 5.10. Training results for above, for output networks generated using the usual seeds.

Seed #	Bit Error	RMS Error	Epochs	Comment
1	1/15	$2.28E-2$	5	1 #7s
2	3/15	$2.04E-2$	5	3 #7s
3	5/15	$2.91E-2$	∞	over gen.
4	5/15	$2.92E-2$	∞	over gen.
5	2/15	$6.25E-2$	∞	over gen.

In tables 5.10 and 5.11, therefore, we present training results for the relations *above* and *left of,* for output networks generated using a range of connection seeds. The comment column at the right of these tables reports whether the bit error is due to a failure to represent some of the longer training sequences, or occurs as a result of substantial over-generalisation. In these, as in all simulations reported in this chapter, we have employed an output network with 200 hidden, context and forward nodes, with copy-back connections between the hidden and context layers, akin to providing

Table 5.11. Training results for left of, for output networks generated using the usual seeds. The use of ∞ in the epoch column indicates that convergence is not achieved, the error describing a low amplitude random walk approximately centred about the figure reported. This regime is, in all cases, established within 5-10 epochs.

Seed #	Bit Error	RMS Error	Epochs	Comment
1	4/18	$2.65E-2$	∞	3 #7s; 1 over gen.
2	6/18	$2.80E-2$	∞	3 over gen.
3	7/18	$2.70E-2$	∞	2 over gen.
4	6/18	$2.60E-2$	∞	3 over gen.
5	7/18	$2.80E-2$	∞	1 over gen.

Table 5.12. Testing on the abv test set. The results show unit activation coupled with a bit indicating error if set. The right-most columns indicate approximate TR and LM position.

Movie #	# Frames	abv(1)	abv(2)	left of	LM	TR
1	5	0.61/0	0.84/0	0.46/0	SW	NE
2	3	0.71/0	0.85/0	0.64/1	SE	NE
3	7	0.61/0	0.27/1	0.48/0	S	NNE
4	3	0.71/0	0.85/0	0.64/1	SSE	NNE
5	5	0.62/1	0.27/0	0.54/0	SE	SW
6	3	0.27/1	0.27/1	0.27/0	SE	N
7	5	0.63/0	0.85/0	0.50/0	S	NNE
8	3	0.27/0	0.27/0	0.27/0	S	W
9	3	0.71/0	0.85/0	0.64/0	S	NNW
10	7	0.27/1	0.27/1	0.27/0	SSE	NNW
11	3	0.27/1	0.27/1	0.27/0	S	NNE
12	7	0.27/0	0.27/0	0.27/1	E	WNW

high-probability, self-recurrent connections. Training is in all cases performed using the *Winnow* rule for $\zeta = 2.0$, with strong weight and threshold bias at the concept layer toward realisation of disjunctions across the forward nodes.

Those configurations which result in a low training error are then used in the generalisation tests of tables 5.12 through to 5.14. Note in particular that the networks distinguish correctly between the target, its antonym and the unrelated alternative in a significant majority of cases (60–70% of test examples). The right hand columns of tables 5.12 through to 5.14 give some indication of the TR and LM positions through the use of appropriate points of the compass.

Table 5.13. Testing on the blw test set. The results show unit activation coupled with a bit indicating error if set. The right-most columns indicate approximate TR and LM position.

Movie #	# Frames	abv(1)	abv(2)	LM	TR
1	3	0.71/1	0.85/1	N	SSW
2	5	0.29/0	0.27/0	NNE	SSE
3	3	0.71/1	0.85/1	NNE	SW
4	7	0.27/0	0.27/0	NNE	ESE
5	3	0.46/0	0.44/0	NNE	SE
6	5	0.70/1	0.82/1	N	SE
7	3	0.27/0	0.27/0	NNE	S
8	5	0.29/0	0.27/0	N	SE
9	5	0.66/1	0.85/1	NNE	SW
10	7	0.27/1	0.27/0	N	SE

Table 5.14. Testing on the right test set. The results show unit activation coupled with a bit indicating error if set. The right-most columns indicate approximate TR and LM position.

Movie #	# Frames	abv(1)	abv(2)	left of	LM	TR
1	5	0.61/0	0.84/0	0.46/0	SW	NNE
2	5	0.27/0	0.61/1	0.27/0	NW	NE
3	5	0.27/0	0.27/0	0.27/0	WSW	E
4	7	0.61/0	0.27/1	0.46/0	SSE	NNE
5	3	0.71/1	0.85/1	0.64/1	SW	NE
6	7	0.70/1	0.27/0	0.56/1	WSW	ENE
7	3	0.44/0	0.41/0	0.43/0	NE	SSE
8	7	0.27/0	0.27/0	0.27/0	WSW	SE
9	3	0.71/0	0.85/0	0.64/1	WSW	NNE
10	5	0.27/0	0.27/0	0.27/0	NW	SSE

5.5.6 Dynamic Relation Simulations

The networks employed for the example dynamic simulations differ little from those of the previous section, with the obvious exception of the motion map component of the saliency mechanism. Tables 5.15 and 5.16 show learning results taken over 5 and 7 frame sequences involving both horizontal trajector motion and parabolic curves of the type shown in Figure 5.12, with table 5.17 providing activations on each movie for the best performed networks.

Table 5.15. Training results for *over* for output networks generated using the usual seeds.

Seed #	Bit Error	RMS Error	Epochs
1	3/8	$3.16E-2$	5
2	3/8	$3.13E-2$	5
3	1/8	$2.42E-2$	5
4	4/8	$3.26E-2$	5
5	3/8	$3.28E-2$	5

Table 5.16. Training results for *under* for output networks generated using the usual seeds.

Seed #	Bit Error	RMS Error	Epochs
1	4/8	$3.21E-2$	5
2	3/8	$3.12E-2$	5
3	4/8	$3.28E-2$	5
4	4/8	$3.27E-2$	5
5	1/8	$2.36E-2$	5

Table 5.17. Network responses for the four over and four under training movies, for the best performed networks of tables 5.15 and 5.16.

Movie	over(3)	under(5)	Comment
0	0.29	0.29	over; parabolic; 5 frames
1	1.00	0.29	over; parabolic; 7 frames
2	1.00	0.42	over; horizontal; 5 frames
3	0.89	0.39	over; horizontal; 7 frames
4	0.29	0.29	under; parabolic; 5 frames
5	0.43	0.93	under; parabolic; 7 frames
6	0.45	1.00	under; horizontal; 5 frames
7	0.29	1.00	under; horizontal; 7 frames

5.6 Facial Expression Semantics

In this section we consider an extension of the L_0 task to the domain of facial expressions, and their association with linguistic descriptions of the prevailing emotional state *denoted* by each expression[17]. Human category judgments are obtained from the classical "Pictures of Facial Affect" study (Ekman and Friesen, 1975), the still images of their set of photographs partitioned by Ekman's subjects into the six classes "Happiness", "Sadness",

[17] This section resports joint work with Michael Norris and Joachim Diederich.

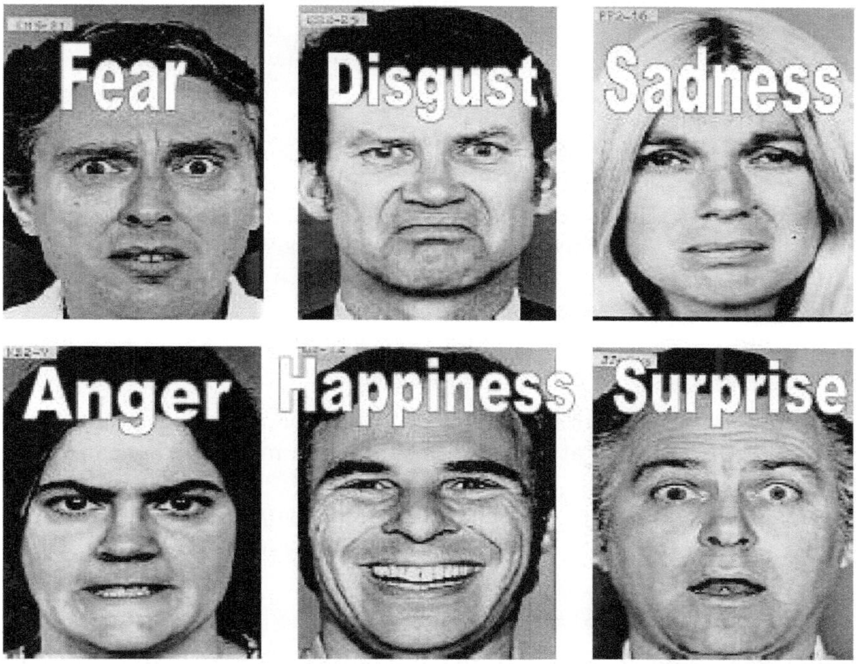

Fig. 5.16. Prototypical examples of the six Ekman and Friesen categories.

"Anger","Fear", "Surprise" and "Disgust" (Figure 5.16). The task of each of our networks is then to recognise one of these categories from training upon positive and implicitly negative examples of that category, these consisting of pre-processed bitmaps derived from scanned versions of the Ekman stills.

As well as using Ekman images unseen by the network for generalisation testing, we also examine network performance on novel images extracted as still pictures from a number of (linear) digital morphs between images derived from the Ekman slides. In this way we are able to identify both the crossover point in the transition 'movie' – the frame at which the final category activation exceeds that for the initial category – and to offer psychologically interesting predictions of perceptual confusion, with some sequences appearing to verge during the transition toward a unrelated category before returning to the appropriate endpoint.

From the cognitive perspective, we argue that this work lies resolutely within the framework of L_0, our task being to establish an association between the semantics of terms such as "happy", and their denotation through the relevant facial expression, which we argue is identifiable through a specific spatial configuration within the image. The literature of face recognition, and even the more specialised area of facial expression processing, is vast, and we shall not attempt to consider it properly in this short section[18]. However it is important to note that verbal descriptions (such as "happy") may be reliably associated with facial expression line drawings by children as young as $4-6$ (Murray and Stanley, 1980), in line with our view of this task as an extension to L_0.

Computationally, our interest lies in the application of a generalised version of the receptive field units described in Section 5.3.1. In these *Domain Gaussian* units (Hogan, Norris, and Diederich, 2000), the intensity at each pixel of a scene is weighted in accordance with its separation from the centre of the map. Formally, we consider an image over a bounded domain $D \subset R^2$ with normalised intensity level:[19]

$$I(x) : D \to [0,1], \tag{5.11}$$

at each point $x \in D$ within the region. Then the activation of the Domain Response Unit centred at c, with transformation matrix T, is defined by:

$$v(I;T,c) = \frac{1}{<G>} \int_D I(x)G(x;T,c)dx. \tag{5.12}$$

Here

$$G(x;T,c) = \exp(-||T(x-c)||_2^2), \tag{5.13}$$

is a Gaussian weight function of $x \in R^n$ centred on $c \in R^n$ and paramterised by the *transformation matrix* $T \in R^{n \times n}$, which determines the shape of the unit's footprint. Usually T is chosen to be diagonal:

$$T = diag(1/r_1, \ldots, 1/r_n), \tag{5.14}$$

r_i being the local radius in direction e_i, the i-th component of the standard Euclidean basis. In the simplest case, $r_i = r, \forall i$, producing a radially symmetric receptive field. The domain unit activation is normalised by its maximum response,

$$<G> = \int_D G(x;T,c)dx. \tag{5.15}$$

In this work we are concerned only with a 2D image map of dimension N^2, each location (in our case a neuron whose activation represents a combination of raw input maps) having identical nominal area, with constant intensity $I_{ij} = I(x_i, y_j)$ across this logical pixel. Equation 5.12 for the domain response is then re-cast to give:

$$v(I;T,c) = \frac{1}{<G>} \sum_{i=0}^{N-1} \sum_{j=0}^{N-1} I_{ij} G(x_i, y_j; T, c), \tag{5.16}$$

[18] See Hogan, Norris, and Diederich (2000) for an extensive review of this material.

[19] This need not be restricted to intensity and may include any suitably normalised scalar combination of image components.

with the corresponding modification to $< G >$. Here (x_i, y_j) is understood to be the centre of the logical pixel (i, j). Our approach departs from the earlier agenda in relying upon gradient descent based weight adjustment. Hogan et al (2000) is best seen as an exploratory study to discover receptive field representations which might be realised through random network components rather than as an attempt to recruit these structures directly.

5.6.1 The Learning Task

In the present study, the 110 images of the "Pictures of Facial Affect" database were scanned from photographic slides to gray-scale pixel images and preprocessed to provide some alignment of facial features between frames, and to eliminate confounding detail. In all cases, the preprocessing operations have an identifiable counterpart within the visual system, and our aim was limited to that of reducing the scope of the facial expression task to that of comparison of images with facial features in gross-level correspondence. Images were scaled (preserving aspect-ratio), cropped and subjected to a localised Gaussian blur and high pass filter The effect is to normalise across complexion, lighting, and hair colour (see Figure 5.17), and the resulting images were reduced to a 'receptive field map' of dimension 30 by 30, a scale at which the facial expressions remain distinguishable by eye.

Images were presented as input to a rectangular array of 64 domain Gaussian units supporting traimable centre and radius parameters. A single concept unit received a weighted sum of the outputs from this array, and targets corresponding to each non-neutral expression were used for training. Due to the paucity of data, the nets are trained on a leave-one-subject out approach, in which *all* of the six images of a single Ekman subject are reserved for testing, each network being trained on the the remaining 104 frames. Receptive field units were observed to quickly 'migrate' across the image to form a small set of highly localised and redundant representations supporting the concept.

5.6.2 Results

Examples of raw generalisation results are shown for unseen subjects NR, MF and MO, in tables 5.18 and 5.19 after 500 and 300 epochs of training respectively. The task provides substantial variation in prototypicality across subjects, with some individual faces being naturally biased – or perhaps biased as a result of the limited technical proficiency of the actor – towards some categories and away from others. The networks of the present study display similar behaviour to the Ekman subjects, with a more pronounced confusion in the 'perception' of Sadness and in the classical confusion of Fear and Surprise. More detailed discussion of the performance is provided in the table captions, and in the additional results presented in Hogan et al (2000).

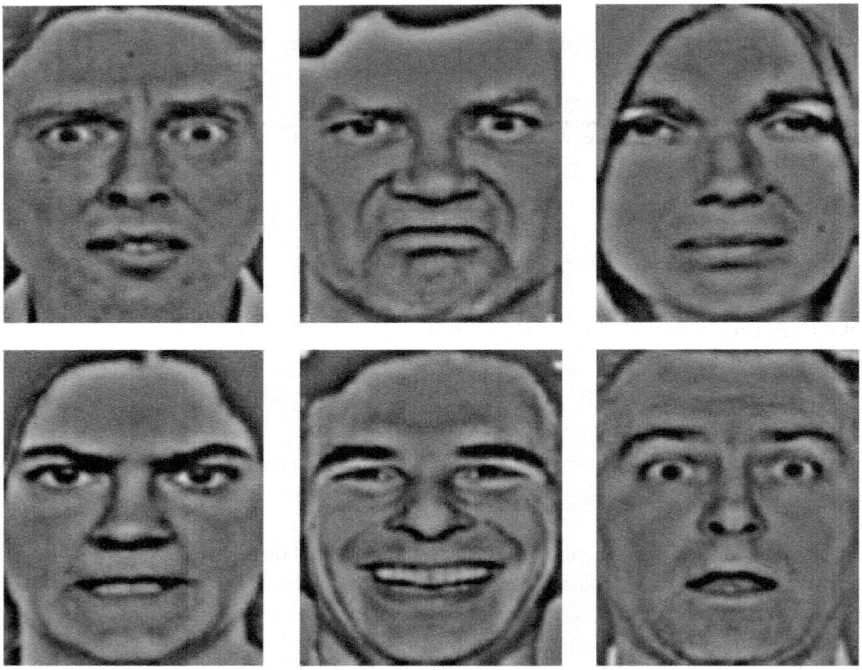

Fig. 5.17. Images derived from the six prototypical examples of Figure 5.16.

The use of digital morphing sofrware allows us to depart from the well settled categories of the Ekman stills to explore the transitions between them, and to make defensible judgments of the category boundaries. For each of the example subjects discussed above, a series of 16 intermediate frames was created, and each frame prepared in the same fashion as before.

Each of the resulting 15 animated sequences (a linear morph between each possible category pairing for a given Ekman subject) is then presented in parallel to all of the model networks. As one might expect, the networks representing the initial and final categories of the morph showed respectively decreasing and increasing activation levels. However, for some of the subjects, the responses of the remaining categories suggest a perceptually confusing trajectory may be followed as the transition takes place, with significant activation of a number of the judgment networks. This is in contrast to our intuitive expectation of a cartesian category space – with transitions constrained to pass linearly through a conceptual origin – although there is more than a hint of this phenomenon in the strong intermediate response of the Sadness network for subjects MF, JM and EM, with Sadness known to be confused with Neutral in some of the Ekman and Friesen experiments.

Table 5.18. Network responses for the unseen subject NR, using judge networks trained for 500 epochs. Network categories run horizontally (Happiness, Sadness, Fear, Anger, Surprise, Disgust) and images vertically, the numbers being those from the Ekman photographs. Each numeric column refers to a category network trained on all other images. Note the correct detection of each category and the strong responses for Disgust (as 'perceived' by the Anger network), and Fear (Surprise).

NR	hp	sd	fr	ag	sr	ds
66: hp	0.60	−0.09	0.26	0.26	0.56	0.16
67: sd	0.30	0.72	0.36	0.54	0.14	0.34
68: fr	0.32	0.17	0.67	0.25	0.50	0.15
69: ag	0.50	0.57	0.13	0.83	−0.15	0.55
70: sr	0.27	0.05	0.46	0.38	0.62	0.45
71: ds	0.49	0.44	0.14	0.77	−0.30	0.86

Table 5.19. Network responses for the unseen subjects MF and MO after training for 300 epochs. These results provide very strong indications of the classical confusions between fear and surprise (note in particular the Fear response to image 54, and the Anger response to image 64). The results are also noteworthy for the extraordinary prototypicality of the MO Happy and Fear images (respectively image number 57 with activation 0.91, and image number 60 with activation 0.96).

MF	Hp	Sd	Fr	Ag	Sr	Ds
48 : Hp	0.75	0.15	0.13	−0.03	0.20	0.23
49 : Sd	0.43	0.64	0.43	0.63	0.55	0.59
50 : Fr	0.33	0.32	0.67	0.38	0.37	−0.17
53 : Ag	0.42	0.40	0.31	0.47	0.43	0.55
54 : Sr	0.26	0.50	0.84	0.51	0.64	0.15
55 : Ds	0.10	0.09	0.01	0.28	0.03	0.64

MO	Hp	Sd	Fr	Ag	Sr	Ds
57 : Hp	0.91	0.23	0.19	0.18	−0.02	0.39
58 : Sd	0.28	0.56	0.30	0.40	0.45	0.35
60 : Fr	0.29	0.14	0.96	0.22	0.44	0.13
61 : Ag	0.25	0.70	0.04	0.67	0.30	0.31
63 : Sr	0.06	0.20	0.46	0.11	0.63	0.22
64 : Ds	0.39	0.48	0.29	0.60	0.28	0.57

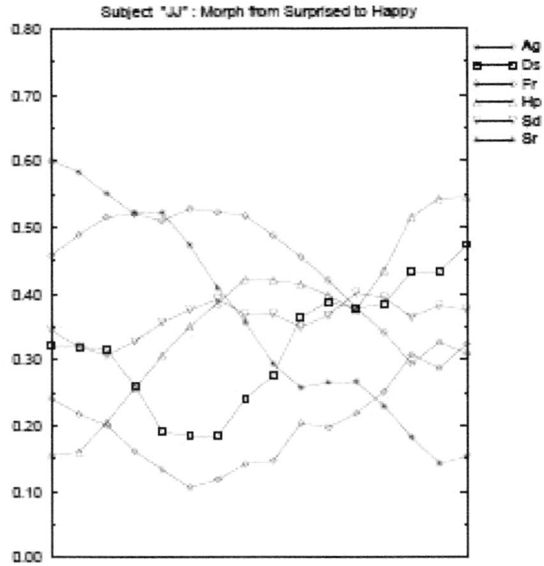

Fig. 5.18. Network responses to the Surprise → Happy transition for subject *JJ*. This figure suggests the existence of a complex transition space between categories, with the prediction of an intermediate perception of *Fear* as the face changes expression between Surprise and Happiness. Moreover, the middle region suggests a substantial break between the often correlated responses of Anger, Disgust and Sadness.

Figures 5.18 through 5.20 show the response of particular category networks to the image transition indicated, with a more detailed discussion of the results being given in the associated caption. Both the static and dynamic results of this section suggesr a category decomposition of the form

$$H_p \vee \{F_r, S_r\} \vee \{A_g, D_g, S_d\}.$$

This assertion is further supported by principal component analysis of the RF unit activations, which suggests that as few as four dimensions may suffice to distinguish the categories. At the highest level, a decomposition into $H_p \vee \neg H_p$ is straightforward, with difficulty increasing as one attempts to distinguish between the Fear-Surprise and Anger-Disgust-Sadness sets, and finally between the elements themselves. In this latter case at least the targets are necessarily probabilistic, and while Ekman's human judges report lower confusion rates than the domain Gaussian nets of this study, they are nevertheless significant within these distinctions.

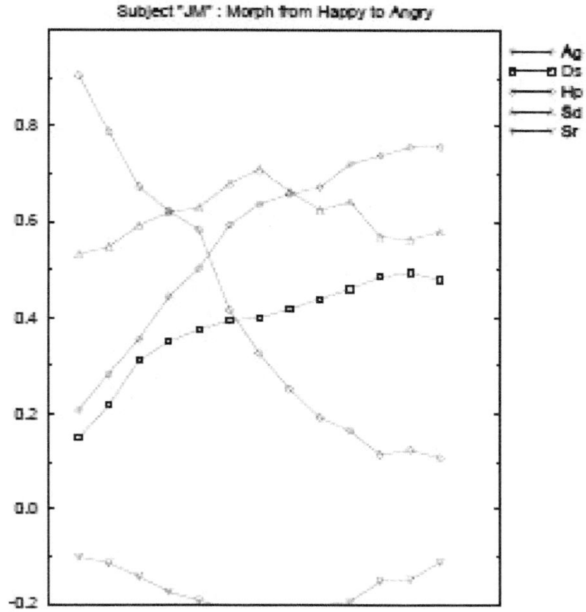

Fig. 5.19. Network responses to the Happiness → Anger transition for subject *JM*. Here we observe an intermediate Sadness peak and moderate correlation between Anger and Disgust. Note the negligible response of the Surprise network.

Such indeterminacy is a fundamental aspect of human categorisation and the results of the present study are in some respects reminiscent of the confusion reported in the 'CUP' study of Labov Labov (1973), cited earlier. At some point, gradual variations are found to give way to hard categorical distinctions, and even the use of controlled linear frame interpolation provides no guarantee that the resulting category judgments will be any more predictable.

5.7 Discussion

In this final section of Part I of this volume, it is appropriate to reflect on the development of recruitment learning in its historical context of discrete time neural systems, and to make some assessment of the currency of these ideas. Unit models such as the LTU (McCulloch and Pitts, 1943) and the perceptron (Rosenblatt, 1961) appear now somewhat incomplete, lacking representational power and computational subtlety. A single real-valued activation parameter is unable to capture the full richness of the spike train, for example,

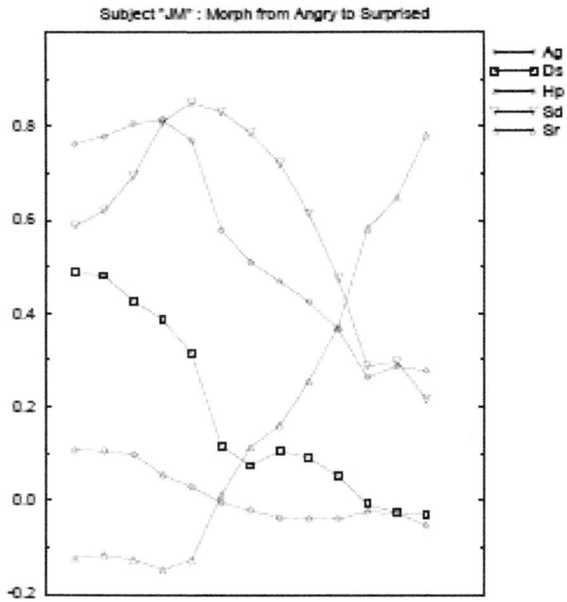

Fig. 5.20. Network responses to the Anger → Surprise transition for subject *JM*. This set of plots shows again an intermediate peak for Sadness, but preserves qualitative similarity between Sadness, Anger and Disgust.

and a single, 'central' transfer function may not adequately model local dendritic summation.

Yet while the discrete time models considered here are giving way to the more realistic spiking neurons of Part II, they remain a very convenient abstraction for investigating cognitive phenomena and bounding the mechanisms which realise them. These chapters have considered the foundations of recruitment learning, both as part of the broader connectionist moverment and as a unique perspective on the rapid acquisition of cognitively interesting tasks. This emphasis was clear from Chapter 1 onwards, with an extensive treatment of the current undertsanding of neural circuitry, which underpinned the connectivity and structural models of Chapter 3. The theme of rapid – even one-shot – learning was developed further in Chapter 2, leading into the analyses and simularions of Chapter 4, which saw the recruitment of whole network structures in support of standard benchmark tasks. In Part II of this volume we shall present a major study of stable recruitment of concept hierarchies within systems of spiking neurons, and in the discussion of Chapter 9 we will explore the unification of these ideas as the central plank of a new generation of recruitment models. For now, it is important to focus on the contributions of the present chapter, and their place in this broader context.

As we have observed, the models of this chapter bear a close familial relarionship to those introduced in Chapter 3, with some specialisation to the tasks at hand. All of these components share the structural primitive of constrained random projections, with decay of connection probability with increasing separation. While binding of disparate representations remains a core task for the system, these networks exhibit sophisticated behaviour, including graded responses to stimuli based on size and position (Section 5.3.1) and acceptance of input sequences as legal strings within regular expression grammars (Section 5.3.2). These representations have the subsidiary effect of challenging traditional notions of learning and generalisation in cognitive tasks, the effect being to partition the input domain into a small set of equivalence classes, each of substantial cardinality. Encoding of a relation at the output network is then fundamentally a matter of memorising a sufficient number of the possible transitions, with generalisation emerging through equivalence over the input examples.

These discussions return us naturally to the theme of rapid learning, and the approach to concept recruitment seen earlier in Chapter 4. Given the scale of the forward layer, and its dependence upon random circuitry, however sparse, it is inevitable that the network should adopt a compendious initial hypothesis. Recruitment learning in sparse neural circuits of this type may thus proceed as much through *elimination* of relations as through their direct acquisition. Learning is therefore extremely rapid – requiring only a single presentation for many static concepts and fewer than 5 epochs for their dynamic equivalents – supporting bursts of knowledge acquisition at the price of some overgeneralisation, and mimicing key aspects of early language acquisition, as demanded by the L_0 manifesto. Recruitmnent learning over sparse neural circuits thus provides an attractive mechanism for exploration of cognitive phenomena, an approach which can only grow in plausibility with the introduction of the spiking neurons of Part II of this volume.

Part II
Recruitment in Continuous-time Neural Networks

Chapter 6
Spiking Neural Networks and Temporal Binding

Cengiz Günay

"Cells that fire together, wire together."
Hebb's rule

6.1 Introduction

Previous chapters of this book gave an overview of connectionist recruitment learning or the *neuroidal* network (Feldman, 1982, 1990; Valiant, 1988, 1994, 1998, 2000b, 2005) as a discrete-time algorithm. By extending recruitment into continuous-time to make it closer to the neuronal networks of the brain, this chapter puts it in the context of spiking neural networks (Hebb, 1949; Von der Malsburg, 1981, 1994; Abeles, 1991; Gerstner, 2001) and shows its similarity to temporal binding (Von der Malsburg, 1981; Singer and Gray, 1995).

The chapter is organized as follows. First, classical neural network research is reviewed to describe the theory of spiking neural network models. Then, the development of concept representation in neural networks and the binding problem is reviewed to present the relationship between recruitment learning and the temporal binding hypothesis. A historical account of these developments is also given.

6.2 Spiking Neurons in Neural Network Research

Artificial neural networks or connectionist networks are mathematical models inspired by biological nerve cells, or neurons. The now 'classical' ANNs were first proposed by McCulloch and Pitts (1943), intended to capture a logical abstraction of the nervous system (see Chapter 1). These networks of simple discrete processing elements were proven to be capable of universal computation. Their processing elements are widely known as McCulloch-Pitts neurons, or linear threshold units (LTUs), seen in Figure 6.1. The inputs and output of the LTU are binary and the function of the gate is to indicate if the weighted sum of its inputs exceeds a threshold. In a recent classification by Maass (1997), the LTU is referred to as a first generation neural network, although earlier abstractions of neurons were made at the beginning of the 20th century. Early machine learning, or Maass's second generation networks, came with the invention of the

J. Diederich, C. Günay, and J.M. Hogan: Recruitment Learning, SCI 303, pp. 183–198.
springerlink.com

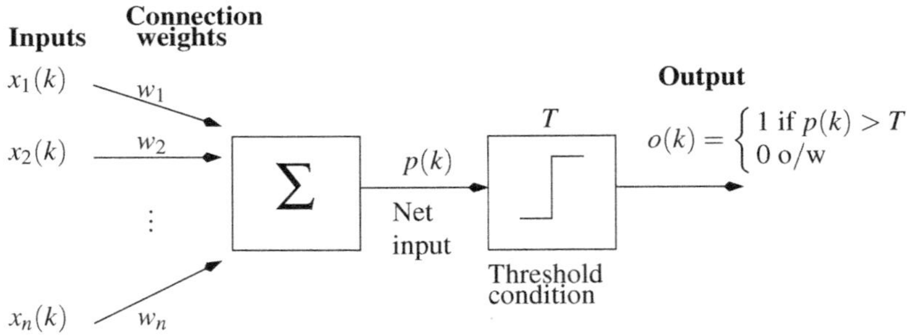

Fig. 6.1. The McCulloch-Pitts neuron, or the LTU.

backpropagation algorithm, which allowed solving linearly non-separable classification problems (Werbos, 1974; Rumelhart et al, 1986a).

More recent advances in neural modeling resulted in – Maass calls these the third generation – spiking neural networks. The essential difference of spiking neurons from McCulloch-Pitts neurons is the interpretation of the activation (output) function. In second generation neural networks, the output of a unit consists of a real value that is hypothesized to represent the average firing rate of a biological neuron at a discrete-time step k (see Figure 6.1). However, in the continuous-time spiking neurons the output at time of firing is the actual spike as seen in Figure 6.2.

Before spiking models, average firing rate models have been commonly used since the 1920's (for a review see Maass, 1997 and Gerstner, 1999). However, the information carried by this rate coding is limited by the natural frequencies of neural firing. For instance, Thorpe and Imbert (1989) observed that certain visual tasks can be completed in about 100 ms by the brain, with the computation involving about 10 stages of propagation across synapses. This leaves enough time for the neurons in each stage to emit only a few spikes, given that highest spiking rate of a neuron is about 100 Hz (Maass, 1997). If neurons were averaging their inputs before

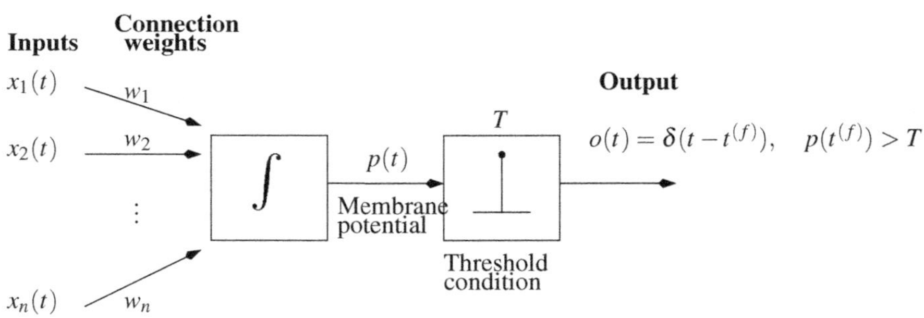

Fig. 6.2. A simplified spiking neuron to compare with the LTU. The schematic does not indicate the reset after spike and the constant decay effects on the membrane potential.

giving an output, this computation could not have been carried out meaningfully with so few spikes. However, to avoid this criticism, it may be argued that the rate coding actually represents the total activity of a neuron population at a time instant (Gerstner, 1999). This approach has other pitfalls by assuming an unrealistic system that has homogeneous populations of neurons connected to points where collective responses are evaluated.

These observations suggested that neurons use a faster code which may be based on the timing of the spikes emitted. Observations indicating that biological systems actually depend on timing of signals have supported this view (Gray et al, 1989; König and Engel, 1995; Singer, 1995; Singer and Gray, 1995; Ritz and Sejnowski, 1997; Engel et al, 1999). A number of alternative approaches to rate encoding have been proposed. These include spike time coding (time-to-first-spike), phase encoding, coding with correlations and synchrony, and stimulus reconstruction with reverse correlations, which all support computation with spiking neurons (Gerstner, 1999).

Von der Malsburg (1981) and Abeles (1991) as well as Hebb (1949) are among those who proposed models empowered with neurons using exact spike timing rather than an average firing rate (Crick, 1984). The following sections discuss why spiking neurons are important for information representation in ANNs.

6.3 Representation, Composition and the Binding Problem

Essential problems in information representation were identified by many researchers in cognitive modeling. As a cognitive architecture similar in spirit to other works (Anderson, 1983; Newell, 1990; Langley et al, 2009), recruitment learning needs to tackle several such problems. We thus consider how recruitment learning employs representational mechanisms adequate for cognitive computation and how it can compose novel concepts as new knowledge is acquired. This section gives an account of the developments leading to these features of representation and composition. We start with the types of representation employed in classical ANNs.

6.3.1 Localist Representations

The *localist* knowledge representation scheme is best described with the extreme view of the single neuron doctrine for perception of Barlow (1972). The doctrine postulates that in principle, one can always find a single neuron in the brain responsible for any decision taken. In time, this definition has extended to include finding single neurons representing any concept perceived or imagined. This implies that a distinct neuron is used to represent each possible combination of these single concepts as well, such as a "blue square" or a "yellow Volkswagen" as shown in Figure 6.3. The localist units that represent such high level features are also called gnostic, or cardinal cells.

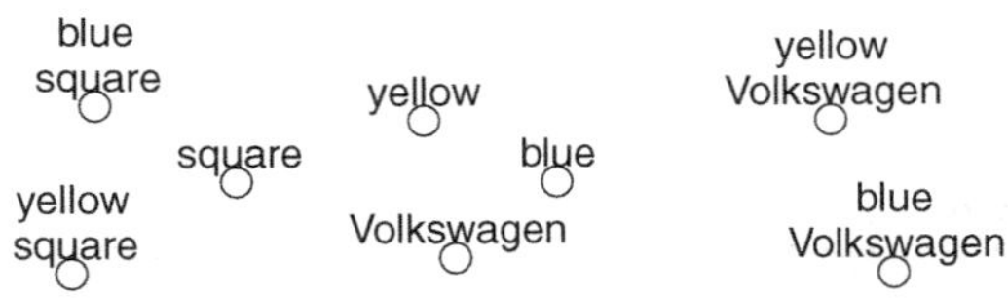

Fig. 6.3. A simple example of an extreme localist representation where each unit represents a single conceptual entity. Notice that while pure concepts such as blue, square, yellow, and Volkswagen take one unit, their conjunctions (e.g., yellow Volkswagen) are also represented by separate units.

The most critical argument against the plausibility of finding a "yellow Volkswagen cell" concerns the storage capacity of the brain, first mentioned by Harris in 1970's (see Weisstein, 1973 and the review of Page, 2000). The argument is that one may need cells to represent general features such as "yellowness" or "Volkswagen," but if a distinct cell is used for an arbitrary combination such as "yellow Volkswagen," then the number of other multi-place combinations that need to be represented is 2^n, which increases exponentially with the number of possible features n. For instance, taking a set of features $S = \{a, b, c\}$, the number of possible combination of features becomes $\bar{S} = \{\emptyset, a, b, c, ab, bc, ac, abc\}$, where the size of the latter is $|\bar{S}| = 8 = 2^3$. These combinations will also unavoidably include never-to-be-used concepts.

One can easily see the kind of problems this combination coding creates, considering gross facts of neuroscience: Taking the 10^6 fibers connecting the retina to cortex, if one uses a purely localist approach to represent pairs of points in the retina (e.g., lines connecting a pair of points), 10^{12} neurons would be needed for all combinations which is more than the 10^{11} available in the whole of cerebral cortex (this example is due to Feldman, 1990).

Another criticism of the extreme localist approach is concerned with its unreliability as a storage system. If the localist scheme was employed in the brain, one would have a "grandmother cell" that fires at the sight of the grandmother (or a celebrity such as Jennifer Aniston or Halle Berry; see Quiroga et al, 2005). Considering that some neurons will always be dying, one would forget a concept completely when the neuron representing it is lost. Obviously, the brain is more fault tolerant than this theory predicts and must have at least a few (or, likely, thousands of) units redundantly representing each high-level concept. For instance, imagine a neuroscientist finding a cell representing an animal's mother, independent of angle of view, or even if the display is an abstraction. If the subject lost the cell permanently and therefore was not able to imagine its own mother, even though it can imagine all other things related to her.[1] This is unlikely because finding this cell does not rule out the existence of other cells representing the same concept. Note

[1] This story originated from Jerome Lettvin, in a talk given in 1969 (see appendix to Barlow, 1995).

that we make a distinction between having units for the constituents of a concept and having a unit that unambigously represent a concept.

This kind of extreme localist view is admitted to be faulty in more recent discussions (Feldman, 1990). Furthermore, contrary to the neuron doctrine for perception, it was found that a single stimulus can activate large populations.

6.3.2 Distributed Representations

The most prevalent alternative to the much criticized single-neuron, extreme localist representation is distributed representation. Distributed representations are purely featural representations that show fault tolerance and have generalization capabilities (Hinton et al, 1986; Page, 2000). They are sometimes called holistic or holographic representations because they involve the whole set of units in the representation of any entity.

Initial motivation for this kind of representation came from the early work of Lashley (1950). He stated that the reason for certain patterns of deficits in lesioned rats can be best explained by looking at the total amount of cortex removed. Since more recent findings show that most of the primary sensory and motor areas have specialized structure, he hypothesized that distributed representations are only used in the higher brain areas whose functional organization is not yet understood. Another motivation for distributed representations was the idea of dynamic cell assemblies of Hebb (1949). Research in these kinds of dynamic cell activity were pursued by the temporal binding proposal of von der Malsburg (1981, 1995) and synfire chains of Abeles (Abeles, 1991, 1994; Abeles et al, 2004), which have some experimental support (Ikegaya et al, 2004; but also see Mokeichev et al, 2007).

In strictly distributed representations, concepts are represented by multiple neurons, and the same neurons can take part in the representations of different concepts, thereby offering economy in representing composite concepts. Furthermore, distributed representations enable representation of general rules that belong to a group of objects that have similar, but not identical features. Since individual instances are not represented separately but by activation of feature groups, then a class of concepts having common features will share neurons in their representations. This enables forming mechanisms that act on behalf of the whole class, therefore offering generalization capabilities. Also, since in distributed representations concepts are represented by the activity of a large number of neurons, they are more tolerant to fault and noise than localist models.

Distributed representations also have their disadvantages. Since a representation of a concept may be distributed over many features, another concept that shares similar features cannot be represented simultaneously, thus preventing parallelism. Another disadvantage of distributed representations is their apparent lack of suitability for representing structured knowledge (Hinton, 1990; Feldman, 1990; von der Malsburg, 1995). There is an exception to this limitation in the recent studies that use distributed cell assemblies, for which there is experimental evidence from the hippocampal regions of the brain (Lin et al, 2005). These assemblies, when put into the context of *formal concept lattices*, can represent whole concept hierarchies (Wennekers, 2009) and can be used to achieve arbitrary cognitive functions (Wennekers and Palm, 2007). However, these studies use lateral inhibition to alleviate the shortcomings found in purely distributed representations.

To illustrate, consider the task of representing the two objects; 'grandmother,' and 'the white house' in a fully distributed fashion. It is straightforward to represent each object with its set of features separately. However, what if we want to represent 'the grandmother in the white house'? A simple superposition of their features will result in crosstalk of the two different objects, yielding a meaningless result. This inability to combine objects in distributed systems is thus named by von der Malsburg (1995) as the *superposition catastrophe*. Other solutions to the problem of representing structured knowledge so far have not yielded satisfactory results for fully distributed representation systems. The difficulty is also because, contrary to localist systems, observing an individual neuron does not reveal the state of the whole system, thus making distributed representations more difficult to interpret.

6.3.3 Representation in the Time Dimension

In recruitment learning, an integral object is represented by correlated activity of its features. This representation method involves the temporal properties of neuronal firings, and is therefore called *temporal binding,* originally proposed by Christoph von der Malsburg (1981) (reprinted as von der Malsburg, 1994). The temporal binding mechanism uses synchronous firing activity to represent features that belong to a single object, and desynchronized activity to distinguish features of different objects (Milner, 1974; von der Malsburg, 1981; von der Malsburg and Schneider, 1986; von der Malsburg, 1995; Abeles, 1991, 1994; König and Engel, 1995; Singer, 1995; Singer and Gray, 1995; Engel et al, 1999).

However, von der Malsburg's proposal was not the original motivation for Valiant's neuroidal network (Valiant, 1994) that implemented recruitment learning. Valiant rather sought a network that can dynamically link its nodes that represent previously known concepts to new nodes that are intended to represent a novel concept. This mechanism is our main subject, recruitment learning, which operates on a network that consists of nodes with sparse random connections, originally proposed by Jerome A. Feldman (1982). The way novel concepts are formed requires simultaneous activation of existing concepts that relate to the novel concept, which is essentially the same mechanism proposed by von der Malsburg. Although similar, there are differences in the approaches of Feldman versus von der Malsburg, mainly in having different motivations. On the one hand, von der Malsburg proposed temporal binding inspired by the synchronized firing activity observed in the brain. Feldman's recruitment learning, on the other hand, focused on issues like avoiding crosstalk and combinatorial explosion in representing binding structures, and additionally specified how to compose novel concepts in connectionist models through learning.

The following discussion describes, in detail, the motivation behind these two parallel approaches and their similarities. Then, an account of contributions from other researchers to Valiant's neuroidal model are given.

We are going to quote, in places, von der Malsburg's account of the history of temporal binding (see review in von der Malsburg, 1995). The issue of binding was identified as a central problem in neural representation earlier by Rosenblatt (1961). The term 'binding,' however, is due to another line of research on the problem of *variable binding* in the knowledge representation subfield of artificial intelligence (AI) which is explained next.[2]

6.3.4 The Variable Binding Problem

The variable binding problem is a special case of the more general binding problem. Variable binding is concerned with finding appropriate bindings for first-order logic variables in rule-based systems. This serves to correctly assess relations between objects. Binding an object to a variable means using the variable to represent the object.

The problem can be demonstrated with an example in the context of a perceptual recognition task as follows (see Figure 6.4). As the knowledge base of a system, take two first-order logic sentences describing the properties of objects in a visual scene. The first sentence specifies the geometric properties of available objects with $\exists a,b\ \mathsf{square}(a) \wedge \mathsf{triangle}(b) \wedge (a \neq b)$ which is interpreted as "there exist a square and a triangle (in the scene)." The second one specifies the location of the available objects by using predicates that can detect the presence of any object in parts of the visual field. This sentence is $\exists c,d\ \mathsf{upper}(c) \wedge \mathsf{lower}(d) \wedge (c \neq d)$ which is interpreted as "there exist two objects such that one is in the upper part and one is in the lower part (of the scene)." Possible input scenes described by the predicates employed are depicted in Figure 6.4a.

Assume one wants to use the available information in the knowledge base to draw inferences. Let us consider the rule "if there is a triangle at the upper part, then assert (something)." This rule can be encoded in first-order logic with $\exists a\ \mathsf{upper}(a) \wedge \mathsf{triangle}(a) \supset \mathsf{something}$. Drawing the inference requires verifying the antecedent. That is, if "there is a triangle in the upper part," the consequent something will be implied as seen in Figure 6.4c. This is not obvious because the relations between the variables a,b and c,d in the sentences of the knowledge base are not known to the system. A search must be conducted by testing possible bindings of variables that correctly represent the scene. Here there are just two possible bindings: $\{a = c, b = d\}$ or $\{a = d, b = c\}$. It is seen that the rule can be applied if the first binding $\{a = c, b = d\}$ holds on the input scene. However, in more general cases this search takes time growing factorially by the number of variables in each sentence and exponentially by the number of sentences in the knowledge base. For researchers in the AI field, the variable binding problem is the time complexity of this search (Valiant, 1998).

[2] The term binding is alternatively called fast links (Bienenstock, 1999).

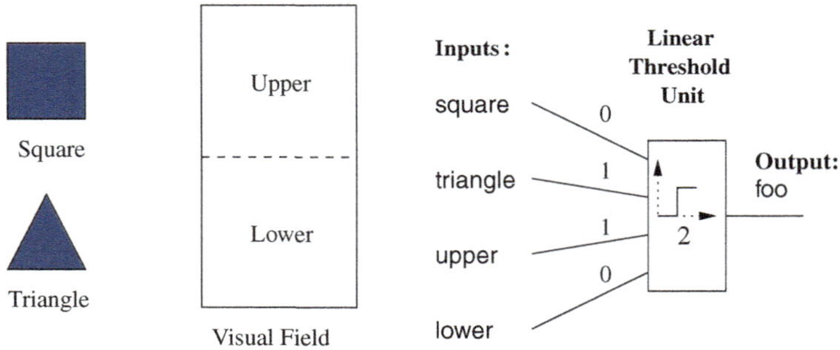

(a) Elements of the scene as predicates.

(b) The network architecture in Rosenblatt's example.

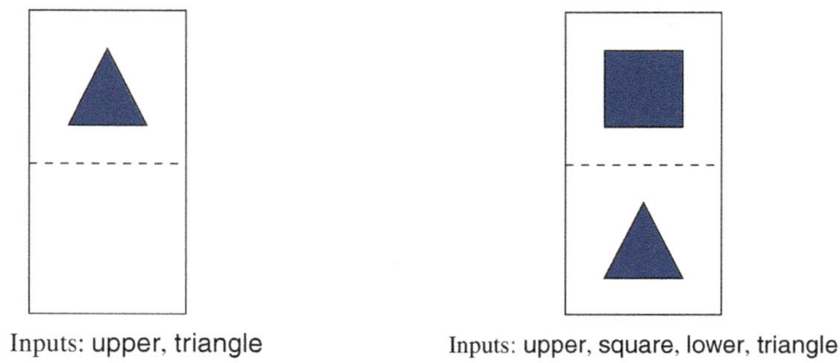

(c) The target scene. Activates the two predicates triangle and upper that can be recognized by a computational mechanism.

(d) The scene that causes the incorrect detection by activating the same inputs as the case on the left, however having a different configuration.

Fig. 6.4. Example for demonstrating the binding problem. The four predicates given in the text are abbreviated as square, triangle, upper and lower for simplicity. See text for details.

The binding issue described above becomes graver if no variables are allowed in the representation. This is the case if propositional logic is employed instead of first-order logic, where each proposition specifies the properties of the totality of the input. The contents of the knowledge base in the above paragraph would then become square-object $\wedge$ triangle-object and object-in-upper-part $\wedge$ object-in-lower-part and the rule would become object-in-upper-part $\wedge$ triangle-object $\Rightarrow$ something. The antecedent of the rule detects the presence of two propositions, that is, a production system needs to test if "there is an object at the upper part" and that "there is a triangle." The system can correctly detect the presence of the triangle at the upper part of the scene depicted in Figure 6.4c because the two propositions required for the rule are available. However, the rule does not assert that it is indeed the triangle which is at the upper part of the scene. The rule works correctly as long as the triangle is the only object in the scene.

In order to show the conflict, another object needs to be added to the scene. Consider the scene depicted in Figure 6.4d where there is a square at the upper part and the triangle is shifted to the lower part. This case activates all four propositions: square-object, triangle-object, object-in-upper-part, and object-in-lower-part. Having more than one object while employing propositional representations creates an ambiguity of which features belong to which object. The result is catastrophic; the two propositions required by the rule are still available, and therefore the rule will succeed by incorrectly detecting a triangle at the upper part even though there is none!

This brings us to the example described originally by Rosenblatt (1961) to illustrate the more general problem of binding. His example is given next to introduce the binding problem. The example is designed to demonstrate the limitations in representational machinery of classical neural networks.

6.3.5 The Binding Problem

Rosenblatt, in describing the binding problem, uses a neural network as seen in Figure 6.4b to recognize the geometric-shapes-at-two-locations task discussed above. Like in propositional logic, inputs carry no binding information in classical neural networks, and active inputs simply indicate objects that exist in the scene. If the network is trained to recognize the stimulus in Figure 6.4c, its objective is to detect the conjunction of the two active inputs in this case: triangle-object and object-in-upper-part. If the network is presented the scene in Figure 6.4d with the square above the triangle, it would receive activity on all four inputs simultaneously and spuriously detect a triangle in the upper part like propositional logic would. Thus, when there is more than one object in the scene, neural networks suffer from the binding problem because of the crosstalk of features from different objects.

This inherent limitation of the classical neural networks has generally been avoided by employing combination-coding schemes for encoding input representations (Rumelhart et al, 1986b). The combination-coding approach is an extreme localist knowledge representation influenced by the neuron doctrine for perceptual psychology of Barlow (1972) that we discussed earlier in §6.3.1, where separate inputs are added for each possible combination of features that may appear simultaneously (e.g. upper-triangle, upper-square, lower-triangle, lower-square). However, this makes the number of inputs grow exponentially with the number of interacting objects, causing a combinatorial explosion when a realistic number of features is considered. Thus, the nature of the binding problem becomes more in representation than in computation unlike variable binding in first-order logic.

Von der Malsburg remarks that the ambiguity arising from the lack of binding information in inputs prevents neural networks from being trained faster. If the binding issue is overcome, he adds, the network should be able to learn input stimuli with a single presentation, instead of being trained over a large number of trials. Otherwise without binding information, the network has to guess appropriate correlations between inputs by observing a large number of samples from the statistical distribution of the data.

Finally after presenting the limitations of different computational schemes, the binding problem of representation can be summarized by the question of how related inputs should be grouped together.[3] The term binding has also alternatively been called *fast links* (Bienenstock, 1999).[4]

6.3.6 Temporal Binding

Von der Malsburg (1981) supports the arguments against the combination-coding approach associated with the localist representations discussed earlier. However, instead of proposing a strictly distributed approach as an alternative, he postulated that the brain uses transient representations for arbitrary conjunctive concepts such as the yellow Volkswagen. These representations are temporary since the brain may only need them during a task, such as analyzing a scene, dispelling the problem of combinatorial explosion.

The composite concepts are formed by dynamically binding distributed groups of neurons that represent their elementary features.[5]By employing distributed representations, temporal binding inherits some of their advantages, such as the ability to generalize. However, these purely dynamical representations are also rather limited if they cannot be transformed into long-term storage, that is if novel concepts cannot be composed out of the binding structures (Feldman, 1990).

Von der Malsburg's proposal is best known as the theory of *temporal binding*, that is binding via synchronous activity of cells, to prevent the problems described.[6] This theory can be briefly described as to temporarily associate entities that are active simultaneously to coherently represent an object. Conversely, explicitly separated, or desynchronized, activity represents different objects. Temporal binding, by using time as coding space, requires only elementary feature units to be present and allows combinations to be formed dynamically via transient potentials at interconnecting synapses. In terms of the number of units needed to represent entities that a cognitive system is exposed to, the magnitude is thus lowered from being exponential to linear with respect to the number of features.

Unlike a classical artificial neural network, the computational system to recognize the scene needs to be augmented with the notion of time. Labeled patterns, forming training examples, in classical neural networks are presented to the network one at a time without any relational information. With temporal binding, an

[3] The bindings in the example are fairly simple with few elements, whereas in real world situations large numbers of bindings are required.

[4] Bienenstock finds the term binding problem "imprecise and stale" which is consistent with the account given here. He speculates on naming the issue "... specific idiosyncratic short-lived functional links. In short, fast links."

[5] In contrast, these simpler features are assumed to have localist representations (von der Malsburg and Schneider, 1986).

[6] The term 'temporal binding' should not be confused with the description of bindings required for maintaining continuity in temporal events (Treisman, 1996).

example is presented to the network for a continuous time duration. The propositions that appear synchronously are bound together and said to represent a single concept or object.

Temporal binding is motivated by the study of the brain. In the context of computations that take place in the brain, functionally and physically separate areas of the sensory cortex (e.g., visual, auditory, somatosensory) analyze different stimulus features in the environment. Temporal binding can explain how these physically distributed feature representations are combined to form coherent unitary percepts. This proposal is further supported by psychological evidence that human subjects exhibit binding errors of detecting *illusory conjunctions* if they are given insufficient time to analyze a scene. This suggests that timing may employ an important part in perceptual processing in the brain (Treisman and Gelade, 1980; Treisman, 1996). Experimental studies of Singer and Gray (1995) provide supporting evidence for temporal binding in their *temporal correlation hypothesis*. Many other experimental (Gray et al, 1989; Singer, 1995; Singer and Gray, 1995; König and Engel, 1995; Ritz and Sejnowski, 1997; Usher and Donnelly, 1998; Engel et al, 1999) and simulation (von der Malsburg and Schneider, 1986; Hummel and Biederman, 1992; Shastri and Ajjanagadde, 1993; Schillen and König, 1994; Lisman and Idiart, 1995; Terman and Wang, 1995; Sougné and French, 1997) studies followed these proposals. There are also critics of the phase-coding approach to the temporal correlation hypothesis (Wennekers and Palm, 1997; Knoblauch and Palm, 2001a,b), which we review in Chapter 7.

Consider again the geometric-shapes-at-two-locations example. The use of temporal binding in Figure 6.5 should solve the ambiguity that is demonstrated earlier. The single-object scene of Figure 6.4c can be represented with the time profile of

(a) Simultaneous activity representing the scene in Figure 6.4c.

(b) Two separate groups of simultaneous activity (ignoring delays) that form the necessary bindings for correctly representing the scene in Figure 6.4d.

Fig. 6.5. Demonstration of temporal binding on the example developed earlier in Figure 6.4 to overcome the binding problem that arise with propositional representations. Time profile of activity in inputs are depicted in the above figures. Impulses indicate activity of the inputs as neuronal spikes.

input activity shown in Figure 6.5a. Simultaneous activity in the inputs triangle-object and object-in-upper-part results in forming the concept of the triangle object taking the role of being in the upper part. Such a binding is not necessary in a single-object scene as discussed before. However, the advantage of temporal binding becomes more obvious with the multiple object scene that created ambiguous inputs to the earlier propositional system. The scene depicted in Figure 6.4d is represented by the activity in the inputs as shown in Figure 6.5b. Here, two groups with desynchronized activity unambiguously represent the two separate objects in the scene. The recognition mechanism is not mistaken any longer by detecting a triangle in the upper part when temporal binding is employed.

6.3.7 Dynamic Connections versus Temporal Binding

At about the same time as von der Malsburg, Feldman (1982) independently proposed the dynamic connection network, which is a equivalent solution to temporal binding, in a neural network (or connectionist modeling) context (see §6.3.1 and Chapter 1). Although Feldman's network was sometimes criticised to be a type of static, combination-coding structure (Shastri and Ajjanagadde, 1993; von der Malsburg, 1995), it is not affected by combinatorial explosion. Both of these two ideas seem to have been influenced by the cell assemblies of Hebb (1949) – see Figure 6.6.

6.3.8 Discussion

As mentioned earlier, both von der Malsburg and Feldman's solutions to the binding problem are analogous, although they were possibly proposed independently. Whereas von der Malsburg approaches the problem from the neurophysiological perspective by emphasizing synchronous firings reported in the brain, Feldman discusses more of the systems and computational aspects. Von der Malsburg formulates the problem successfully since his work has been most influential (Crick, 1984; Singer and Gray, 1995). Feldman's proposal has mostly been criticized for combinatorial explosion and unreliability due to the localist representation. Only later, Feldman (1990) defends his view against each of these criticisms.

Feldman's stance was at neither end of the localist-distributed dilemma reviewed back in §6.3.1. The representations with $c \log N$ redundant units is a middle point between localist and distributed extremes, essentially still advocating a localist approach. The redundancy and randomness, as Feldman (1990) argues, counters arguments made against localist representations. This mode of representation is also called *distributed localist* or *modularly distributed* (Browne and Sun, 2001).

Even though in both approaches the mechanisms are triggered by synchronous activity, in Feldman's approach this is not necessarily temporal synchrony. In Feldman's units, the output reflects the average frequency of firing and not a single firing event. This puts Feldman's network into the category of second generation neural networks discussed earlier. It was Valiant who later pursued Feldman's approach to

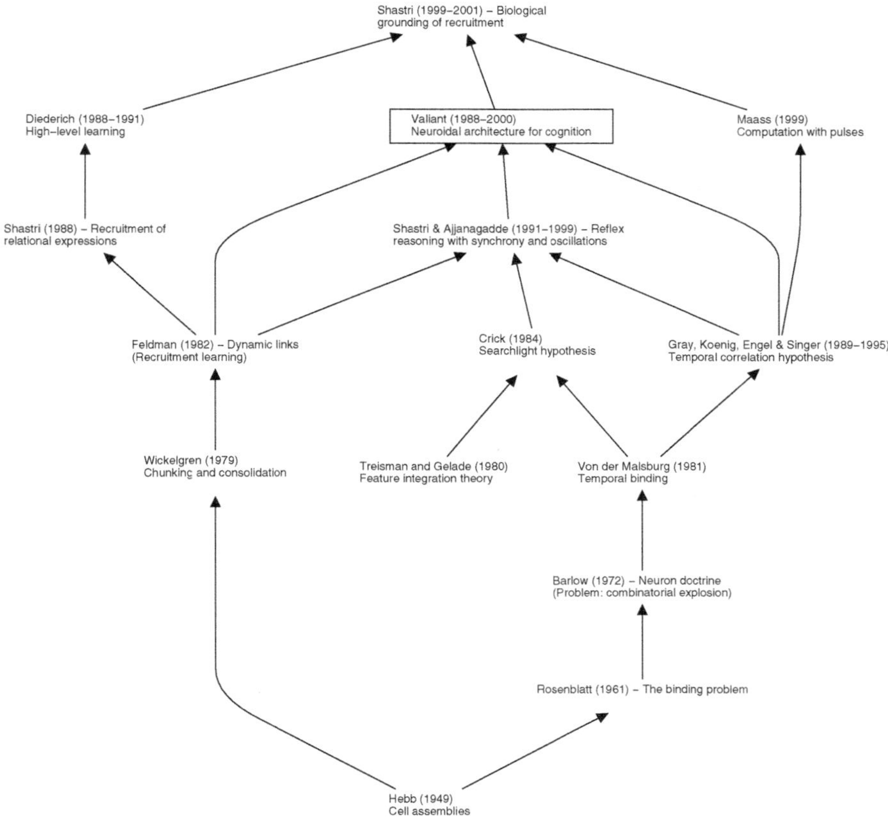

Fig. 6.6. Historical progress described by a citation graph leading to the proposal of recruitment learning. The graph shows ideas of researchers contributing to the notion of concept representation and composition in Valiant's neuroidal model. Feldman and von der Malsburg apparently proposed similar approaches independently. For simplicity in the graph, most direct influences from hierarchically lower items are not repeated if there is an indirect path connecting them.

change the interpretation of its outputs to single firings, even though in a simplified binary form (Maass (1997) acknowledges that this type of network conforms better to the third generation of neural networks). This leads to the unification of the two theories of temporal binding and recruitment learning, both of which so far had been developed separately. Even Shastri (2001a, 1999b), who has worked on synchronous firings in semantic networks, did not put recruitment learning into the picture until later. See Figure 6.6 depicting the historical development of the above discussed ideas.

To reiterate, the similarities and distinctions between the temporal binding and the most recent proposed recruitment learning approaches can be given as follows:

1. Both use synchronous activity to indicate tuples to be bound;
2. Temporal binding is purely dynamical and simply associates the tuples together, whereas recruitment composes a novel concept representing the binding; and
3. Temporal binding does not specify an interconnection scheme, some implementations use full or local excitatory and global inhibitory connectivity, whereas recruitment learning uses properties of random graphs with redundant representations to achieve a feasible concept production scheme without causing a combinatorial explosion.

Von der Malsburg and Feldman's proposals influenced many researchers in different fields. One of Feldman's students, Shastri (1988, 1999b, 2001a), adopted temporal binding for representing relations between objects in a neural network as a logical inference system. This network employed temporal binding, and therefore used *phase-coding,* for representing features of objects. A number of significant methods were proposed within this framework, including a method for representing relational expressions in a neural network context and drawing inferences via reflexive reasoning. The model incorporated synchronous firing and oscillatory activity similar to that of the brain, with neurally-inspired units and arbitrary connection topologies crafted for solving various logic problems (Shastri, 1988; Shastri and Ajjanagadde, 1993; Shastri, 1999a). However, the biological plausibility of this system was criticized (Dawson and Berkeley, 1993). Other applications of temporal binding include invariant shape recognition (Hummel and Biederman, 1992), and binding in multiple feature domains (Schillen and König, 1994). Later studies followed, once the connection between recruitment learning and temporal binding was established with spiking neurons.

6.4 Spiking Neurons for Recruitment Learning

Even without employing spiking neurons, recruitment learning has been biologically motivated (Wickelgren, 1979; Feldman, 1982; Valiant, 1994). It is also compatible with more recently discovered phenomenon of neurogenesis as reviewed in Chapter 1, which explains how new neurons can be recruited into a network. However, even in the absence of the support from neurogenesis, the algorithms described in the next two chapters are consistent with biology when they employ more realistic spiking neuron models. For instance, recent experimental evidence from monkeys showed that synchronized firing of groups of neurons in the motor cortices correlates with bindings of combinations of sensory and motor parameters (Stark et al, 2008). Synchronized firing is also what drives the finite state machine (FSM) algorithms that we use to switch the state of neurons or synapses in Chapter 7. These algorithms are consistent with spike-timing dependent Hebbian and anti-Hebbian plasticity rules (Bi and Poo, 1998, 2001; Leslie et al, 2001; Lau and Bi, 2005) that result from the experience-dependent spine formation of neocortical synapses (Turrigiano et al, 1998; Turrigiano and Nelson, 2004;

Nelson and Turrigiano, 2008). In experiments, the spike-timing dependent Hebbian plasticity correlated with formation of new synaptic spines that were stable for hours at a time (Turrigiano et al, 1998) and were actively preserved by homeostatic mechanisms (Rabinowitch and Segev, 2006). This phenomenon is simulated by the *memorized* state in recruitment learning. Conversely, the anti-Hebbian rule scales all synaptic weights to normalize them for learning new knowledge on top of previously memorized synaptic patterns (Leslie et al, 2001). These observed phenomena can be simulated in recruitment learning if it is extended to a continuous-time algorithm by employing spiking neuron models with realistic synapses.

Spiking neurons were first proposed for recruitment learning by Shastri (2001a). The integrate-and-fire (I/F) spiking neuron model featured a realistic synaptic response profile and spiking and bursting modes of activity. This model demonstrated the biologically-grounded properties of long term potentiation (LTP) and depression (LTD). In this and following works, Shastri (2001b, 2002a,b) used this spiking neuron recruitment framework to model temporal binding based on observed biological parameters of the hippocampal regions of the brain. A more complete and coherent picture of biological learning mechanisms in the context of binding has been reviewed more recently by Huyck (2009), who showed that stateless and transient bindings (e.g., formed by synchrony) can be simulated with biologically grounded short and long term potentiation algorithms to demonstrate a robust performance in keeping the persistence of bindings, achieving a large capacity and still providing the speed of one-shot learning. Huyck (2009), however, did not address the synchrony and stability issues we are interested in spiking neurons.

Historically, Shastri's work was followed by Günay and Maida (2006a,b) that we address below in Chapters 7 and 8, and by James and Maida (2009) that also employed spiking neurons. In the latter, some of these neurons were also delay units, which repeat previously presented inputs, to memorize temporal sequences of concepts in a hierarchical recruitment network. The network achieved this by using the biologically-inspired features of winner-take-all (WTA) and heterosynaptic long-term depression (LTD). It was shown that the LTD mechanism is necessary to stop the same units to be recruited more than once, and that the WTA mechanism is needed to control the size of the recruited unit set, similar to what we demonstrate in Chapter 8. These biologically-realistic mechanisms therefore were sufficient, without the finite-state machines employed in the vicinal algorithms of Valiant (1994), to demonstrate recruitment learning for sequence learning. James and Maida (2009) demonstrated coding of temporal sequences of 4 digits as an XOR problem with a 4-layer network, and then presented a 5-layer network that can memorize 7-digit phone number sequences. This work is an important extension of the approach described in the following chapters, generalising the results to more elaborate network structures and providing a continuous time counterpart to the studies of Hogan encountered in chapter 4.

However, in all the networks of James and Maida (2009), the perfect synchrony required for temporal binding was achieved by the equidistant feedforward connections between layers and well-timed delay units, for which a biological grounding was not given. These and other issues that arise with spiking neurons for recruitment are addressed in the following chapters (Günay and Maida, 2001; Günay, 2003; Günay and Maida, 2006a,b).

6.4.1 Asymmetric Delays and Stability Problems in Recruitment Learning with Spiking Neurons

As a starting point, we investigate the effect of delayed lines on converging direct versus indirect pathways on the synchronization requirement of proposed mechanisms in Chapter 7. Delays are employed since they are artifacts of any kind of realistic theory or implementation. We believe that it would be a significant refinement if properties of signals that indicate temporal binding are maintained as they travel a multiple-stage, asymmetrical cortical path. This may also improve the limitations of other architectures: for instance, the SHRUTI architecture was reported to be capable of drawing inferences of only a few stages of propagation deep before losing synchronization (Shastri and Ajjanagadde, 1993; Shastri, 1999a).

To investigate the effects of delays, a more biologically realistic neuroidal model is employed, by translating the discrete-time recruitment model to continuous-time, in order to profit from research and data with spiking neural models. We aimed to assess the theoretical bases of maintaining temporal binding in a minimal multi-stage architecture that exhibits a principal threat to synchronization (Günay and Maida, 2001, 2006b). Then, simulations were run to test the hypotheses put forth (Günay and Maida, 2003b,c, 2006b). These simulations feature Valiant's neuroidal model augmented with mechanisms and neural algorithms.

Our second major concern is improving the stability of the recruitment learning procedure (Günay and Maida, 2006a). Recruitment learning becomes unstable when it is used to recruit a chain of concepts in cascade. The size of recruited sets becomes increasingly perturbed due to the statistical variance of its expected value. We proposed a mathematical boost-and-limit model to improve the stability of recruitment, and verified the applicability of this method with a rudimentary software model in a spiking neuroidal net simulator (Günay and Maida, 2006b).

In Chapter 8, we then describe a biologically supported mechanism that may serve to implement the previously proposed boost-and-limit method in neural hardware (Günay and Maida, 2006a). We exploit the existing variable delays between neurons to our advantage in this model. It also offers the advantage of recruiting multi-place conjunctions that were difficult with the original recruitment learning mechanism. In this context, we attempt to provide solutions using the properties of such models, using for instance, inherent delays, inhibitory effects, and other mechanisms.

Chapter 7
Synchronized Recruitment in Cortical Hierarchies

Cengiz Günay

"You may delay, but time will not."
Benjamin Franklin

7.1 Introduction

In the brain, functionally and physically separate areas of cortex (e.g., visual, auditory, somatosensory) analyze different features of environmental objects. This raises the question of how physically distributed feature representations are combined to form coherent unitary percepts. This question of neural representation was first identified by Rosenblatt (1961) as the binding problem. A solution to this problem is *temporal binding* (von der Malsburg, 1981) that we showed in the previous chapter that it is consistent with recruitment learning when spiking neurons are employed. This solution has become known as the *temporal correlation hypothesis* (TCH), which posits that binding of disparate feature representations is achieved by synchronized neural firing across cortical areas (von der Malsburg, 1981, 1995; Gray et al, 1989; Singer, 1995; Singer and Gray, 1995; König and Engel, 1995; Ritz and Sejnowski, 1997; Engel et al, 1999; Treisman, 1996) based on what has more recently been termed the study of the *functional connectivity* of the brain (Baeg et al, 2007). The viability of TCH depends on maintaining the synchrony of neural spikes coding features from the same object and desynchrony of spikes coding features from different objects.

Our previous work (Günay and Maida, 2001) noted that some cortical connection topologies pose challenges for maintaining synchrony. These topologies consist of variable-length pathways that converge onto some destination area. An example of such a topology is the converging direct/indirect pathways in Figure 7.1. Here, varying delays are obtained with varying the number of synapses crossed by each pathway. If the sources are synchronized, one would expect that the destination readouts would have a way to detect the synchrony at the source. Analytical and simulation studies of Diesmann et al (1999) show that, if appropriate connectivity is employed, spike volleys propagating across neuronal pools can become more synchronized. The neurons favor synchrony allowing synchronous volleys to propagate protected, filtering out uncorrelated noise. Diesmann et al. did not address the problem of direct/indirect converging pathways. Our previous work proposed constraints on the

J. Diederich, C. Günay, and J.M. Hogan: Recruitment Learning, SCI 303, pp. 199–242.
springerlink.com

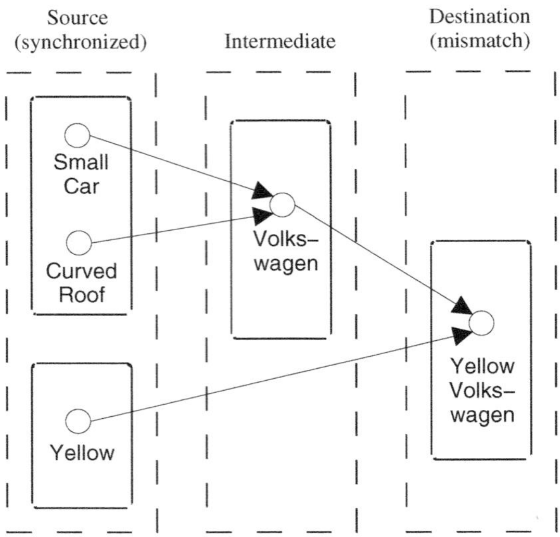

Fig. 7.1. Connection topology showing possible direct and indirect pathways from an initially synchronized source of activity converging at a destination. Dashed boxes indicate the hypothesized stages of processing, in which each individual solid box indicates the localized processing for a single feature.

timing and integration properties of these circuits in order to address the problem of synchronization.

In this chapter, we review the work of Günay and Maida (2006b) that proposes: 1) to conduct simulation studies that explore the effectiveness of the proposed constraints; and, 2) to place the studies in a broader context of synchrony-driven recruitment learning. These intend to help establish a biologically plausible substrate for *structured connectionist models* (pioneered by Feldman, Shastri, and others) that are capable of reasoning and other cognitive functions (Fanty, 1988; Shastri, 1988; Shastri and Ajjanagadde, 1993; Valiant, 1994; Feldman and Bailey, 2000; Valiant, 2000a; Shastri, 2000; Valiant, 2005; Feldman and Valiant, 2009).

7.1.1 Spike Timing: Tolerance and Segregation

The viability of synchrony-driven recruitment learning depends on the accuracy of neuron spike timing. Figure 7.1 shows a neural connection topology that naturally emerges in a hierarchical recruitment scenario. Spike synchrony is difficult to preserve in this topology because, by virtue of different signal transmission delays, initially synchronous activity may dissipate after passing through separate direct/indirect converging pathways.

The figure illustrates a hypothetical circuit formed by recruited neurons for recognizing a "yellow Volkswagen" object. In the leftmost part of the figure, the neural firings representing the primitive object properties occur synchronously as the subject focuses attention. This is consistent with stimulus-dependent synchrony behavior (Singer and Gray, 1995). It is reasonable to require an additional level of processing for the potentially more complex shape properties than the color property of the object. This allows the circuit to use features such as "small car" and "curved roof" as elements to form the intermediate concept of a Volkswagen independent of color. If temporal binding is employed, the signals at the destination need to be synchronous to represent the same object. However, when signal transmission times are considered, the degree of synchrony at the final destination is degraded due to signals crossing varying length pathways with varying delays.[1]

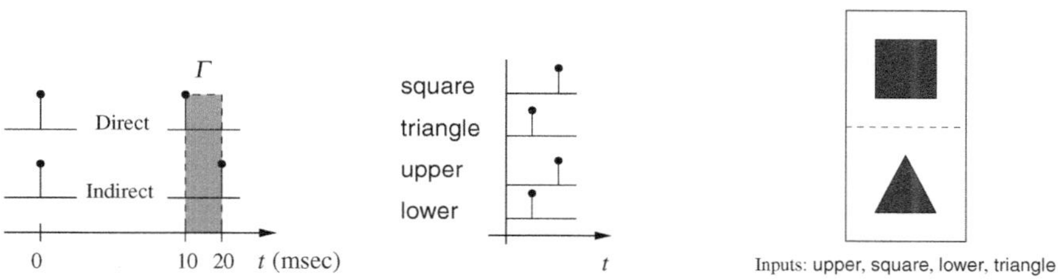

(a) Tolerance window. (b) Phase segregation. (c) Scene with two objects.

Fig. 7.2. (a) The tolerance window Γ required to integrate inputs such as shown in Figure 7.1. Initially on the left, spikes corresponding to primitive object properties are synchronized. After traveling over separate pathways with different delays, their synchrony is degraded, as seen on the right side. By defining this tolerance window the spikes can still be treated as synchronous. Timings are chosen arbitrarily for illustration. (b) Activity in separate phase windows for each object in a scene. (c) The scene contains a triangle in the lower part of the visual field and a square in the upper part.

This work addresses two related aspects of spike timing in direct/indirect connection topologies. These are tolerating delays and preventing crosstalk. Figure 7.2a shows how a *tolerance window* can be used to integrate signals with varying delays into the same response. This window specifies the maximum time allowed between two incoming spikes that can both contribute to generate a particular action potential (Koch et al, 1996; Shastri, 1993, 1999b, 2001a). Its purpose is to implement feature binding. In contrast to signals that correspond to the same object, signals that correspond to different objects must remain separate, or desynchronized, in order to avoid mismatching features from different objects as seen in Figure 7.2b. Desynchrony can be obtained by assigning each object to a different tolerance window. This causes object representations to be segregated into separate *phase* windows

[1] Similar examples can be formed with multi-modal sensory stimuli. Regarding formation of object representations, since reaction time for auditory stimuli is faster than for visual stimuli, the signals that account for the sound of an object are processed faster than the signals representing the image of the object.

(Schillen and König, 1994; Lisman and Idiart, 1995; Shastri and Ajjanagadde, 1993; Shastri, 1999a; Terman and Wang, 1995; Jensen and Lisman, 1996; Campbell et al, 1999). Its purpose is to implement feature segmentation.

This phase-coding approach is one possible method for desynchronizing responses. This method puts neural activity pertaining to each object in a separate phase window (of a large oscillatory period). This prevents interference as the activity propagates to deeper structures, preserving its integrity with respect to temporal binding. Many theoretical and simulation studies show that segregation between object representations, or desynchronization, can be obtained by inhibitory projections. Some propose using global inhibitory projections (Lisman and Idiart, 1995; Terman and Wang, 1995), while others propose using lateral recurrent inhibitory connections (Schillen and König, 1994; Jensen and Lisman, 1996).

Other support for phase windows arises as a solution to the problem of *multiple instantiation* in connectionist models (Mani and Shastri, 1992; Sougné, 1998a,b). These studies offer a biologically plausible way to represent multiple instances of objects by placing them in phase windows. This approach, in contrast to symbolic systems, models more closely the defects in performance observed in psychological studies. In symbolic systems, it is possible to instantiate an arbitrary number of representations of the same object. On the other hand, in neural models an object representation is usually associated with the same units. Therefore, there is a problem if the object needs to appear more than once in a relation. Thus, phase-coding is an appropriate solution to this problem.

The phase-coding approach also has its skeptics. Knoblauch and Palm (2001a,b) suggested that results of some electrophysiological recording experiments are inconsistent with phase coding. In particular, recordings of activity pertaining to multiple separate stimuli (e.g., bars moving in opposite directions) cause a flat correlogram, whereas phase-coding predicts a non-zero time lag of the central peak. Knoblauch and Palm suggested that the TCH can be interpreted differently. In their model, circuits represent bound entities with fast, synchronized oscillations and unbound entities with slow unorganized firing. Circuits can quickly switch between these two states to represent dynamic bindings. Another alternative to phase coding is synchronization among synfire chains for representing dynamic bindings (Abeles et al, 2004).

The present work extends the study of phase segregation by investigating how to preserve synchrony and desynchrony in the presence of delays involving direct/indirect converging pathways. Here, the relevance of recurrent inhibitory circuits is secondary, since we are interested in the effect of delays on the synchronous activity coming from upstream areas, converging to a single destination site. We study the constraints on tolerance and segregation parameters required to maintain the phase-coding at the destination read-out site and its ability to control recruitment learning.

We have chosen the connection topology in Figure 7.1 for studying the effects of delays. It arises naturally in a recruitment learning framework and more complex connection topologies can be transformed to this case by hierarchical reduction. For this topology, we found constraints on the tolerance and segregation measures that satisfies the conditions for temporal binding. For each of the measures, we calculated a lower bound, that is a minimal value.

The need to define measures of tolerance and segregation has been addressed before. In forming hierarchical learning structures, Valiant (1994) suggested that higher levels of processing need to operate in slower time scales than lower levels. In slower operation, the higher level units may integrate information from the faster lower levels of processing. For instance, high-level units representing a scene may need to integrate all information about the contents of the scene presented sequentially before giving a response. This implies that the scene unit has a tolerance duration many times longer than the lower level. This kind of time scale hierarchy, which may range from milliseconds to years, has been known to allow neural processing of various cognitive behaviors (Newell, 1990; Mauk and Buonomano, 2004).

VanRullen and Koch (2003) identified relevant experiments in reviewing the theory of discrete perception. It is known that, if two successive events follow one another too closely, they are perceived as a single event. The critical time interval for this phenomenon to occur is 20–50 ms. This interval is consistent with the time required on each item to prevent perceiving illusory conjunctions in visual search experiments (Shiffrin and Schneider, 1977; Treisman, 1996). VanRullen and Koch (2003) reported that, in composing a higher level representation of a multi-object scene, the perceived number of objects is limited to ~10–12 items/s. Overall, a perceptual time frame of ~100 ms is proposed while stimuli are grouped to form a single event. Note that, the purpose of these experiments is to establish the time frame of a single event in perceiving timing of moving scenes. Even though these experiments support our idea of multiple timing levels, the actual measurements may not apply directly to our current purpose, which is to establish the timing while building hierarchical representations of static scenes.

Moreover, it was found that this critical time interval also applies to the initial acquisition and maintenance of these representations. If, in the first few hundred milliseconds of an object's presentation, its temporal contiguity is compromised, the neurons in the inferio-temporal (IT) cortex that initially respond to the object become less selective (Li and DiCarlo, 2008). With a longer exposure to the compromised stimulus the neurons can even reverse their object selectivity, learning a novel preference. These experiments suggest temporal tolerance is essential in learning these location-invariant object representations in high-order cortical areas of the brain, which also supports employing tolerance windows in recruitment learning.

7.1.2 Psychological and Neuroscientific Theory and Evidence

We previously discussed evidence for the type of scenario depicted in Figure 7.1 (Günay and Maida, 2001). Here, we give a brief overview of the theory and evidence. Such direct-indirect connection topologies are found frequently when neocortical interareal connections are considered (see Figure 7.3). The figure shows

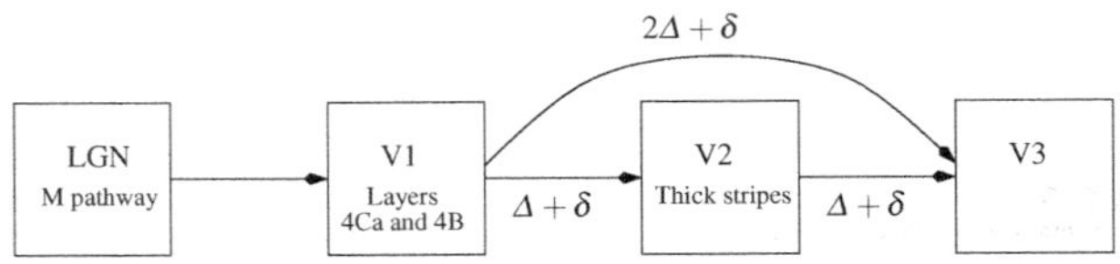

Fig. 7.3. Simple example of direct/indirect connections in visual cortex possibly leading to mismatched arrival times of spikes. Δ stands for axonal propagation delay and δ stands for synaptic transmission delay and integration time.

the M-pathway of the interareal connection between visual areas V1, V2 and V3 (Livingstone and Hubel, 1988; Boussaoud et al, 1990).

Many experiments are consistent with the TCH. Psychological studies identified visual search tasks where attentional mechanisms serially scan multiple objects in a scene (Shiffrin and Schneider, 1977; Treisman, 1996). The pattern of binding errors (illusory conjunctions) at high search rates is consistent with the hypothesis that attentional mechanisms bind object features together. There is also evidence that attention shifts initiate stimulus-dependent synchronous activity in the primary visual cortex (V1) (Singer and Gray, 1995).

If temporal binding is employed, it is reasonable to assume that synchronous activity representing an object should be sustained as it propagates through various functional areas until the computation terminates. In support of this hypothesis, synchrony has been experimentally observed in different cortical areas (Abeles, 1994; Singer and Gray, 1995).

7.1.3 Direct/Indirect Connection Topologies

Assuming a phase-coding approach, timing is crucial in preserving the integrity of signals with respect to temporal binding, especially when synchronized spike volleys must meet after taking alternate cortical paths. This is similar to the hypothetical example discussed above in Figure 7.1. There are likely many cortical connection topologies that exhibit direct/indirect converging pathways (Figure 7.3) The propagation pathways in the figure can be confirmed by consulting the visual response latencies given in Table 7.1 (Nowak and Bullier, 1997; Lamme and Roelfsema, 2000).

Table 7.1. Response latencies in ms in visual areas taken from Lamme and Roelfsema (2000). The early response in MT is due to the connection received from superior colliculus (SC) which we disregard in this study.

Area	Earliest	Mean
V1	35	72
V2	54	84
V3	50	77
MT	39	76

Consider a synchronous response that originates in area V1 and which then propagates first to area V2 and next to V3. Area V1 has direct connections to V3. Thus, V3 will receive two synchronized spike volleys caused by the same stimulus: one directly from V1, and one through V2. Even though we are not aware of direct evidence supporting that meeting pathways actually converge at the cellular level, it is reasonable to believe that the arriving signals interact since local cortical circuits are highly interconnected.

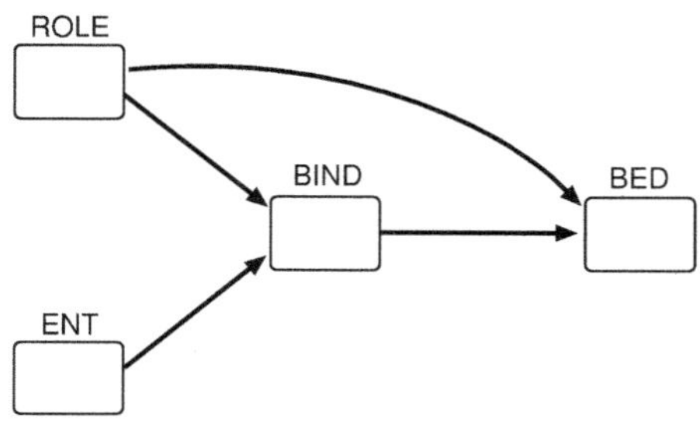

Fig. 7.4. A direct/indirect converging pathway circuit used by Shastri (1999b).

Shastri (1999b) uses a direct/indirect connection topology shown in Figure 7.4. Shastri (1999b, 2001a) describes the recruitment of binding detector cells (BIND) and binding-error detection (BED) cells. Although Shastri is aware of the issue of signal propagation delays as evidenced in Shastri and Ajjanagadde (1993) and that his model neurons (cf. Shastri, 2002a) include signal transmission delays, he does not provide a general solution to handle timing issues in this direct/indirect problem. Figure 7.4 shows an area named ROLE *directly* projecting to an area named BED. There is also an *indirect* projection from ROLE to BED via BIND. The purpose of BED is to recruit binding error detector cells based on associative LTP. This requires synchronized input from ROLE and BIND. Shastri (2001b) says that, once recruited, a BIND cell takes up to 25 ms to respond (and 100 ms to be recruited if not already recruited). However, his tolerance window for coincidence detection is 5 ms. He does not discuss how BIND outputs are synchronized with ROLE outputs. Shastri reports quantitative results pertaining to number of nodes recruited, but his analyses only use the connectivity of the underlying graph structure and do not make reference to signal transmission timing properties within the structure. But they should *also depend* on preserving the integrity of synchrony. Despite the fact that his model reportedly includes transmission delays, and that theoretically he subscribes to a temporal correlation framework with phase-segregated integration windows, he does not report exactly how he gets SMRITI to function. Similar criticisms apply to (Shastri, 2002a).

Defining a lower bound on the segregation between phases in hierarchical structures with direct/indirect converging pathways allows predicting their maximal frequencies of activity. Conversely, from the observed range of brain activity frequencies, the maximum depth of this type of recruitment hierarchies can be predicted.

For the parameter values estimated from the visual cortex, and under the assumption of homogeneous conduction speeds, we find that only two-level deep hierarchies can be supported for gamma-band (20–70 Hz) frequency range. For slower activity in the alpha-band (8–12 Hz) range, hierarchies up to six levels can be formed. Based on this, we hypothesize that small hierarchies are formed at local areas producing gamma-band activity, and larger hierarchies form over larger areas of cortex producing alpha-band activity. This is consistent with the observation of local gamma-band activity and global alpha-band activity (cf. VanRullen and Koch, 2003). In their interpretation, VanRullen and Koch associated the slow alpha-band activity across areas with the "context" and the faster gamma-band activity with the "content" of neural representations.

7.1.4 Overview

The main objective of this chapter can be summarized as follows. We review our earlier investigation of the tolerance and segregation parameter constraints for performing temporal binding across varying-delay converging pathways (Günay and Maida, 2006b). We present formal and simulation results that confirm earlier timing hypotheses (Günay and Maida, 2001). Our result verifies that the tolerance window for keeping coherent representations and the amount of phase segregation to prevent crosstalk for such a given topology can be calculated. This enables both interpretation of biological circuits and design of artificial networks. We also improve the recruitment stability in deep hierarchies. This allows the use of recruitment in limited size structures suitable for simulation studies.

Our work is consistent with previous work on recruitment learning (Shastri, 1988, 2001a; Fanty, 1988; Diederich, 1991; Feldman and Bailey, 2000) and provides an interesting counterpoint to the simulations of chapter 4 and the sequential binding of chapter 5. We augment Valiant's neuroidal model to continuous-time by using the *spike response model* (SRM) of Gerstner (1999). We prefer using spiking neurons because they are:

- Richer and have more features (e.g., spikes allow using temporal binding, modulators and ion channels serve different purposes not found in ANNs, etc.)
- Amenable to be implemented in parallel analog hardware since electrical equivalent circuits can be given and neurons are independent.
- Better suited to modeling neuroscientific, biological, and psychological data.

There are also disadvantages of using spiking neuron models, as discussed earlier in section 5.7. They are more complex compared with simplified neural units. Thus they require more sophisticated simulation environments, and they are computationally expensive. The SRM is more efficient than other spiking models, such as the integrate-and-fire (I/F) model, because it allows larger step sizes during simulation. This is because SRM uses time responses of activity functions that do not require integration. Other methods that require integrating differential equations often need very fine step sizes to maintain accuracy. Complementing other studies on recruitment which mainly provided analytical calculations and

Monte Carlo simulations (Valiant, 1994; Gerbessiotis, 1993, 2003; Shastri, 2001a; Feldman and Bailey, 2000), our work extends by implementing a neural simulator that employs recruitment learning and investigates its practical applicability (but also see Feldman and Valiant, 2009). Our work parallels that of Part I in its passage from theoretical analysis to actual implementation, but our paths diverge slightly. Hogan's work exploits the more convenient abstraction of the perceptron to address more difficult applications. Here, we adopt a more sophisticated approach based on spiking neurons, but this greater realism comes at a price: we have uncovered many issues during this simulation study that have required our attention. Among the most important of these has been the instability of the recruitment method.

Our results verify that the tolerance window for keeping coherent representations and the amount of phase segregation to prevent crosstalk for such a given topology can be predicted. This enables both interpretation of biological circuits and design of artificial networks. We also improve the stability of the recruitment method in deep hierarchies which allows using recruitment in limited size structures suitable for computer simulations. The stability mechanism proposed also illustrates a plausible biological mechanisms for maintaining stable signal propagation in the cortex.

The organization of the rest of this chapter is as follows. First, in §7.2, the specific recruitment algorithms employed are described. In §7.3, timing issues in using recruitment in certain problematic conditions are explored. Then, measures of tolerance and segregation for maintaining coherence of temporal binding in direct/indirect connection topologies are defined in §7.4. The methods for testing the proposed hypotheses, followed by the simulation results are given in §7.5. Finally, conclusions and future work is given.

7.2 The Recruitment Learning Testbed

This section describes the spiking recruitment learning testbed we employed in this chapter. The following subsections progressively build the context of the recruitment learning simulation for later sections. §7.2.1 introduces Valiant's neuroidal architecture employing recruitment learning. §7.2.2 describes how we organize our network differently from Valiant's simple random network to test tolerance and segregation. §7.2.3 describes the type of neural representation that is employed. §7.2.4 makes the relation to temporal binding explicit. §7.2.5 gives a simple example to illustrate recruitment. This example is revisited in §7.2.6 to analyze the inner workings of the network during recruitment.

7.2.1 A Neuroidal Architecture

Valiant (1994) describes recruitment learning in the framework of his neuroidal architecture (see Chapter 1 for details). In its simplest form, the neuroidal network is formed by a simple random interconnection network as seen in Figure 7.5(a).

Each node represents a *neuroid* which is the elementary building block of the network. Apart from being a *linear threshold unit* (LTU), the neuroid is also a *finite*

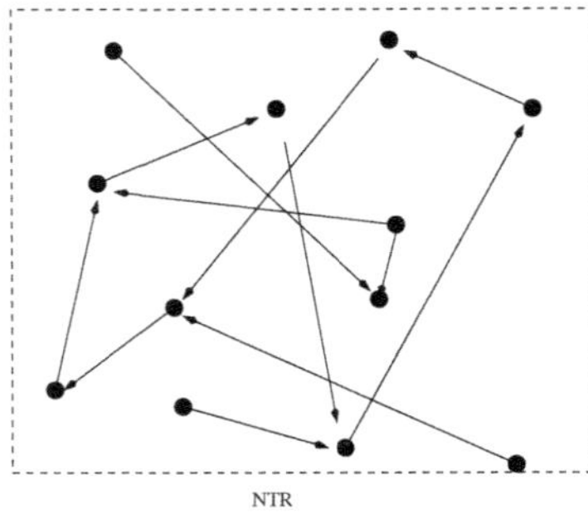

(a) Initially blank random interconnection network of neuroids. Also known as the *neuroidal tabula rasa* (NTR).

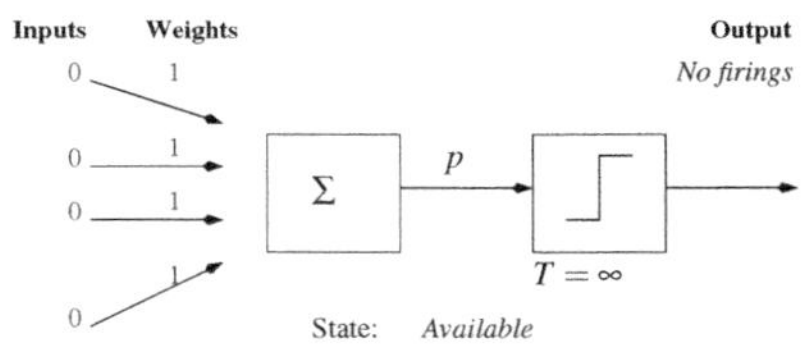

(b) The linear threshold unit of the neuroid.

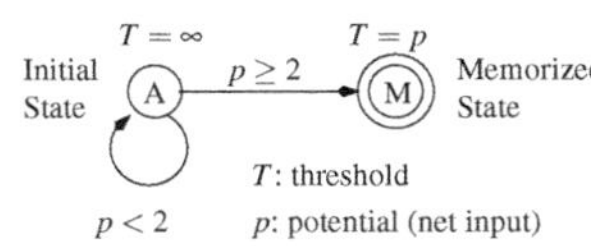

(c) State machine of the neuroid.

Fig. 7.5. Valiant's neuroidal network.

state machine (FSM) that controls the parameters of its LTU (See Figure 7.5(b)–(c)). In the neuroidal network the FSM detects coincidences among the neuroid's inputs. It starts at the initial *available* (A) state and later changes into the *memorized* (M) state, according to its membrane potential (p).

Recruitment learning requires that the connection density between the neuroids to be high enough. Given the random connection scheme of the network, it is possible that projections from existing concept sets have no intersecting neuroids to be recruited for representing the conjunction. Nevertheless, it can be shown that the probability of failure is sufficiently small if the network parameters are adjusted properly. Valiant proposes that the connection probability between any two neuroids should be chosen as

$$p = \frac{1}{\sqrt{rN}} \tag{7.1}$$

to ensure the recruitment property, where N is the total number of neuroids, r is the replication factor, that is the number of neuroids required to form a concept set. This allows the size of the recruited set for a concept to have expectation E and variance σ,where

$$E = \sigma = r - O(\frac{1}{\sqrt{N}}), \tag{7.2}$$

approximately, where O(·) indicates the big-O notation. Intuitively, the probability of connection in (7.1) is designed to yield r neuroids to represent the conjunction of two concepts represented by sets of r neuroids each. The details of the concept sets are discussed in §7.2.3.

7.2.2 Structural Organization

In order to construct models of varying-length pathways, we organized the network into subparts rather than grouping all neuroids in a single random interconnection network. Consistent with Valiant (1994), we group neuroids in abstract sets influenced by cortical areas to form a multipartite graph structure. A multipartite graph contains connections only between its separate partitions. The random interconnections that enable recruitment are located between areas rather than intraareally as seen in Figure 7.6a.

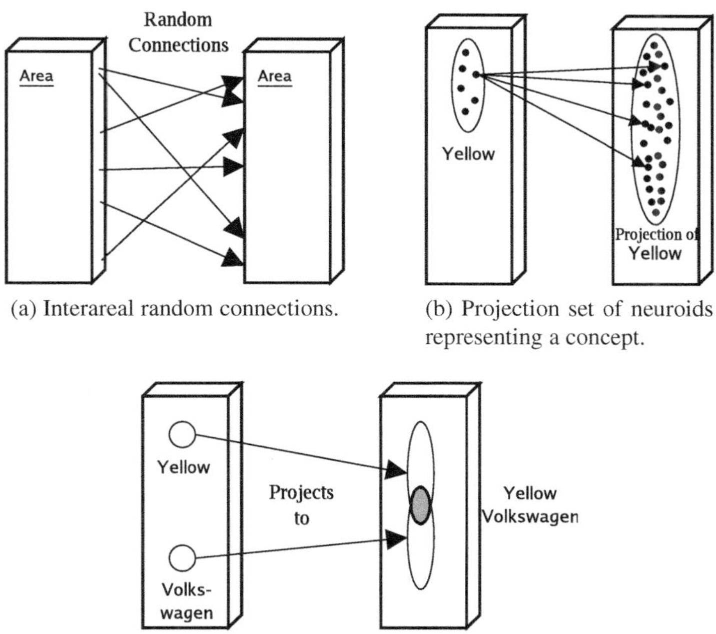

(a) Interareal random connections.

(b) Projection set of neuroids representing a concept.

(c) Temporal binding activates the intersection set.

Fig. 7.6. Network structure.

The probability of connection, p, for recruitment learning in random multipartite graphs can be calculated similarly to simple random graphs. We can show that choosing the probability as

$$p_{1,2} = \frac{1}{r_1}\sqrt{\frac{r_2}{N_2}} \tag{7.3}$$

for connections from area A_1 to A_2 allows us to have the same expectation and variance for the size of recruited set as in Eq. (7.2) for simple random graphs. Here r_i and N_i are the replication factor and total number of neuroids, respectively, for an area A_i. If we take $r = r_i$ and $N = N_i$ constant over areas, the multipartite probability calculation becomes the same as Valiant's formula in (7.1). The deduction of the probability formula in (7.3) is given in § A.1.1 on page 275.

7.2.3 Redundant Localist Representations

The neural representations that we employ can be described as redundant localist representations. Concepts are represented by a small redundant number ($r \sim 10$) of randomly distributed units within the network. This number r is known as the *replication factor*. Since each set represents a single concept, the representation is in essence localist. Redundancy ensures robustness in case some units are lost. This type of representation is also called *modularly distributed* or *distributed localist* (Browne and Sun, 2001). Each of these neuroids projects to a large number of neuroids in connected areas, again randomly, seen in Figure 7.6b.

7.2.4 Recruitment Depends on Temporal Binding

When projections of two separate concept sets target the same area, common units are included in both projection sets due to the statistical properties of the network. The units that receive input from both sources are recruited upon coincident firing of the two concept sets, as seen in Figure 7.6c. The theory of temporal binding suggests that in the brain, features of objects are bound together by synchronous activity. That is, synchronized activity represents features of a single object, whereas desynchronized activity indicates separate objects. Therefore, it can be said that recruitment depends on the principle of temporal binding. Coincident, or synchronous, activity triggers the memorization mechanism for recruitment, caused by activity above a preset threshold in units at the intersection set of the two projections. These units change their internal states and weights to permanently memorize the feature combination. This set of recruited units represents the high-level concept of the conjunction of the two primitive concepts. The premise of recruitment learning is that a suitable connection density can be chosen such that the recruited set will be expected to contain as many units as the initial concepts.

7.2.5 How Recruitment Works

The following simple example illustrates recruitment. Details of the state machine are postponed until the next section. Figure 7.7a shows a target neuroid d with three binary inputs. The neuroid is initially in the “Available” (A) state with all connection weights set to 1 (cf. Figure 7.5c). Consider the following time profile of spikes for the given inputs as seen in Figure 7.7b.

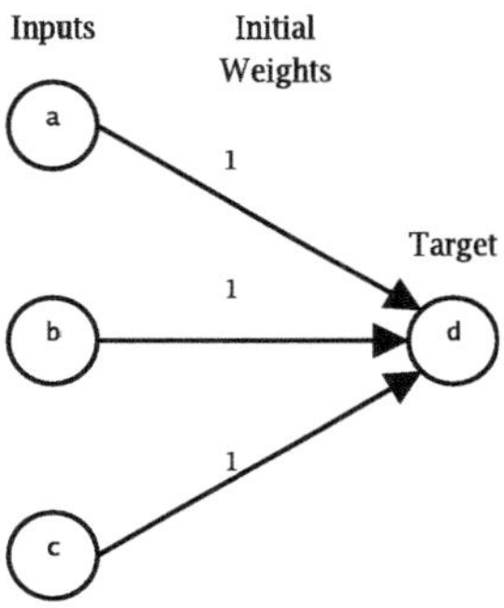

(a) Target neuroid for the recruitment example.

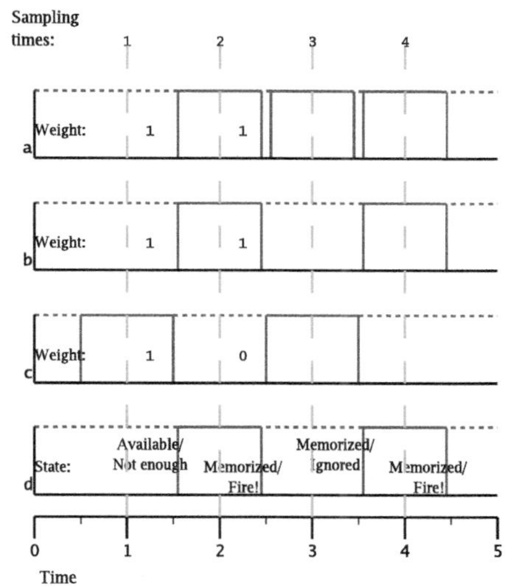

(b) Time profile of activity in inputs and target. The square pulses represent activity timing for each unit. The connection weights or states are indicated inside the activity pulse. The vertical dashed lines only indicate sampling times.

Fig. 7.7. A simple recruitment example.

- At time $t = 1$, input c is active, however, this does not affect the state of the target neuroid. This is because the net input is insufficient to cause the state transition (See Figure 7.5c).
- At $t = 2$, a coincidence has occurred. Inputs a and b are active simultaneously. The net input is above the transition threshold, so the target neuroid changes to the *memorized* (M) state. As seen from the weight changes indicated in Figure 7.7b, while weights from a and b retain their original value of 1, the weight from c decreases to 0. Furthermore, the threshold T is set to the current net input $p = 2$. As a result, the target neuroid emits a spike. Therefore, the neuroid has specialized and memorized to recognize a and b together, ignoring the rest of its inputs.
- At $t = 3$, inputs a and c are active. However, this event will be ignored because the weight from c has been pruned during memorization. Since the neuroid has already specialized, this particular coincidence will be counted as a spurious event, neither causing the target neuroid to change state, nor to fire.

- At $t = 4$, inputs a and b are active, being the memorized event, triggers a spike at the target neuroid.

Note that, the state machine explained so far is simplistic, built for illustration and for fast computer simulations. It minimally consists of two states, only allowing permanent memorization. This may seem to be a coarse model of cognitive memory functions. Human memory is much more dynamic, forgetting unnecessary items if they are accessed with low frequency. The state machine here can be improved to model more realistic memory behavior. For instance, a third state can be introduced for the case when the neuroid enters a "forgetting" phase. Gradual weight adjustment mechanisms can also be built.

7.2.6 Tracing State Transitions

This section describes the details of the state transitions that caused the behavior we briefly explained in previous section. Recall that the state machine consists of two states: an initial Available (A) state and a final Memorized (M) state (See state machine in Figure 7.5c).

At initial state A, as seen from the LTU on the left of Figure 7.8, no inputs are available $\forall i I_i = 0$, weights are $\forall i w_i = 1$, and the firing threshold is $T = \infty$. The neuroid will stay in this state until $p \geq 2$ which indicates a cumulative sum of more than one simultaneous input.

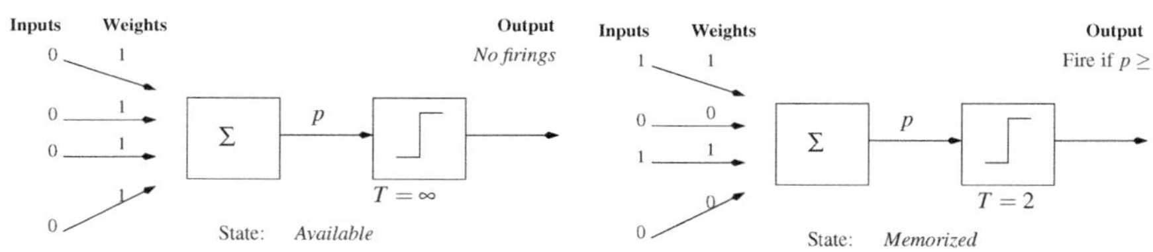

(a) In initial state. (b) State after memorization.

Fig. 7.8. FSM states before and after memorization.

The LTU on the right in Figure 7.8 shows the triggering situation where two inputs are simultaneously active, causing the reset of all but the active connection weights, and the threshold being set to the net input $T = p$ to ensure the neuroid will only respond to the inputs just memorized.

The actual state machine used in this work is slightly more complex. It contains three states, where the intermediate state is for gradual learning of concepts. However, we allow only a single learning step to yield fast, one-shot learning. The weight update rule consists of a one-time application of $w(t+1) = 1.5w(t)$ for active inputs, and $w(t+1) = 0.5w(t)$ for inactive inputs. This learning rule is a multiplicative weight adjustment method consistent with Hebbian learning, and inspired by the Winnow algorithm (Littlestone, 1991) (For details of the algorithm, see Chapter 4).

7.2.7 *Limitations*

Recruitment has been proposed mainly for only forming two-way conjunctions (Feldman, 1982; Valiant, 1994; Shastri, 2001a, 2002a). We believe that this constitutes a starting point for the work in recruitment learning. We predict that n-way conjunctions can be formed via recruitment learning with the help of stabilizing lateral inhibitory connections (cf. Günay, 2003; Günay and Maida, 2006a and the *conjunction* studies of section 4.7.2 and Chapter 8). Therefore, we believe that there is no theoretical limit for forming conjunctions with more than two inputs. For simplicity, we use two-way conjunctions in this work because the problem of direct/indirect converging pathways is independent of this limitation of the present form of recruitment learning used.

The units in our model are restricted to representing a single concept, since it is based on Valiant's model (Valiant, 1994). Nevertheless, hierarchical representation implies that concepts take part in representing other concepts. This allows constituent concepts to take part in formation of multiple complex concepts. Therefore units will always take part in representation of multiple concepts. Biological plausibility of this type of representation has been discussed in several places (Feldman, 1990; Valiant, 1994, 2005; Huyck, 2009).

7.3 Timing Issues in Recruitment

So far we explained the simplest recruitment learning algorithm. We showed that this algorithm depends implicitly on temporal binding. Some problems with timing arise when certain learning conditions are used with temporal binding, such as hierarchical learning and learning in direct/indirect connection topologies with delays.

7.3.1 *Timing in Hierarchical Learning*

The phase-coding approach for temporal binding requires that activity pertaining to different objects occurs in separate phase windows. From the aspect of hierarchical learning, evaluating a higher-level concept composed of the conjunction of two lower-level objects that occur in different phases requires detecting the conjunction of temporally separate events. This is consistent with experiments that show perceiving a multi-object scene as a single event takes ~100 ms (VanRullen and Koch, 2003). This forces us to extend the meaning of temporal binding.

Consider the example shown in Figure 7.9. On the left, a hypothetical connection circuit in the brain is shown. The right hand side shows the timing activity in the circuit nodes. The circuit represents the neuroidal substrate that integrates the features in a visual scene. There are two objects in this scene: A blue square and a green circle. Since we assume temporal binding, both objects are attended to by the brain during separate phases. The first object attended to at phase $t = 1$ has blue and square active simultaneously. The blue square detector unit (b&s) is also active in the same phase on the time profile on the right. At phase $t = 2$, the same happens for the green circle object.

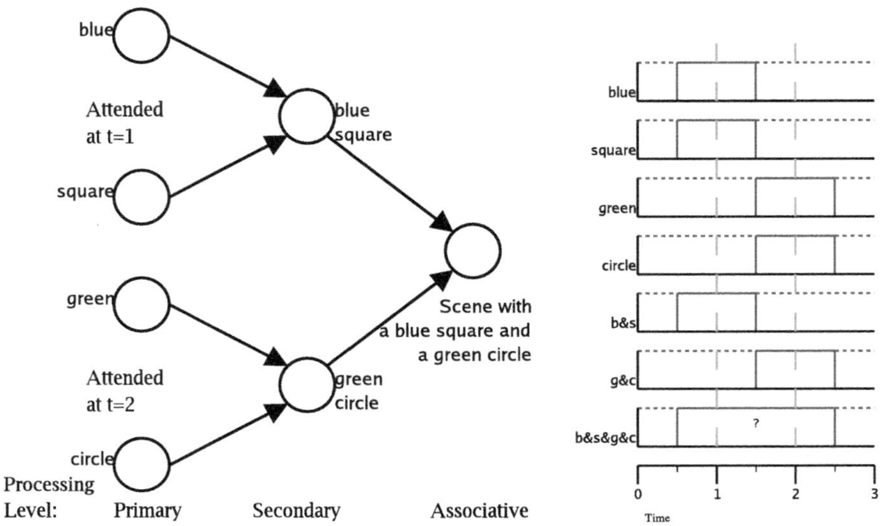

Fig. 7.9. Hierarchical learning example. (Left) Structure of concepts. (Right) Time profile of activity.

Having constructed the lower-level primitives of the perceived scene, assume that a concept that represents the whole scene is needed. That is, a node should represent "the scene with a blue square and a green circle." Notice from the figure that such a node cannot depend on detecting simultaneous activity at its inputs because the b&s and g&c units are active at separate phases, with no overlapping activity. Intuitively as a solution, it can be suggested that the scene detector unit observes activity in adjacent phases during an interval and gives a decision. This can be justified by claiming that different levels of cognitive activity work at different speeds. In this case, while a primitive feature binding happens quickly, the scene can only be perceived after all objects are scanned.

The idea here is motivated by and consistent with other proposals (Valiant, 1994, Ch. 11). Such a scenario is addressed in Valiant's architecture by detecting *timed conjunctions*. These are conjunctions of inputs that appear at different times.

Valiant examined forming hierarchical structures when a recruitment-style learning algorithm is employed. In particular, each level of hierarchy is shown to be slower than the previous one by a factor of the number of objects allowed. In our case, there were two objects which imposes that the scene recognition unit waits for both objects. In general, the operation of the scene unit has to accommodate as many objects that need to be represented.

Valiant showed that, for m possible objects, and a base τ_0 time window for synchrony at the lowest level, a new level needs a $\tau_1 = m\tau_0$ time duration to integrate results from the previous level. In general, this becomes $\tau_i = m\tau_{i-1} = m^i\tau_0$ for a level i. Others have also proposed the use of multiple time scales to explain different cognitive phenomena across levels (Newell, 1990; Mauk and Buonomano, 2004).

7.3.2 Delays in Converging Direct/Indirect Pathways

We proceed to explore further cases which require timed conjunctions. So far, we ignored delays in our examples for simplicity. However, a realistic system should at least consider, or maybe depend on, unavoidable delays in its components and transmission lines.

We treat delays caused by hierarchical learning and delays caused by unequal transmission lines similarly. We study them by focusing on one example case containing varying length converging pathways as shown in Figure 7.10. In this case, a tolerant way of conjoining is needed for integrating signals covering different length pathways before coming to the final destination due to varying delays (Günay and Maida, 2001).

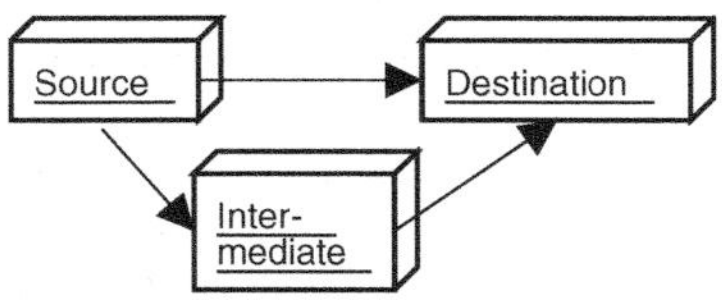

Fig. 7.10. Direct/indirect pathways from a source converging at a destination.

This specific situation is addressed in (Valiant, 1994, Ch. 5), where the following approaches were discussed:

- Ignoring this case by assuming that all paths converging from a source to a destination are of equal length, or
- Having peripheral systems which can provide persistent firings until computations terminate (also see Fransen et al, 2006).

Our work follows the second proposal by defining the peripheral systems that help computations in these structures.

7.3.3 Detecting Temporally Separate Coincidences

The hierarchical learning scenario shows the difficulty in passing information across time scales. The difficulty is in detecting activities that appear for a duration shorter than the higher-level time scale, and which do not overlap in time. This is similar to the case in Chapter 5, where movies of object relations in a scene had to be presented in successive, non-overlapping frames. The no-overlap problem also appears in the case of direct/indirect converging pathways with delays. Without simultaneous activity, the net input of a neuroid cannot provide the information for detecting the desired timed conjunctions. That is, without memory of the past, the state machine we proposed for recruitment learning in §7.2.6 cannot handle timed conjunctions.

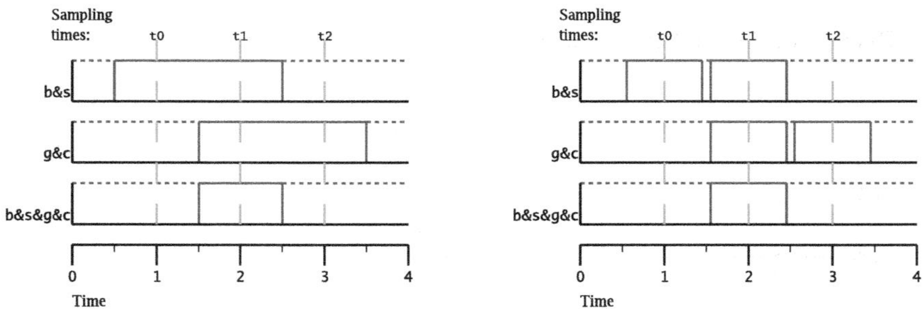

(a) By increasing spread of activity for each spike.

(b) By persistent firing with a fixed spread of activity.

Fig. 7.11. Possible solutions for tolerant conjoining by changing behavior of high-level concept representations.

Two possible solutions can be proposed to the problem of *tolerant conjoining* temporally separate but related activity, or timed conjunctions. The first, seen in Figure 7.11(a), is by increasing the spread of the activation on the postsynaptic membrane caused by inputs at the unit representing the high level concept. Since the activity on the membrane is long enough to overlap in time, the unit can detect simultaneous activity to learn the features of the scene. The second method, seen in Figure 7.11(b), is by persistent or repetitive firing of units representing the low-level concepts. In this case, the inputs need to fire repetitively until an overlapping effect is caused on the destination higher-level concept. Both solutions are based on the same principle of causing overlapping activity to be detected at the target unit, only differing in implementation.

In this work, we only consider the first method of increasing the spread of activation at the destination membrane. We assume the default effect of spikes on the membrane gives the desired behavior at no extra computational cost. Implementing the second method would require an external controller.

7.3.4 Defining Peripherals for Timing with Delays

When delays are considered, tolerant conjoining is needed for simple conjunctions even within the same time scale. Consider the example shown in Figure 7.12 with two objects, similar to the previous example. Each object now has three features;

shape, color, and movement type. We assume that the shape information needs an additional intermediate level of processing in comparison to movement. Even though this lacks direct biological evidence, it is known that the movement is perceived faster through the magnocellular pathways in the visual system. The increase in the transmission speed in the magnocellular pathway is due to the larger axons. A direct connection line may model this increased speed in transmitting movement information.

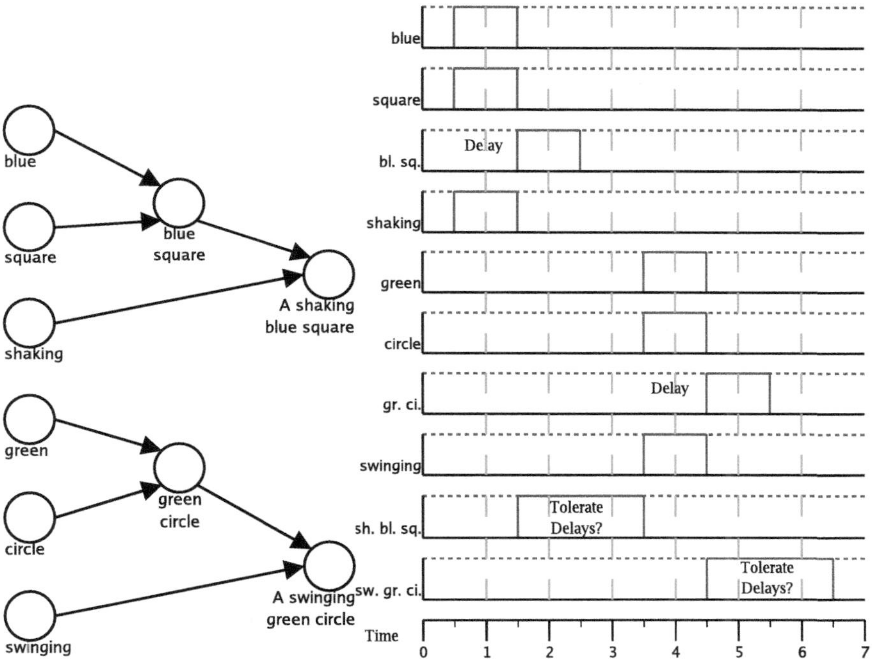

Fig. 7.12. Recruitment example with delays and direct/indirect connections.

When the subject attends to the shaking blue square object, all three inputs are available simultaneously at $t = 1$. The blue square detector will only become active after a delay from this time (a unit delay for simplicity), thus requiring tolerant conjoining with the input shaking.

The main aim of this example is to introduce another measure, the phase segregation parameter. The segregation parameter is defined as the time from the beginning of an object's tolerance window until the beginning of the next object's tolerance window. Tolerance should be limited to avoid crosstalk of activity representing features of one object, with the features of the next object attended. To prevent occurrence of crosstalk, it is necessary for the segregation parameter value to be equal or greater than the tolerance parameter. We call this the *phase segregation constraint*.

On one hand, if tolerance is too long, features of successively attended objects will crosstalk. On the other hand, if tolerance is too short, the system will suffer from limited speed of processing due to inefficient use. An optimal setting is possibly maintained dynamically by the brain. Therefore in the figure, shaking blue square (sh. bl. sq.) should not tolerate any signals arriving after $t = 2$, or it will result in spurious bindings with the features of the swinging green circle object.

7.3.5 Limitations

Tolerant conjoining assumes activity of the constituent parts of the hierarchy is presented in adjacent phases. If, for instance, a phase sequence contains constituents of two hierarchies interleaved, then the time we compute for tolerant conjoining will be insufficient. However, this invalidates the temporal binding assumption, since the parts of a concept are not synchronous, or in this case, close in time. Thus, ordering of constituents in adjacent phases is a requirement for synchrony-driven recruitment learning, and for tolerant conjoining to work. This identifies the *phase ordering* problem. Placing constituents of each hierarchy in adjacent phases is a complex problem. We assume attentional mechanisms are responsible for this task. This may require recurrent circuits and reverberating activity (i.e., multiple passes over the scene) before finding the correct phase ordering. The correct hierarchies must be also influenced by the way attentional mechanisms work. Thus, the way hierarchies are formed should be according to logical grouping of features into objects (e.g., all objects at nearby locations or with similar features are automatically grouped together).

Next, we formalize the constraints on *tolerance* and *segregation* parameters for overcoming the problems described here.

7.4 Defining Measures of Tolerance and Segregation

This section reviews the formal constraints on the tolerance and phase segregation parameters presented earlier for maintaining object coherency with respect to temporal binding (Günay, 2003; Günay and Maida, 2001, 2006b). Readers who are not interested in the mathematical details can skip to Section 7.5 without loss of continuity. For a given connection topology with converging inputs over pathways of varying delays, two requirements are studied: one for the duration of the tolerance window Γ, and a second for the duration of the phase segregation parameter Φ. We first consider the tolerance constraints.

Definition 7.1. *A* tolerance window *Γ_i, for a neuronal unit x_i, is the longest time duration, or interval, during which the unit can integrate a train of incoming spikes that all contribute to emit a single action potential (AP). The unit is said to* delay-tolerantly conjoin *the inputs received during this interval.*

In this work, Γ_i does not change for different units. Therefore, we simply use Γ.

Definition 7.2. *A neuronal unit x, is said to* cover *a set of distributed sources only in case all synchronized spikes from those sources arrive at x within the interval Γ.*

Lemma 7.3. *Let neuronal unit x receive incoming signals from a set of distributed sources at varying distances. x covers this set of sources if*

$$\Gamma \geq d_{\max} - d_{\min}\,, \tag{7.4}$$

where $d_{\max}$ and $d_{\min}$ are the longest and shortest transmission delays from the sources, respectively.

We first employ a simple model, where the effect of an incoming spike on the postsynaptic membrane potential is a discrete pulse with constant magnitude.

Theorem 7.4. *Delay-tolerant conjoining of two disparate spikes from sources covered by a neuroidal unit x_i is possible, if each spike causes a Γ-long constant potential of magnitude P on the postsynaptic membrane p_i.*

Proof. Let the first spike reach x_i at time t_1 and second spike at time t_2, where $t_2 - t_1 \leq \Gamma$ according to Def. 7.2. Potentials caused by the disparate spikes on the membrane p_i would overlap for some interval $0 < I \leq \Gamma$. Since x_i is an LTU, the net input p_i will reach $2P$ during the interval I. A threshold T can be set to detect the sum of the overlap value and ignore non-overlapping spikes by choosing $P < T < 2P$. Crossing the threshold can cause an AP and trigger the recruitment of a concept representing the conjunction, satisfying Def. 7.1.

7.4.1 Implementing with a Continuous-Time Model

Thus far, we used a discrete-time model, consistent with Valiant's recruitment learning framework. This was also the case in the previous examples used for illustration, where we employed unit delays. However, considering timing dynamics of the brain, unit delays and discrete models can only provide a coarse approximation. A simple integrate-and-fire (I/F) spiking neuron model is more appropriate. We use the *spike response model* (SRM) of Gerstner (1999, 2001). This model naturally allows delay-tolerant conjoining of separate signals by generalizing Thm. 7.4, relying on increasing the temporal spread of higher-level concept activations discussed in §7.3.3 (see Figures 7.11a and 7.13).

Using the SRM, a tolerance window for direct/indirect converging pathways can be implemented by adjusting the length of the EPSP on the postsynaptic membrane. Figure 7.13b shows how this allows lagged spikes to have an overlapping effect on the postsynaptic membrane that can be detected with a threshold device. Here we attempt to find constraints for the SRM EPSP, in order to assess its parts which are significant for tolerance or segregation.[2] This allows us to make formal claims.

[2] This is similar to the study of Shastri (2001a) to discretize an EPSP to three regions; a rising part, a plateau, and a decaying part.

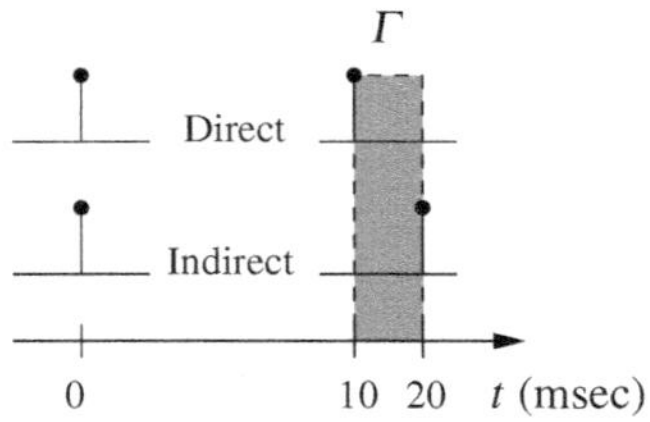

(a) Initially synchronous signals cross direct and indirect pathways, and arrive early and late, respectively. The $\Gamma = 10\,\text{ms}$-long shaded window duration needs to be tolerated.

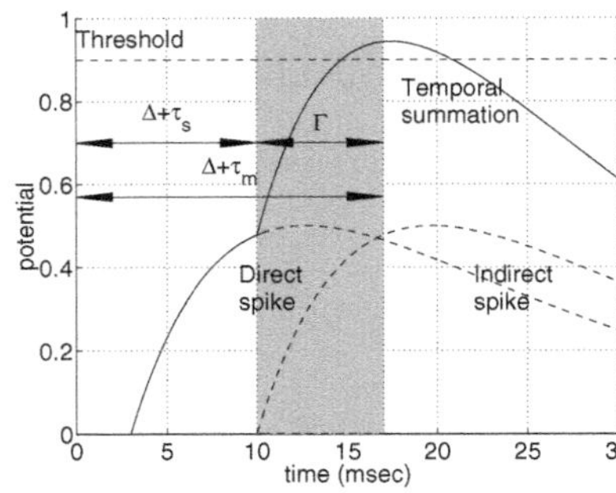

(b) In the Γ-window, early and late arriving spikes cause overlapping *excitatory post synaptic potentials* (EPSPs) that sum to exceed a detection threshold. Non-overlapping EPSPs do not reach this threshold.

Fig. 7.13. SRM helps tolerate delayed signals.

In the SRM, an EPSP caused by a single presynaptic spike i at $t = 0$ is given by the kernel,

$$\begin{aligned}\varepsilon_i(t, \tau_s, \tau_m, \Delta_i^{\text{ax}}) = \frac{1}{1 - \tau_s/\tau_m} \Bigg[&\exp\left(-\frac{t - \Delta_i^{\text{ax}}}{\tau_m}\right) \\ &- \exp\left(-\frac{t - \Delta_i^{\text{ax}}}{\tau_s}\right)\Bigg] \mathscr{H}(t - \Delta_i^{\text{ax}}),\end{aligned} \tag{7.5}$$

where $\mathscr{H}(\cdot)$ is the Heaviside step function, τ_s is the synaptic rise time constant, τ_m is the postsynaptic membrane time constant, and Δ_i^{ax} is the presynaptic axonal delay associated with the spike. For simplicity, we refer to (7.5) as $\varepsilon_i(t)$ when $\tau_s, \tau_m, \Delta_i^{\text{ax}}$ are kept constant. The rise and decay behavior of (7.5) depend respectively on the τ_s and τ_m parameters (Gerstner, 1999) as shown in Figure 7.14 .

First, we establish that there is only one maximum of $\varepsilon_i(t)$, which depends on the parameter selection of τ_s and τ_m. This maximum is written as

$$\hat{\varepsilon}_i^{\tau_s, \tau_m} = \varepsilon(t_{\max}, \tau_s, \tau_m, 0) = \max_{t'} \varepsilon_i(t', \tau_s, \tau_m, 0)\,.$$

Δ_i^{ax} is omitted since it does not affect the shape of the EPSP (see Appendix A.3 for derivation of $t_{\max}$).

We now propose parameters for implementing the tolerance window. For delay-tolerant conjoining, a neuroid should emit an AP when the effective parts of two EPSPs caused by separate incoming spikes overlap, whereas the EPSP caused by a single spike must not cause an AP. The following definition is intended to restrict the SRM to satisfy these necessary conditions.

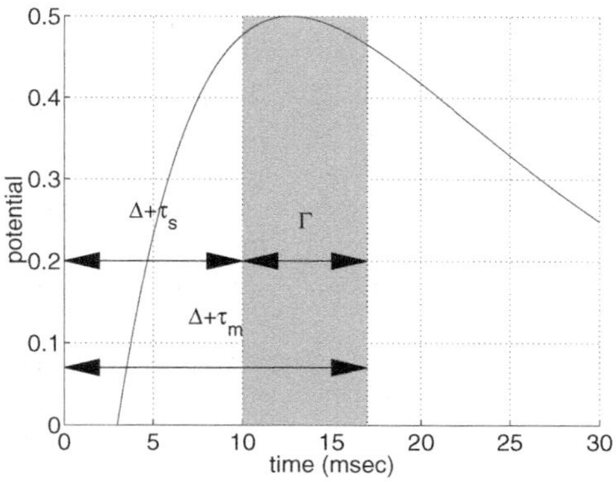

Fig. 7.14. Shape of a SRM EPSP, $\varepsilon_i(t)$.

Definition 7.5. *An* effective window ω *is a time range* $t \in \omega$ *within the EPSP, where*

$$\varepsilon_i(t) > 0.5\hat{\varepsilon}_i^{\tau_s,\tau_m}. \tag{7.6}$$

This definition indicates that during the effective window ω, $\varepsilon_i(t)$ must be larger than its half-maximal value.

Theorem 7.6. *Delay-tolerant conjoining of two disparate spikes is possible with a neuroidal unit* x_i *employing the SRM, if a* Γ*-long effective window* ω *can be chosen.*

Proof. Proof of Thm. 7.4 applies when $P = \hat{\varepsilon}_i^{\tau_s,\tau_m}$. The effective windows from the two spikes would overlap during an interval I. Thus, a threshold T can be chosen to detect the overlap value with $P < T < 2P$.

The part of the SRM EPSP effective for tolerant conjoining can be given with a time range defined by a pair of lower and upper bounds (see the shaded area in Figure 7.14).

Theorem 7.7. *The region* $t \in \omega$, $\omega = [\Delta_i^{\text{ax}} + \tau_s, \Delta_i^{\text{ax}} + \tau_m]$ *is an effective window of* $\varepsilon_i(t)$.

Proof. Since Δ_i^{ax} does not affect the shape of the EPSP, we take $\Delta_i^{\text{ax}} = 0$ in this proof. Then, for $\omega = [\tau_s, \tau_m]$ to be an effective window, the condition in (7.6) must be satisfied within the region. Since the peak $\hat{\varepsilon}_i^{\tau_s,\tau_m}$ always falls inside the region, $\varepsilon_i(t)$ monotonically increases and decreases at the lower and upper boundaries of the region, respectively (see Appendix A.3 for proof). Thus, it is sufficient to show that the condition holds when t is at the boundaries of the region. Furthermore, we can show that the behavior of $\varepsilon_i(t, \tau_s, \tau_m, \Delta_i^{\text{ax}})$ when t is at the boundaries depend only on the ratio τ_s/τ_m (see Appendix A.3). Numerical analysis of change in $\varepsilon_i(t)$ at the boundaries when τ_s/τ_m is varied is given in Figure 7.15. The solid line showing $\varepsilon_i(\tau_s)/\hat{\varepsilon}_i^{\tau_s,\tau_m}$ indicates that the condition (7.6) is satisfied independent of the parameter ratio τ_s/τ_m (it is always above 0.5). However, for the dashed line showing $\varepsilon_i(\tau_m)/\hat{\varepsilon}_i^{\tau_s,\tau_m}$, the condition can only be guaranteed when $\tau_s/\tau_m > 0.1$.

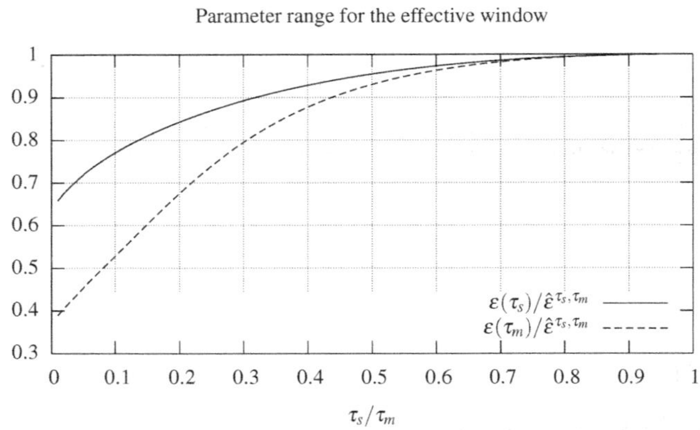

Fig. 7.15. Change of $\varepsilon(t)$ with τ_s/τ_m at the boundaries of the region $t \in [\tau_s, \tau_m]$.

This theorem can be interpreted requiring that τ_m needs to be long enough to include the rising time τ_s before the effective magnitude of the EPSP is reached (see Fig. 7.13). Assuming that the rise time τ_s is constant, and the membrane time constant τ_m can be varied by biological processes that modify the membrane conductance, we offer the next corollary.

Corollary 7.8. *Delay-tolerant conjoining can be achieved if the membrane time constant*[3] *is chosen as*

$$\tau_m = \tau_s + \Gamma . \tag{7.7}$$

7.4.2 *Phase Segregation*

The second requirement for temporal binding concerns the phase segregation measure for separating activity pertaining to different objects.

Definition 7.9. *Phase segregation Φ is the time separation between the synchronized activity pertaining to two different objects represented successively.*

We first employ a discrete model as in Thm. 7.4 for delay-tolerant conjoining, for asserting that tolerance windows are exclusive to each object.

Theorem 7.10. *Segregation should obey $\Phi > 2\Gamma$, to prevent crosstalk between elementary features of different objects at a neural unit x covering the sensory sites for these features.*

Proof. Let S_1, S_2 be the sets of spike timings pertaining two successively presented objects o_1, o_2, respectively. Assume, for $t_0, t_1 \in S_1$, the earliest spike arrives at

[3] An refinement is made to the earlier paper (Günay and Maida, 2001) where it was proposed that $\tau_m = \Gamma$ is sufficient.

$t_0 = \Delta_1$, and the latest spike arrives at $t_1 = \Delta_1 + \Gamma$. According to Thm. 7.4, the effect of a spike arriving at t_1 will cause a constant potential until $t_1 + \Gamma$. Thus, the earliest $t_2 \in S_2$ can arrive is at $t_2 > \Delta_1 + 2\Gamma$. According to Def. 7.9, the segregation is the difference between the originating times of spikes pertaining to each object, yielding $\Phi = (t_2 - \Delta_1) - (t_0 - \Delta_1) > 2\Gamma$.

Informally, if each spike has a Γ-long spread on the destination membrane, then the latest arriving spike at the end of the tolerance window will be effective for another Γ amount. The next tolerance window cannot start until the effect from this previous spike has ended.

After defining requirements for segregation in the discrete model, we generalize for the spiking model. Tolerant conjoining should apply to spikes emitted during a tolerance window, but not spikes emitted in two different tolerance windows. We again restrict the SRM EPSP to satisfy these conditions to give a reciprocal of Thm. 7.6.

Lemma 7.11. *Delay-tolerant conjoining of separate groups of spikes pertaining to different objects is possible at a neuroidal unit x_i employing the SRM, if no effective windows of EPSPs from spikes pertaining to different objects overlap in time.*

The part of the SRM EPSP excluded from tolerant conjoining can be given with a time range.

Theorem 7.12. *There is no effective window in the region of $\varepsilon_i(t)$ outside the time range $t \in [\Delta_i^{\mathrm{ax}}, \Delta_i^{\mathrm{ax}} + 2\tau_m]$.*

Proof. Since Δ_i^{ax} does not affect the shape of the EPSP, we take $\Delta_i^{\mathrm{ax}} = 0$ in this proof. Then, for not finding an effective window outside $\omega = [0, 2\tau_m]$, the condition in (7.6) must fail outside the region ω. On the left side of the region, $\varepsilon_i(t) = 0, t < 0$

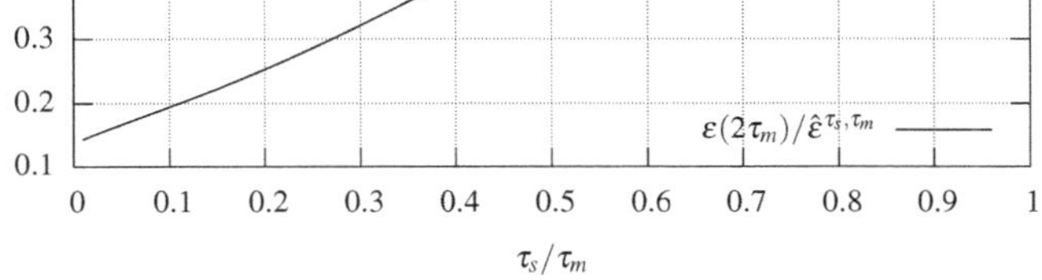

Fig. 7.16. Numerical analysis of change of $\varepsilon(2\tau_m, \tau_s, \tau_m, \Delta_i^{\mathrm{ax}})$ with τ_s/τ_m.

fails the condition. Since the peak $t_{\max} \in [\tau_s, \tau_m]$ (see Appendix A.3), the EPSP is monotonically decreasing at $t = 2\tau_m$. It is sufficient to show that if the condition fails at this boundary, there cannot be another effective window. Figure 7.16 shows the change in $\varepsilon_i(2\tau_m)/\hat{\varepsilon}^{\tau_s,\tau_m}{}_i$ when τ_s/τ_m is varied. The data in the figure shows that the condition fails when $\tau_s/\tau_m < 0.5$.

Combining the SRM parameter constraints found in this proof with the one in the proof of Thm. 7.7, we obtain $0.1 < \tau_s/\tau_m < 0.5$. Notice that the above theorem can be further optimized. We then give the segregation measure adopted for use with the SRM to achieve the effect depicted in Figure 7.17.

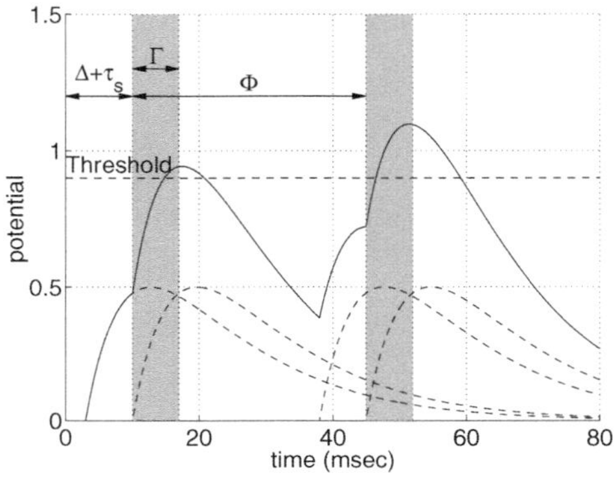

Fig. 7.17. Phase segregation of SRM EPSPs. Two sets of spikes pertaining to two separate objects are depicted. Shaded areas show the effective windows of the EPSPs that form the tolerance windows.

Theorem 7.13. *The segregation for the SRM should obey*

$$\Phi_{\mathrm{SRM}} > \Gamma + 2\tau_m = 3\Gamma + 2\tau_s . \tag{7.8}$$

Proof. Using $t_0, t_1 \in S_1$ from proof of Thm. 7.10, the arrival time of the earliest spike $t_2 \in S_2$ should be outside any effective windows pertaining to o_1. According to Thm. 7.12, $t_2 > t_1 + 2\tau_m$ should be satisfied in order to avoid overlaps. This yields $\Phi_{\mathrm{SRM}} = (t_2 - \Delta_1) - (t_0 - \Delta_1) > t_1 + 2\tau_m - t_0 = \Delta_1 + \Gamma + 2\tau_m - \Delta_1$.

Corollary 7.14. *The value of the segregation parameter limits the maximal firing frequency of the destination neuroid by $f < 1/\Phi_{\mathrm{SRM}}$.*

Phase segregation or desynchronization can be implemented by having a globally inhibitory projection (Schillen and König, 1994; Lisman and Idiart, 1995; Terman and Wang, 1995; Günay and Maida, 2001), which suppresses the source units for the duration of an inhibitory time constant $\tau_i = \Phi_{\mathrm{SRM}}$.

7.4.3 The Spike Response Model

The SRM we employ is based on an integrate-and-fire (I/F) model. The SRM is equivalent to the standard I/F model with appropriate parameter selections (Gerstner, 1999). In the model employed here, a synapse is modeled as a low pass filter and the membrane as an *RC* couple as seen in Figure 7.18.

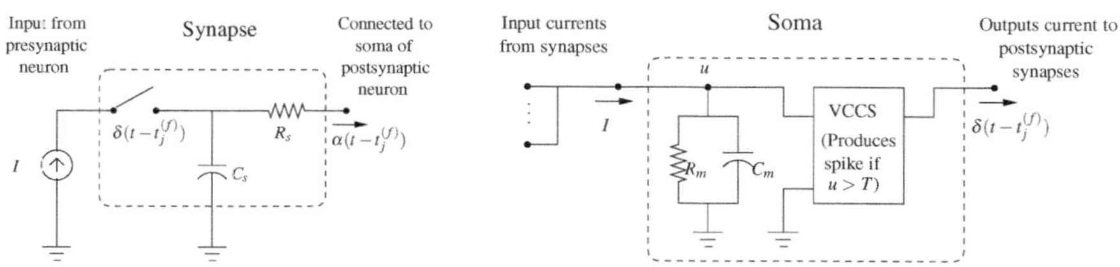

Fig. 7.18. Circuit equivalents for the SRM components. (Left) A synapse, and (Right) the membrane. VCCS stands for a *voltage controlled current source.*

The approximate time response of the membrane potential is

$$p_i(t) = \eta_i(t - t_i^{(f)}) + \sum_{j \in \Gamma_i} \sum_{t_j^{(f)} \in \mathscr{F}_j} w_{ji} \varepsilon_{ji}(t - t_j^{(f)}) \tag{7.9}$$

where $\eta(\cdot)$ is the refractory kernel and $\varepsilon(\cdot)$ is the excitatory synaptic kernel, given in Eq. (7.5). The refractory kernel is defined as

$$\eta(t) = \vartheta \exp(-t/\tau_r)\mathscr{H}(t),$$

where ϑ is the magnitude of the refractory effect[4] and τ_r is the refractory time constant.

Notice that there are no differential equations to integrate in using the SRM kernels. This is because time response kernels are used instead of a system of differential equations, and hence the name "response model." The time response for the membrane potential in (7.9) is obtained by using the exact solution of the differential equations for the effects of single spikes and approximating the interaction inside and between neurons.

7.4.4 State Machine for Continuous-Time Neuroids

The recruitment learning algorithms we discussed earlier are based on discrete sampling times. It turns out that finding a simple way to upgrade the existing discrete-time algorithm is difficult. This is because in a continuous-time system there is no fixed sampling time but a changing continuous value.

[4] Taken as unitary for the normalized potential values in this work.

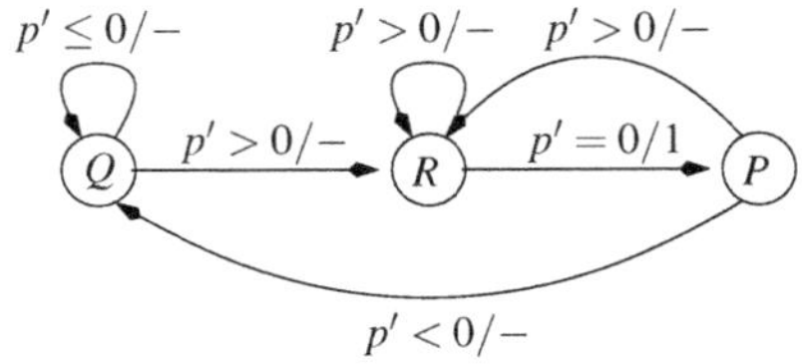

Fig. 7.19. Continuous-time state machine working in conjunction with the discrete-time state machine for recruitment. Arcs indicate state transitions. The transition condition and machine output are displayed atop the arc, separated by a slash (/). The three states are; Quiescent (*Q*) for no activity on potential, Rising (*R*) when potential is increasing, and Plateau (*P*) when a local maximum is reached. Sampling time is set by the transition $R \longrightarrow P$.

We decided to use another state machine, in addition to the one in Figure 7.5c, that operates on continuous parameters seen in Figure 7.19. This machine determines the sampling time for the existing discrete-time recruitment algorithm. This proves to be a very elegant and simple way to upgrade the existing discrete-time model, without obscuring it with continuous-time parameters. This state machine detects peaks (local maxima) of neuroid's membrane potential p according to its derivative p', thereby providing the sampling time for the discrete-time learning algorithm (see Figure 7.13). In this way, we have a simple addition to the system and we can use the previously defined discrete-time machine without modification. In summary, the transition $R \longrightarrow P$ in the continuous state machine in Figure 7.19 triggers the discrete-time state machine in Figure 7.5c. Using a system with discrete and continuous parts together is sometimes termed a *hybrid* approach.

Using a state machine based on differential equations is similar to the *qualitative reasoning* approach (Kuipers, 1994). Qualitative reasoning approximates infinitely detailed physical systems using an incomplete representation of a few selected characteristics. Differential equations are translated into a form that enables reasoning about the system. Discrete states can be obtained from special conditions (such as $p' \leq 0$, $p' = 0$, and $p' > 0$ in our case). Here, the $p' = 0$ condition, in transition $R \longrightarrow P$, indicates that the the membrane potential reached a local maximum, and therefore must be checked against the threshold to emulate firing behavior.

7.5 Methods and Results

This section presents the results of simulation studies whose purpose is to study the predictions of the bounds calculations in the previous section. These studies simulate the recruitment learning task in the presence of direct/indirect converging topologies. The architectures studied are feedforward, multipartite random graphs that vary in the number of layers and the amount of indirection as illustrated in Figure 7.20.[5] Each architecture, or testbed, consists of an equal number of input areas, I_i, and intermediate or middle areas, M_i, where $i = 1, \ldots, k$, and $k \geq 2$. The variable

[5] All timings indicated are simulated milliseconds, not actual time measurements.

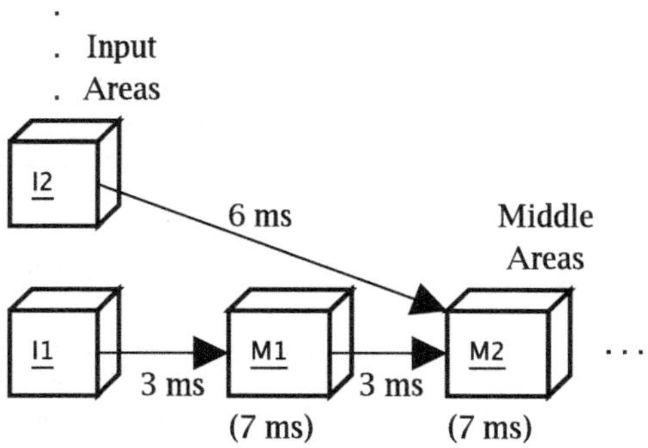

Fig. 7.20. Example testbed used to measure the effects of manipulating the tolerance and segregation parameters. The middle areas serve to create indirect pathways with more synapses than the direct pathways (compare the path $I_2 \rightarrow M_2$ with $I_1 \rightarrow M_2$). The number of input and middle areas is varied for testing. Total axonal delays for both the direct and indirect pathways are 6 ms. This is a two-layer testbed.

number of areas k allows testing the bounds calculations which are dependent on the connection topology.

A simulated attentional controller presents objects to the input areas decomposed into their elementary features (e.g., a blue square is decomposed into blue and square). Each elementary feature for a particular object is represented in a distinct input area by a set of $r = 10$ neuroids (replication factor discussed in §7.2.3). To present an object to the network, the attentional controller causes synchronous spikes to occur in all of the units representing the elementary features for that object. Note that input to a network consists only of the elementary features that were prompted by the attentional controller. To simulate attention to more than one object (say, in a scene), the attentional controller presents each object to the network, segregated by the interval Φ_{SRM}, calculated according to Eq. (7.8). Indirect pathways are created by having signals cross the middle areas.

Successful recruitment depends on both the timings within the testbed and the statistical properties of projections across the areas. To test the capacity of the testbed, which depends on its statistical properties, a control experiment is performed by changing only the number of objects presented. Then, to test the timing hypotheses, the tolerance and segregation parameters are varied between runs. These experiments are repeated for different testbeds when $k \in \{2\ldots4\}$. Simulation performance is evaluated by observing changes in the network's internal organization. We expect to find, at the final converging area, units recruited to represent the attended objects. Each of the recruited concepts for the intermediate layers, and for the objects at the final layer, is expected to have roughly r neuroids. The following defines the maximal capacity property of recruitment in an area:

Definition 7.15. *If all neuroid sets representing concepts contain r neuroids, and each neuroid represents at most one concept, then the maximum capacity of an area is N/r concepts, where N is the number of units in that area of the network.*

Thus, for these networks, the maximal capacity is 10 concepts, when there are $N = 100$ neuroids in the final area and $r = 10$.

To test the performance of the binding scheme, the results of binding errors are examined. Binding errors cause *illusory conjunctions* of features belonging to separate objects and, therefore, result in spurious concepts being recruited in the network. The quantitative performance measure for a network is given by the composite quality of both the correct and the spurious concepts that exist at the end of the simulation. The relative magnitudes of these quality measures determine if a readout mechanism, such as a threshold, can distinguish the valid concepts from the spurious ones. For instance, the readout can be in the form of a winner-take-all mechanism. We assume this is a necessary condition for the network to perform correctly.

To assess the quality of representations formed for a single object that has been presented to the simulation, we give the following definition.

Definition 7.16. *Let the* quality *measure for an object i represented in the network be*

$$q_i = \frac{c_i}{r},$$

where c_i is the number of units allocated for the object. The quality q_i of an object i is maximally 1, if r units are allocated to represent it, and minimally 0 if no units represent it.

In order to evaluate the set of correct or spurious concepts in a simulation, we define the *overall quality* of representations formed for a set of objects as follows.

Definition 7.17. *Let the* overall quality *of an object set O be*

$$Q = \frac{1}{|O|} \sum_{i \in O} q_i . \tag{7.10}$$

We can now use (7.10) to calculate the overall qualities Q_c and Q_s using the object sets O_c and O_s, for correct and spurious objects, respectively.

Several simulations are run to collect statistical data for each test, since both the connections within the network, as well as the feature combinations for objects are chosen from a uniform random distribution in each simulation. Details of the methods are given in subsequent sections for each of the specific tests. The structure of the input representations is given in §7.5.2.

7.5.1 Testbed Details

Only one synapse is required to cross an area. Axonal delays are homogeneous, and increase linearly with distance in the testbed. Therefore, the difference in transmission time (cf. Lem. 7.3) is caused only by synaptic delays. We vary the number of input and middle areas, k, to create larger differences in delays. If $k = 2$, this two-layer topology creates a two-synapse indirect pathway compared to a single-synapse direct pathway to the destination. Therefore, for a given topology the tolerance window can be chosen as

$$\Gamma = (k-1)\tau_s \,, \tag{7.11}$$

since the shortest path always contains a single synapse. Inserting this into Eq. (7.8) we get the required segregation as

$$\Phi = (3k-1)\tau_s \,. \tag{7.12}$$

Since we choose to manipulate τ_m for implementing the tolerance window Γ, we combine (7.7) and (7.11) to get

$$\tau_m = k\tau_s \,. \tag{7.13}$$

Model parameters are chosen according to timing data from visual cortex (Nowak and Bullier, 1997; Lamme and Roelfsema, 2000). The time it takes to cross an area is assumed to be ~ 10 ms. In turn, axonal delays are estimated to be about $\delta = 3$ ms according to their diameter and physical distance. This leaves $\tau_s = 7$ ms for the synaptic rise time (including the dendritic delays), which is a slower process than axonal transmission. Employing these parameters, when segregation values are calculated from Eq. (7.12) for a 2-layer topology similar to the circuit formed by cortical areas V1, V2 and V3, we get 35 ms for segregation. The segregation value predicts the maximum oscillation frequency in this circuit to be 28 Hz, which falls within the gamma band (20–70 Hz). The activity in the gamma frequency band is suggested to be used for object representations (see §7.6.1 for more discussion on oscillatory rhythms).

For all the simulations, each area contains $N = 100$ neuroids with the replication factor $r = 10$ for representing concepts. The connection probability of two neuroids from connected areas is given by $p = \sqrt{\mu/rN}$. This probability is calculated by extending the methodology described in (Valiant, 1994) for simple random graphs. The parameter μ stands for the amplification factor. We employ $\mu = 6$ for increasing the expectation of the set of recruited neuroids for stability reasons discussed in Chapter 8. This value is determined empirically to yield satisfactory recruitment in deep hierarchies. If it is increased further, it creates interference between objects and therefore causes more spurious concepts.

Other parameters used in the simulation include the refractory reset after each spike with a 10 ms time constant. The spike threshold for the middle areas, $T = 1.5P$, is chosen according to the proof of Thm. 7.6.

7.5.2 *Behavior of the Inputs and Concepts*

Inputs to the network are formed by pre-allocated sensory concepts represented by sets of r neuroids within input areas. The sensory concepts are located in the input areas of the network. Each input area provides for the representation of a primitive sensory feature type. The sensory concepts within an area represent different values on the dimension of the specific feature type. For instance, a feature value, such as square versus circle, will be represented by sensory concepts in the shape input area. These sensory concepts are named numerically as S_i^j, where j is the concept number in area I_i. To model the system as it attends to a particular object whose shape, say, is circular, attentional controllers cause the circle sensory concept to be activated. When a sensory concept is activated, each of the neuroids in the set representing the concept emits a single synchronous spike.

Attention to multiple objects is modeled by activating a sensory concept from each input area synchronously for each perceptual object. As an exception to this rule, two concepts need to be chosen from input area I_1, since middle area M_1 is only connected to I_1, and the recruitment mechanism requires simultaneous activation by two separate concept sets. In a multi-object scene, separate objects are attended to at different times, separated by the amount of the segregation parameter Φ_{SRM} given in (7.12).

The total number of unique objects that can be represented using this scheme can be calculated by $\binom{n}{2} n^{k-1}$, where n is the number of sensory concepts in each input area, and k is the total number of input areas. For the simulations in this work, $n = 4$ sensory concepts are allocated in each input area I_i, where $i = 1, \ldots, k$, and $2 \leq k \leq 4$. Thus, there are 24 and 96 possible objects to choose from, for 2- and 3-layer topologies, respectively.

Neuroid sets representing concepts are recruited in the middle areas upon activation of sensory concepts. A new concept is labeled according to the sensory concepts that caused its recruitment. For instance, the concept recruited in middle area M_1 upon simultaneous activation of the sensory concepts S_1^0 and S_1^1 is labeled as $S_1^0 \wedge S_1^1$.

A simulation consists of presenting a sequence of multiple perceptual objects to the network, segregated in time. At the end of the simulation, the set of concepts created in the network is analyzed. Recruitment of the concepts and the weight update rule is the same as described earlier in Section 7.2. Correct concepts are the conjunctions of the originally presented sensory concepts for each perceptual object. Spurious concepts are all the concepts created, except the correct concepts and the anticipated intermediate concepts recruited in the middle areas.

7.5.3 Intuitions on Tolerance Window Parameter from the Simulations

Here, we describe a simulation run to illustrate delay-tolerant conjoining in the neuroids of the network. The simulation was successful when the tolerance parameter Γ was selected according to the criteria proposed in (7.4) and accordingly selecting the membrane time constant τ_m as in (7.7).

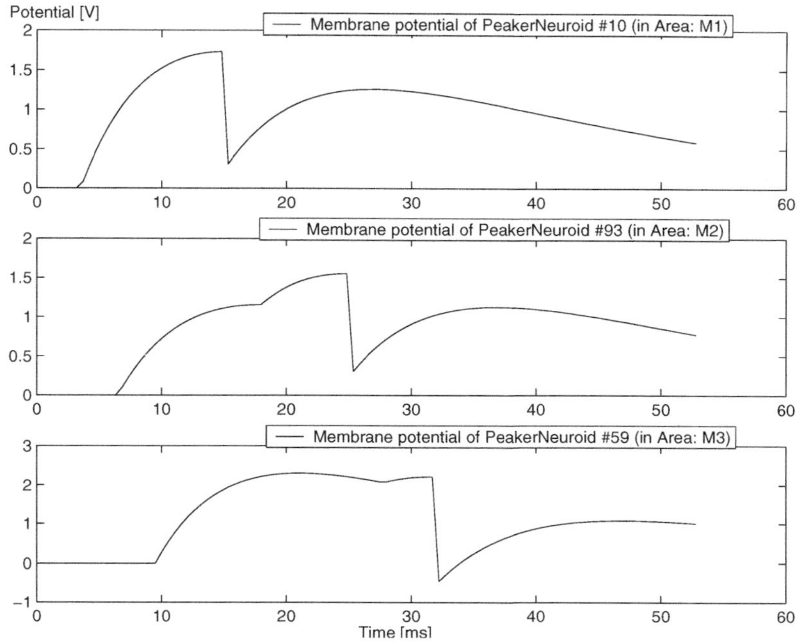

Fig. 7.21. Membrane potentials from a selected neuroid from each middle area, from M_1 to M_3, shown from top to bottom, respectively. The resets on the membrane potential show the time of spikes emitted. The action potentials are not depicted since they are ideal Dirac delta functions in the SRM.

The three profiles in Figure 7.21 show the membrane potential time profiles of a selected recruited neuroid from each middle area. This shows the progress of signals from sources I_i, $i = 1, \ldots, 3$ to the destination M_3:

1. Initially the neuroid in M_1 (profile at top of figure) receives a signal from the input area I_1. After a $\delta = 3$ ms onset delay due to axon transmission, and $\tau_s = 7$ ms synaptic rise and dendritic delay, the neuroid fires at $t \sim 15$ ms since the membrane potential $p > T = 1.5$, the threshold.
2. A recruitment candidate in M_2 (profile at middle of figure) receives a signal from the input area I_2 with a 6 ms onset delay and another signal from the recruited neuroids in M_1 after a 3 ms transmission delay, at $t \sim 18$ ms. The cumulative effects of both these signals makes the neuroid fire at $t \sim 25$ ms. Notice that the first spike's effect is not sufficient to recruit the M_2 candidates, even though a local maximum that triggers the state machine is reached on the way (see figure and the state machine for recruitment in §7.4.4).
3. An effect similar to M_2 is obtained in M_3 (profile at bottom of figure). This time the signals originating in input area I_3 arrive after a 9 ms delay. However, the signals from M_2 arrive at $t \sim 28$ ms and neuroids are recruited and fired at $t \sim 32$ ms.

However, there is an anomaly in M_3 worth mentioning. On close examination of the membrane potential plot, one can see that the signal coming from M_2 does not raise the potential above a value higher than what has been reached by input I_3. This may seem contradictory to the tolerant conjoining described so far, which distinguished the effect of multiple inputs from the effect of an individual input, by using the potential level to discriminate between them. In this case, this may indicate that the input signal from I_3 is sufficient to cause recruitment without waiting for the results of the computation coming from M_2. The reason that recruitment and spiking are not observed is because there is a recruitment limit in our simulation for maintaining the stability of recruitment. Since there are already neuroids in M_2, representing active inputs in I_3, no neuroids in M_3 are allowed to join the recruitment. This can be interpreted as an inhibitory effect raising the threshold. The artificial recruitment limit imposed by the boost-and-limit mechanism is discussed in Chapter 8. Due to the current implementation of the limiting mechanism, if a concept is allocated in multiple areas, then the inhibition spans globally to those areas.

7.5.4 *Quantitative Results*

In the following subsections, results are given for observing the effects of different values for the number of objects presented, the tolerance window, and the segregation parameters. For each of these parameters, figures are given with simulations on varying-size network architectures. These networks are constructed with varying levels of indirect pathways as described in §7.5.1. An architecture with a two-level indirect pathway is depicted in Figure 7.20.

The performance measure is the overall quality of object representations given by Definition 7.17, simply referred to as *quality* hereafter. In each figure, qualities for correct and spurious concepts are plotted while some parameter is varied. The graph plots the average quality value over a number of trials indicated on each figure. Each data point represents the mean quality value of ten simulations. The errorbars represent the maximum and minimum quality values from the trials.

In the figures, an x-axis value for the parameter varied may be marked by a dash-dotted vertical limit bar for the calculated bound. Other network parameters employed are included in the figure legends and captions.

As stated at the beginning of §7.5, a network is successful if the correct concept quality can be distinguished from spurious concept quality via a threshold value.

7.5.4.1 Network Object Capacity

Figure 7.22 displays the overall object-representation quality as the number of objects presented to the network is varied. Parts (a), (b), and (c) of the figure show the results for a two, three, and four-layer testbed, respectively. Recall, that a k-layer testbed has an indirect pathway of k synapses. In all cases, $N = 100$ and $r = 10$, so the maximal capacity of these networks is 10 (see Definition 7.15). The vertical dash-dotted line at the point where the number of objects is five indicates the point of half-maximal capacity. In all of the graphs, the solid plot shows the overall quality

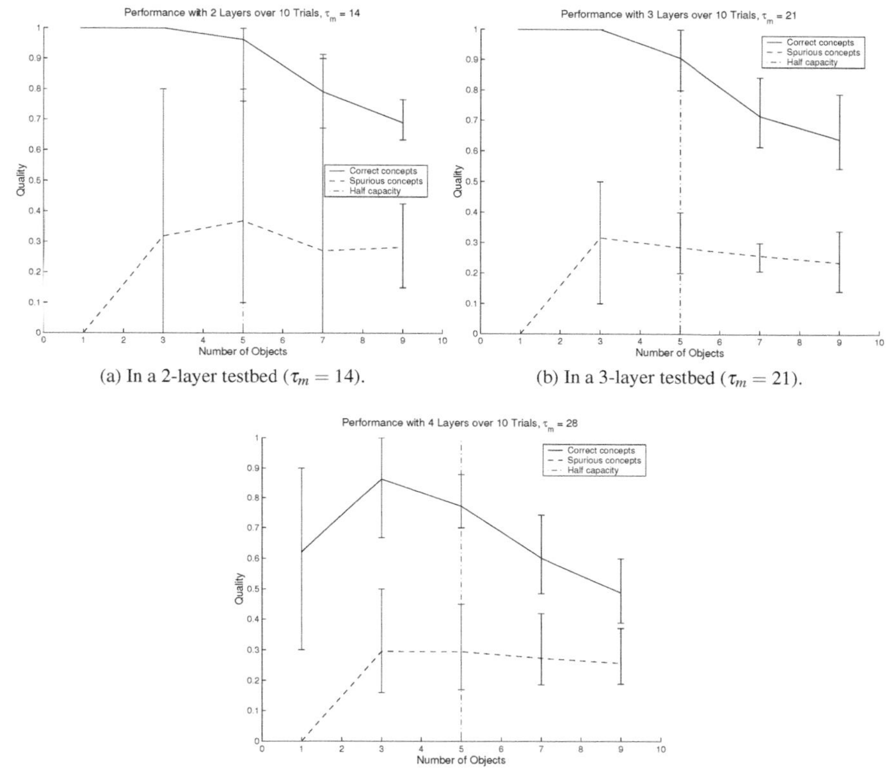

(a) In a 2-layer testbed ($\tau_m = 14$).

(b) In a 3-layer testbed ($\tau_m = 21$).

(c) In a 4-layer testbed ($\tau_m = 28$).

Fig. 7.22. Variation in concept quality as a function of the number of objects presented to the network. Plots show the robustness on the expected capacity of the network. The dash-dotted line represents the half-maximal capacity value for the number of objects. See §7.5.4 for reading the plots.

for the correct concepts while the dashed plot shows the quality for the spurious concepts. The most important feature of the graph is the threshold distinguishability of the correct from the spurious concepts. The error bars indicate maximum and minimum values over ten runs. As long as there is no overlap in the two plots then threshold discriminability is guaranteed.

All three testbeds successfully represent all objects at the half-maximal capacity of five. Furthermore, the network behaves gracefully as the maximal capacity is approached. However, as the number of indirect pathways increase, the performance slightly degrades. The maximal capacity is achieved for topologies with a low number of indirect pathways, such as the ones shown here. The quality of correct concepts is still distinguishable from spurious ones even for capacities at the maximal limit (see Figure 7.22). Theoretically, the maximal capacity of the network cannot be achieved easily. As the network is populated with concept sets, fewer neuroids will be available for recruitment for new concepts. Thus, the probability of finding random connections to the few available neuroids left will be lower. Nevertheless, simulations indicate that if the assumption about recruiting r neuroids for each final concept does not hold (less than r neuroids are allocated for some concepts), the network capacity can become higher than expected.

7.5.4.2 Tolerance Window Parameter

Figure 7.23 gives the quality of concepts in the network as the tolerance parameter is varied, for simulation testbeds of two, three and four indirect layers. The tolerance bound of τ_m, indicated with the dash-dotted vertical line, is calculated by (7.13) for each topology. As with the results on network capacity, the method is successful

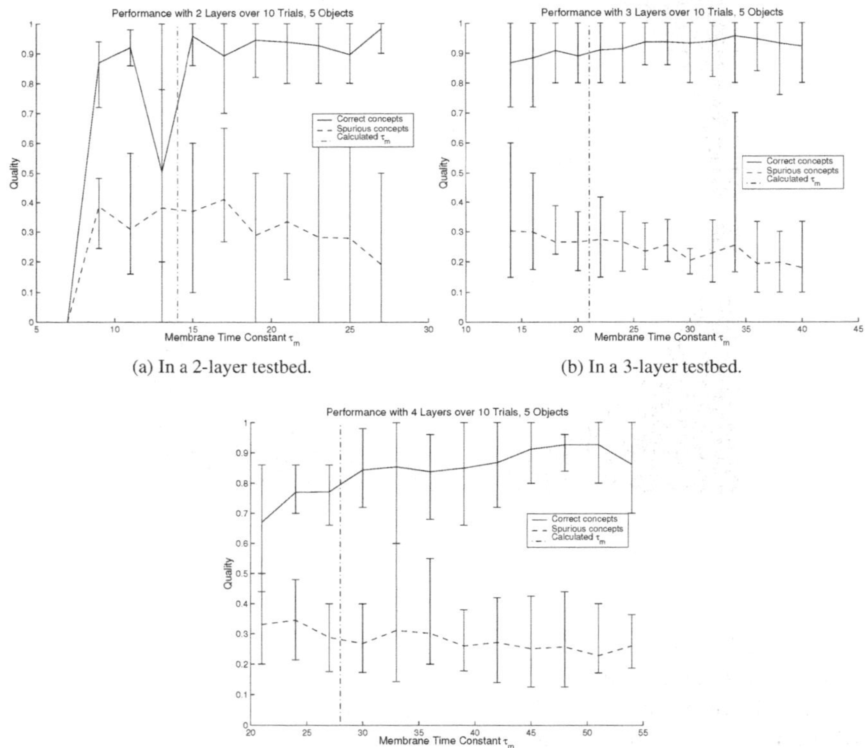

(a) In a 2-layer testbed.

(b) In a 3-layer testbed.

(c) In a 4-layer testbed.

Fig. 7.23. Variance in concept quality as a function of the membrane time constant τ_m. In the simulations, the tolerance window Γ is varied, implying that τ_m and segregation between activity Φ is calculated according to Eqs. (7.7) and (7.8), respectively. The calculated operating value of τ_m, given by (7.13), is shown with a dash-dotted vertical line. Five objects were presented to the network. See §7.5.4 for reading the plots.

with the architectures employed, even though the performance slightly degrades as the number of indirect pathways is increased.

The reason is not apparent for the sudden trough on the correct concept quality of the two-layer testbed in the figure. Since the same behavior can later be seen in the two-layer network results of the phase segregation plots, we claim that it does not depend on any one of the parameters, but maybe depends on both of them. In other words, this temporary loss of quality may be due to a resonance behavior between the tolerance and segregation parameters which becomes only significant in the two-layer architecture.

7.5.4.3 Phase Segregation Parameter

Figure 7.24 shows the variation in the quality of concepts in the network as the segregation parameter is varied, for simulation testbeds of two, three and four indirect

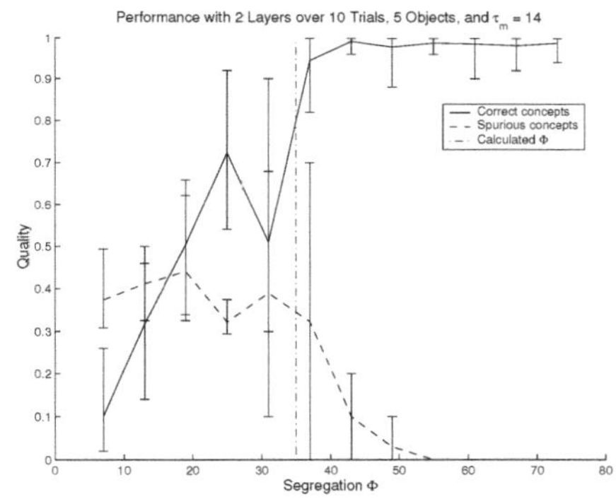

(a) In a 2-layer testbed ($\tau_m = 14$).

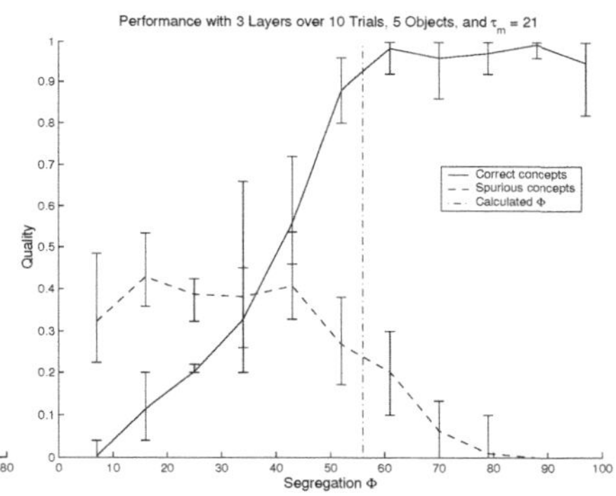

(b) In a 3-layer testbed ($\tau_m = 21$).

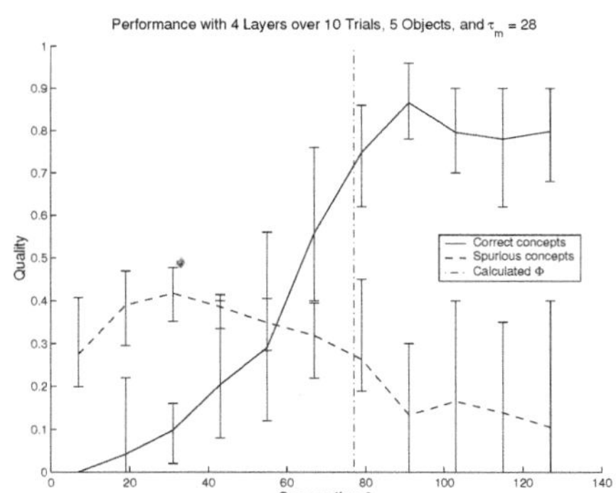

(c) In a 4-layer testbed ($\tau_m = 28$).

Fig. 7.24. Variance in concept quality as a function of the segregation amount Φ. In the simulations, the segregation Φ is varied, while the tolerance τ_m is kept constant at a value calculated according to (7.13). The segregation bound of Φ, shown with a dash-dotted vertical bar in the graph, is calculated by (7.12). Five objects were presented to the network. See §7.5.4 for reading the plots.

layers. The segregation bound Φ, indicated by the dash-dotted vertical line in the graph, is calculated by (7.12) for each topology.

Our calculated segregation limit seems to be a proper lower bound for the architectures tested. We expect the network to be more stable as the segregation is increased. However, the quality of correct concepts initially increases and then plateaus in the four-layer testbed (see Figure 7.24(c)). While this decrease is an undesirable effect due to a deficiency in the model, possibly due to selection of SRM parameters, it is not significant. The decrease in the quality possibly occurs because the tolerance value is kept fixed while the segregation is increased. This reduces the interference between adjacent phases as the segregation becomes large. Apparently, for a certain range, the interference causes a subthreshold activation between adjacent phases, that increases concept correctness without causing disruption. We report that this kind of interference can sometimes be useful rather than disruptive, even though this was not our intention. Since we cannot guarantee when this interference is going to be useful, we do not depend on this feature.

7.6 Discussion

7.6.1 Maximal Number of Hierarchical Levels Possible in the Brain

As the number levels in direct/indirect converging topologies increases, the minimal required segregation value also increases (see Eq. (7.12)). According to Corollary 7.14, segregation limits the frequency of periodic activity. This allows us to use the observed brain activity frequencies to predict maximum levels of hierarchies that can exist.

Section 7.5 showed how the tolerance and segregation measures depend on the topological parameters according to Eqs. (7.11) and (7.12), respectively. The segregation depends on both the number of synapses in the longest indirect pathway k and the synaptic rise time constant τ_s. These equations assume that τ_s and conduction velocities are homogeneous throughout the topology. In the following predictions, the limits only apply on the levels of direct/indirect converging pathways. Diverging pathways need not be synchronous and thus do not need to be segregated.

Using Eq. (7.12), the connection topology formed by the visual areas 1–3 shown in Fig. 7.3 requires $\Phi = 35$ ms. The maximal periodic activity becomes 28 Hz in this case, which falls into the gamma frequency band (20–70 Hz). The number of levels that gamma-band activity can support is limited by $k \leq 2$. From this we can hypothesize that for small hierarchies of direct/indirect topologies can only be supported locally in the gamma-band frequency range. For topologies with more levels, we assume they span larger areas of cortex and that their activity falls into the alpha frequency band (8–12 Hz). The number of hierarchical levels supported by the alpha-band is $k \leq 6$. This hypothesis is consistent with gamma-band activity being observed at local structures while alpha-band activity occurs over large structures or globally (cf. VanRullen and Koch, 2003).

7.6.2 Spurious Concepts

Although it seems that the amount of spurious activity in our network increases with the number of indirect layers, this failure to scale is not because of our calculation of the tolerance and segregation parameters, but it is an artifact of the recruitment learning method. When recruiting concepts in a cascade, if some spurious concepts appear at one stage, they cause the recruitment of additional spurious concepts at further stages, based on the conjunction of the spurious concept with other legitimate concepts.

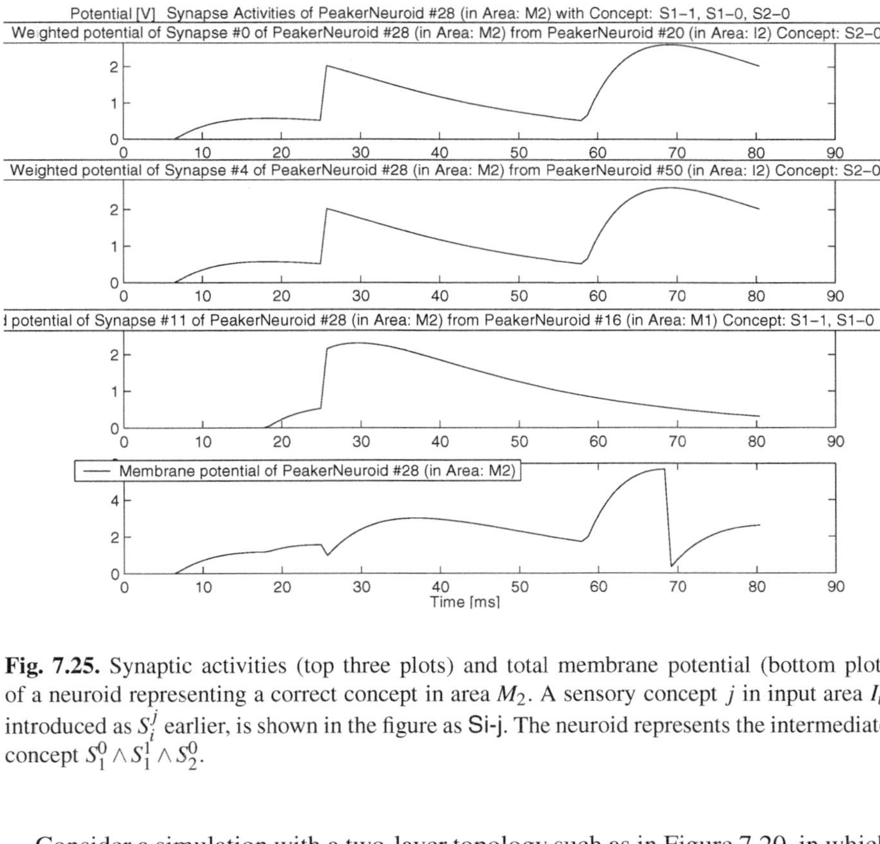

Fig. 7.25. Synaptic activities (top three plots) and total membrane potential (bottom plot) of a neuroid representing a correct concept in area M_2. A sensory concept j in input area I_i, introduced as S_i^j earlier, is shown in the figure as Si-j. The neuroid represents the intermediate concept $S_1^0 \wedge S_1^1 \wedge S_2^0$.

Consider a simulation with a two-layer topology such as in Figure 7.20, in which two hypothetical perceptual objects are presented at separate times. The first object is represented by the sensory concept conjunction $S_1^0 \wedge S_1^1 \wedge S_2^0$, and the second by $S_1^1 \wedge S_1^2 \wedge S_2^0$ (each have two concepts in input area I_1 and one concept in input area I_2). Figure 7.25 gives the synaptic activities and the total membrane potential for a neuroid that belongs to the set of a correct concept $S_1^0 \wedge S_1^1 \wedge S_2^0$ in area M_2. The figure shows three incoming synapses from neuroids in areas I_2 and M_1. These synaptic neuroids belong to the sets of concepts $S_1^0 \wedge S_1^1$, and S_2^0. Note that there are two synapses from the set for S_2^0, but only a single synapse from the set for $S_1^0 \wedge S_1^1$. The neuroid is recruited for $S_1^0 \wedge S_1^1 \wedge S_2^0$ at $t \simeq 25$, when three synchronous spikes, received from synapses 0, 4, and 11, approach their maxima. The sudden increase in the synaptic potentials reflects the change in the weight values. The sudden decrease in the membrane potential at this time (bottom plot in figure, time $t \simeq 25$), however, is due to the reset after the neuroid fires.

When the second object $S_1^1 \wedge S_1^2 \wedge S_2^0$ is presented to the network at $t \simeq 55$, we expect this neuroid to stay silent. However, the neuroid produces an action potential seen from the reset at $t \simeq 70$. The reason for this erroneous action is that the combined effect from two strong synapses from set S_2^0 produces enough activation to cross the threshold calculated for the recruited concept. The culprit is both the learning algorithm for not weakening the synapses enough, and the uneven distribution of synapses from different concept sets. Neither of these issues are resolved in this work since they are artifacts of the theory behind recruitment learning and not the tolerance and segregation parameters. We are working on making the network more noise tolerant rather than tweaking the parameters to suppress this kind of natural outcome. One possible solution is to employ a probabilistic approach to interpreting the concept boundaries (Feldman and Valiant, 2009), so that spurious activity with low number of units can be rejected.

The important consequence of this erroneous activity is that the spike caused by this neuroid is going to cause more spurious effects in the downstream areas (see Fig. 7.26). The postsynaptic neuroid will assume that it received a spike from a synapse representing the concept $S_1^0 \wedge S_1^1 \wedge S_2^0$, even when the second object does not represent the concept S_1^0. In the simulator, recruited concepts are labeled at the time of their recruitment according to their incoming neuroids. Afterwards, the simulator reads the previously assigned label, rather than actually observing the neuroid's activity. As a solution, it may be argued that, since the neuroid fired in phase with the second object, it should represent an intermediate concept for the second object. Another possibility is to dynamically change the concept to which the neuroid belongs. This implies using a more advanced learning algorithm that allows gradual adjustment of weights after the initial memorization, or n-shot learning.

7.6.3 Implementation of the Tolerance Window

We suggested that lower bounds on the tolerance and segregation parameters can be calculated for a direct/indirect connection topology. The results confirm that the network performs better as the tolerance increases. Excessive stability is not desirable since there is a trade-off with processing speed. We prefer to have the lowest tolerance value to achieve the fastest speed without compromising network correctness. For this purpose, the values chosen for tolerance seem to be appropriate, since correct concept quality values can be distinguished from spurious ones. In this respect, there even seems to be room for further optimization of the tolerance parameter.

We proposed that the membrane time constant can be dynamically adjusted (possibly by biological processes that vary the membrane resistance), to accommodate

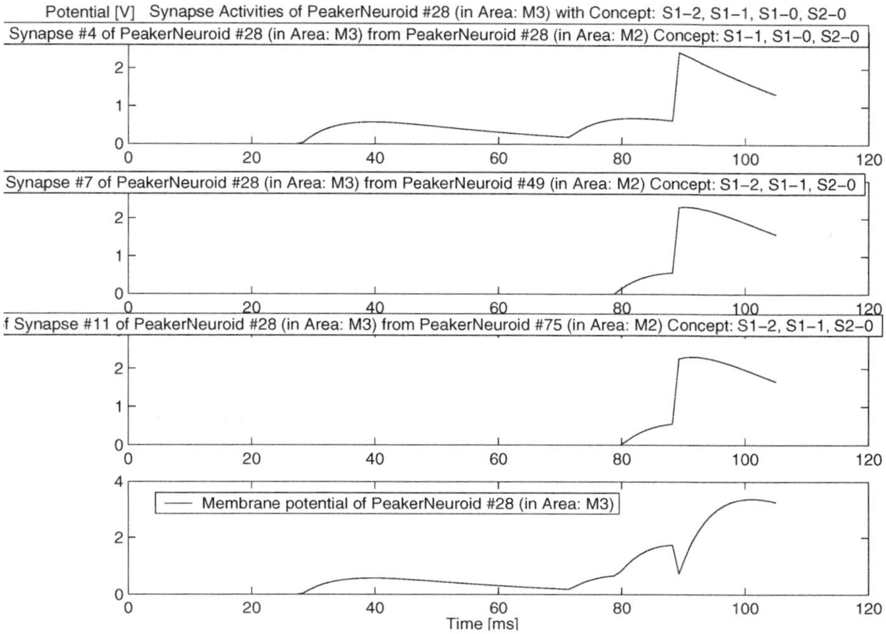

Fig. 7.26. Synaptic activities (top three plots) and total membrane potential (bottom plot) of a neuroid representing a spurious concept in area M_3. This concept $S_1^0 \wedge S_1^1 \wedge S_1^2 \wedge S_2^0$ is caused by the correct concept $S_1^0 \wedge S_1^1 \wedge S_2^0$ in Fig. 7.25 firing in the wrong phase. A sensory concept S_i^j is written as Si-j.

the calculated tolerance window (see §7.3.3). Another modeling study also showed that synaptic dynamics with different time constants can help maintain memories in networks (Volman et al, 2007). An alternative to varying the membrane or synaptic time constant is utilizing the persistent firing of inputs occurring at separate times, for creating an overlapping effect on the destination read-out site (Cannon et al, 2002; Günay and Maida, 2003d; Koene and Hasselmo, 2007). Persistent or reverberating activity has been observed in experiments of *in vivo* (Funahashi et al, 1989) and of dissociated cortical culture preparations (Lau and Bi, 2005). Yet another solution for adjusting tolerance is by dynamically adjusting the threshold (excitability) of the destination unit.

However, there are other approaches to the problem of variable delays. In particular, it was proposed that introducing synapse-specific delays and integration times adopted during development can accommodate for differences in delays (Senn et al, 2002; Steuber and Willshaw, 2004). If cortical circuits can adapt to varying delays, this may solve the problem with the direct/indirect connection topologies, as well.[6]

---

[6] Due to personal communication with Benjamin Rowland and comments from an anonymous referee.

7.6.4 Implementation of Segregation

The calculated segregation, however, needs to be applied at the initial source (possibly by attentional mechanisms). Therefore, feedback connections from the destination site should inhibit the source areas for the desired segregation amount. For instance, the dense feedback connections from visual area V1 back to the lateral geniculate nucleus (LGN) may be responsible for this kind of modulation (see the direct/indirect connection topology in Figure 7.3). However, it is difficult to assume that there is a direct feedback connection to the initial source in all such topologies. Instead a more complex attentional mechanism may be responsible for segregating signals. The segregation also predicts the maximum firing frequency in the local circuit.

The field of signals and systems has also contributed to the theory and application of timing issues in interconnected circuits. In particular, the industry for fabricating *integrated circuits* (ICs) nowadays gives high importance to timing properties of circuits with the need to produce faster computers. Some theory from this field may apply to the issues we discuss in this work.

The problem of synchronizing varying-length or varying-delayed paths is especially important in fabricating ICs. Three mainstream approaches can be identified in the current literature as a solution to the problem (Chandrakasan et al, 2001, Chs. 9,11). The first solution is achieved by using a central global clock signal to synchronize events at different parts of the circuit. The clock signal controls the time when the computational units start processing their inputs. This requires buffering devices to store and synchronize the inputs arriving at various times for each unit. For structures with varying delays between the source and destination stages, synchronization can be achieved if the clock period is made large enough to tolerate the maximally-delayed signals.[7] This solution is equivalent to the approach we take to calculate the tolerance window Γ with (7.4). A major disadvantage to this approach is that even faster computations need to wait for this longer duration.

The second approach proposes using circuits without a global clock signal. These circuits are called *asynchronous*, where each unit produces an output as it completes its computation. Here, a special effort must be made for ensuring that the varying-delay pathways do not appear. To achieve this, paths between different stages of computations are shortened and unified. The major disadvantage of this approach is that this type of fine tuning is expensive and susceptible to errors caused by noise or slight variations in component properties due to fabrication artifacts.

The third approach attempts to combine the strengths of both previous approaches. Each interconnected processing stage consists of interacting components. Results of a computation from a stage are only transmitted to the next stage after receiving a release signal. This approach is most interesting for our purposes because it is easier to model it with biological circuits. This method does not require a global

[7] The clock is assumed to control the source and destination stages of the computation. Intermediate stages of computation between the source and destination need to be controlled by an independent and faster clock signal or function asynchronously.

clock and the nature of connections are more localized. Modeling this approach into our system is left as future work.

7.6.5 Conclusions

Our previous work proposed lower bounds for the tolerance window Γ and the phase segregation Φ parameters. Here, we presented results from later work (Günay and Maida, 2006b), where we improve these hypotheses in (7.7) and (7.8) and show that the SRM parameters must also obey $0.1 < \tau_s/\tau_m < 0.5$. We apply these constraints to direct/indirect converging topologies and run simulations to test their viability. We ran simulations on networks with varying size direct/indirect connection topologies. We tested for binding errors among multiple hypothetical objects presented, while the tolerance and segregation parameters were changed over a range including the predicted values. We conclude that appropriately chosen tolerance and segregation parameters enable use of temporal binding for recruitment learning in direct/indirect connection topologies. Furthermore, using a spiking neuron model is appropriate for recruitment learning, which was originally proposed with simpler discrete-time neuron models. Recent studies are consistent with our view (Shastri, 2001a, 2002a; James and Maida, 2009).

We also improved the stability of recruitment with aid of a stabilizing mechanism (Günay and Maida, 2006a) that we present next in Chapter 8. As a result, our simulations indicate that up to roughly half of the predicted capacity can be achieved with reasonable performance. The statistical variance inherent in the recruitment method prevents recruiting a chain of concepts in cascade. This problem is especially apparent in smaller network sizes, such as we employ here with a low number of neuroids ($N \sim 10^2$) for each area. Earlier work on the stability of the recruitment method for larger network sizes such as $N \sim 10^8$ (Valiant, 1994), and under asymptotic conditions $N \rightarrow \infty$ (Gerbessiotis, 1993, 1998, 2003) indicates recruitment can be used up to four levels deep. Feldman and Valiant (2009) showed that high capacity can be obtained by recruitment learning in a two layer network with a probabilistic approach in counting units that are part of concepts, overlapping concept representations and more complex learning rules. Our stabilizing method can potentially be applied to these larger networks.

7.6.6 Future Work

Our results here point to the need to design neural circuits that adaptively adjust the tolerance and segregation parameters, rather than calculating and setting them to fixed values according to each topology. Since cortical circuits are known to change, tolerance and segregation should be managed dynamically according to changing conditions. For managing tolerance, it can be shown that, if only the membrane resistance is externally manipulated to vary the membrane time constant, the desired effect can be achieved without increasing the time to spike (see Appendix A.2).

Another mechanism that deserves further work is the neural circuits that may be responsible for managing the proposed stabilizing machinery for hierarchical recruitment. There are a number of neural circuits that can be proposed for realizing this boost-and-limit function. For instance, it can be proposed that global inhibition by itself, or local lateral inhibition with noisy delays to trigger an inhibitory circuit to shut off all activity after sufficient recruitment is reached, can be used to control recruitment. This is the subject of the next chapter.

Chapter 8
The Stability of Recruited Concepts

Cengiz Günay

"Even the largest avalanche is triggered by small things."
Vernor Vinge

8.1 Introduction

Recruitment learning is prone to instability when a chain of concepts is recruited in cascade as seen in Figure 8.1. The statistical variance inherent in the recruitment method causes increasing perturbations to the recruited set size, and thus instability (Valiant, 1994). We previously proposed a boost-and-limit algorithm to improve recruitment stability, and verified the applicability of this method with a software

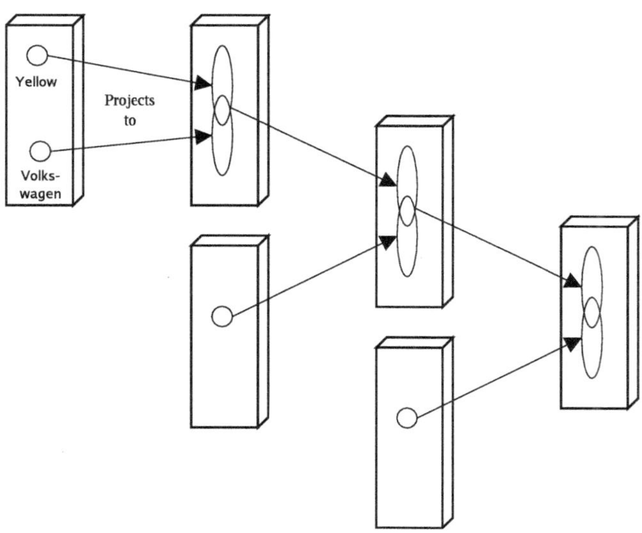

Fig. 8.1. A hierarchical recruitment scenario. The circles indicate the set of neuroids that represent each concept, whereas the large ellipses indicate projection sets of these neuroids. Number of neuroids in intersections of projections vary in an unstable manner when recruitment is used repetitively.

J. Diederich, C. Günay, and J.M. Hogan: Recruitment Learning, SCI 303, pp. 243–274.
springerlink.com

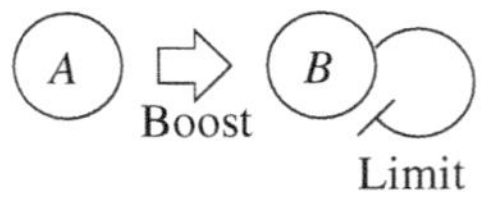

Fig. 8.2. Basic structure of the boost-and-limit mechanism. Boosting signifies increased connectivity between A and B. Limiting applies to the size of the recruited set via negative feedback (possibly lateral inhibition).

model in a spiking neuroidal net simulator (Günay and Maida, 2001, 2006b), which was overviewed in Chapter 7. In that model, excess recruitment candidates were rejected to enforce a stable recruitment level. This chapter describes the improvements of the previous boost-and-limit method to use a more biologically realistic neural mechanism (Günay and Maida, 2006a).

The boost-and-limit method, sketched in Figure 8.2, works by first increasing the statistical expectation of the recruited set size. This limits the probability of under-recruitment. Then, to control this increase, negative feedback is applied by setting a hard limit on the size of the recruitment set. This limits the possibility of over-recruitment. We propose a biological model with similar function, using both the variable delays inherent in cortical networks, and lateral inhibitory effects between principal neurons as the negative feedback. In this model, the initially synchronized spike volley intended to cause recruitment is assumed to be subject to varying delays in the individual spikes. The delays are caused by spikes travelling through axons of slightly varying lengths and by the varying dendritic placement of synapses. The background noise in cortical networks interacts with membrane charge time and thereby can affect the timing of a spike emitted by the postsynaptic neuron. The time to charge up the membrane to the threshold depends on the current membrane voltage achieved by the background noise. This adds another source of uncertainty to the spike time. These varying delays in the spike arrival times cause the destination neuroids to fire and become recruited in a temporally dispersed sequence. During this process, we propose using the local lateral inhibition as a mechanism that saturates to fully inhibiting the localized area after enough neuroids are recruited. This is possible if each recruited neuroid emits a delayed lateral inhibitory signal within a properly connected structure. In other words, recruitment causes the neuroid to fire (as proposed by Valiant) and emit a lateral inhibitory spike (our proposal), thereby slowing down further recruitment. In this work, we assume that neuroids are capable of having both excitatory and inhibitory synapses.

In Section 8.3, we describe the stochastic population approach that we employ to study the properties of the proposed boost-and-limit mechanism within the context of an integrate-and-fire (I/F) neuroidal network. We are interested in variations of the expected population size of recruited neuroids with respect to perturbations to the input population size. We first introduce a feedback control system where the recruitment is modeled with a monopolar excitatory synaptic population. The expected size of the recruited neuroid set has an equilibrium point r_o. In Section 8.3.1 we confirm that in the original open-loop recruitment system, the equilibrium point is unstable. That is, for perturbations in the size of the input set, the size of the recruited set diverges from the equilibrium point. We also verify the population model results with Monte Carlo simulations.

In the closed-loop system, for an ideal instantaneous feedback condition, the equilibrium point becomes stable, but the model is prone to oscillations. However, since we encounter inconsistencies in the steady-state behavior of this model, we switch to a more detailed model using dipolar excitatory-inhibitory synaptic populations. This model shows an improved convergence rate to the stable equilibrium point for the less restricted uniform-delay feedback condition. Section 8.4.5 describes the low pass filter that is required at the output of both these models to prevent the unwanted oscillations in the activity level, these being an artifact of the population model. Monte Carlo simulations, in Section 8.4.6, verify that the feedback model is stable and exhibits no oscillations. Using Monte Carlo simulations, the effect of changing the feedback delay on stability results is analyzed in Section 8.4.7. In Section 8.5, we extend the recruitment process to allow an arbitrary number l of synapses, instead of just two, for improved biological realism.

The final model allows one to choose the desired recruitment size, r_o, for representing concepts, and the number of neuroids per localized area N, arbitrarily. Another free parameter of the model is the feedforward excitatory connection density. This is calculated for a given r_o and N according to the definition of recruitment learning. The connection density can also be adjusted with a positive gain constant λ. The choice of λ affects the rate of convergence to the stable equilibrium point. Given these parameters, we can calculate the required lateral inhibitory projection density for stable recruitment in hierarchies.

8.2 Previous Work Related to Recruitment Stability

The problem of recruitment instability was originally explored by Valiant (1994) in the so-called vicinal algorithms in randomly connected graphs. Valiant proposed that hierarchical recruitment with a depth of three to four levels can be achieved if the parameters of the system are chosen appropriately, based on the work of Gerbessiotis (1993, 2003). This study assumed a replication factor of $r = 50$ neuroids for representing each concept. It was also assumed that the total number of neuroids in the system was large, approaching infinity, which is reasonable given the large number of principal neurons (pyramidal cells) in the cerebral cortex. Gerbessiotis (2003) provided a rigorous formalism of the expected recruited set size in random graphs. Gerbessiotis (1998) also showed that a constructed graph can guarantee the replication factor to be a constant r only for a graph with $3r$ vertices (neuroids), and not for a graph with an arbitrary size.

Our earlier work (Günay and Maida, 2006b) suggested that the instability becomes graver when the total number of neuroids in the network is low (e.g., on the order of hundreds). In our case, networks that are divided into localized areas with a small number of neuroids are interesting because they are better suited for computer simulation. Even though the mammalian brain contains a large number of neurons in total, there are smaller substructures where our analysis could be applied. For instance, possible candidates are cortical areas and microcolumns.

Levy (1996) presented a hippocampal-like model for sequence prediction which used a recurrent network having random asymmetric connectivity. They analyzed parameters statistically, in a manner similar to our work, to find appropriate neural threshold and weight parameters for maintaining stable activity levels (Minai and Levy, 1993, 1994). Their model differs from ours in having excitatory feedback connections, and employing a rate model with discrete time steps, unlike the continuous spiking model used in the present work. Yet, our model lacks effects of spike rates and variable thresholds since these are of secondary importance in our framework (Günay and Maida, 2003b). Previous work using statistical analysis on network parameters goes back to Amari (1974). This kind of analysis is closely related to mean-field methods in statistical physics.

Shastri (2001a) modeled recruitment learning based on the biological phenomena of long-term potentiation (LTP) and long-term depression (LTD) with idealized I/F neurons. Then, assuming that recruitment learning is employed in the interactions between the enthorinal cortex of the medial temporal lobe, and the dentate gyrus of the hippocampal formation, he calculated the probability of finding a sufficient number of recruitment candidates according to the anatomical properties of these structures and a suitable selection of parameters. Shastri also extended the recruitment learning method to: 1) allow multiple redundant connections from each of the input concepts, which makes the method more robust; and, 2) allow a recruited neuron to take part in representing other concepts, which increases the concept capacity of a network containing a finite number of units.

Diesmann, Gewaltig, and Aertsen (1999) and Tetzlaff, Geisel, and Diesmann (2002) have analyzed the boundary conditions, for which propagation of synchronized spike packets within a feedforward subgraph is possible. These architectures are composed of feedforward layers of I/F neurons with each neuron having convergent inputs from the previous layer. Diesmann et al found that for a synchronized spike packet to propagate undisturbed, there are lower bounds for the size of the packet and the connection density. Tetzlaff et al found that there are also upper bounds, above which, meaningful propagation of activity is no longer possible. Their results were shown for fully connected feedforward networks with stationary background activity. Tetzlaff et al also point out that feedback from inhibitory populations could potentially stabilize the network to enable controlled synchrony over a larger range of network parameters. Our networks use inhibitory feedback and do not have background activity.

Other research relevant to the controlled propagation of synchronous activity is by van Rossum, Turrigiano, and Nelson (2002) and Litvak, Sompolinsky, Segev, and Abeles (2003). The primary issue addressed by this research is whether rate code, coupled with a high-input regime and associated background activity is a viable representation of the neural code. Van Rossum et al, extending earlier results of Shadlen and Newsome (1994, 1998), provide simulations showing that, in the presence of noisy background current, firing rates propagate rapidly and linearly through a deeply layered feedforward network. Furthermore, they find that background noise is essential but does not lead to deterioration of the propagated activity. Litvak et al are skeptical that rate code in the presence of background activity is possible at all. They point out that van Rossum et al had to fine-tune their noise parameters to obtain propagation of rate code, and that each neuron fired in an almost periodic manner. Such firing patterns are not commonly observed in biological cortical activity. For the purposes of the present paper, the relevant issue is whether controlled propagation is possible, and on this point the answer is in the affirmative.

8.2.1 Relation to Winner-Take-All Mechanisms

In the brain, neural firings result in stereotypical action potentials (APs) with constant magnitude. However, the firing times and spike rates carry important information (Gerstner, 1999). In our model, since all neuroids are assumed to fire once or only a few times during the period of analysis, the time-to-first-spike is the most significant variable. In this sense, our model can be considered as a winner-take-all (WTA) mechanism (Feldman and Ballard, 1982) if the winners are chosen according to temporal precedence, similar to the work of Indiveri (2000). Specifically, our model is a k-WTA, because it allows roughly k winners to be fired and recruited, where k is the number of neuroids redundantly representing a concept. It is also a *soft* WTA, which sorts k real valued outputs according to the magnitude of the corresponding real valued inputs, in contrast to a *hard* WTA whose outputs are binary (Maass, 2000). Regarding the computational power of WTA networks, Maass (2000) showed that networks that use lateral inhibition as a WTA mechanism have the same universal computing power as layered feedforward networks. Using WTA networks in a biological context goes back to Elias and Grossberg (1975).

Shastri (2001a) suggested that the set of recruited neurons should be controlled by a soft WTA mechanism, without actually implementing it. James and Maida (2009) used a WTA with instantaneous global inhibition for learning of sequences in hierarchical recruitment. Knoblauch and Palm (2001a) and Wennekers and Palm (2007) use a terminating inhibition similar to Shastri's model for ensuring the retrieval of only a single item from a spiking associative network. Wennekers (2009) extended this idea to retrieve variable size fully-connected assemblies, or *cliques*, in bidirectional associative networks by using a similar feedback inhibition component. In a related work, Gao and Hammerstrom (2003) provide performance comparisons for the Palm network (Palm et al, 1997) which is an earlier neural associative memory model that also features a k-WTA component. However, the latter study uses digital circuits to calculate the output of the k-WTA, separating it from our approach.

There is much previous work on WTA networks, in the context of associative attractor networks with symmetrical connections studied by Hopfield (1982). Recent work includes WTA networks (e.g., Tymoshchuk and Kaszkurewicz, 2003), and k-WTA networks (e.g., Calvert and Marinov, 2000) which feature stable equilibrium points, based on other results on global stability in neural networks

(Kaszkurewicz and Bhaya, 1994; Arik, 2002). These differ from our approach because they employ full-recurrent-connection networks and use sigmoidal rate-coded neurons. They also require iteration until convergence to a stable solution is achieved and some may suffer from local minima. WTAs built with competitive learning networks are superior to Hopfield-type networks, because they do not have the local minima problem (Urahama and Nagao, 1995). There are efforts to analyze and implement various types of WTA networks (Indiveri, 2000; Calvert and Marinov, 2000; Badel et al, 2003; Ganapathy and Titus, 2003).

8.3 Components of the Model Framework

We start with a simple model of the recruitment process at a destination area B, caused by inputs from an initial area A, where A and B are disjoint sets of neuroids. The reason for choosing this two-area model is to assess the stability in the size of the recruited set when the process is repeated. The input R_A, which represents a concept of interest, is some subset of neuroids in area A that projects to area B. Inputs from the neuroids in set R_A in area $A^{(1)}$, which project to area $B^{(1)}$, cause a set $R_{B^{(1)}}$ of neuroids to be recruited (see Figure 8.3b). This process can be repeated by assuming the set $R_{B^{(1)}}$ is in an area $A^{(2)}$ and it is used to recruit a new $R_{B^{(2)}}$ set in the next area $B^{(2)}$. We wish to show that the variance of $|R_{B^{(k)}}|$ is sufficiently small after k iterations using our proposed method. We call this the *variance problem*.

Generalized Recruitment

Notice that we employ a generalized recruitment learning method to generate an output set from a single input set (see Figure 8.3b). This general solution can later be transformed into specific networks requiring multiple inputs. Recruitment learning was originally designed to require two input sets to be activated to create an output

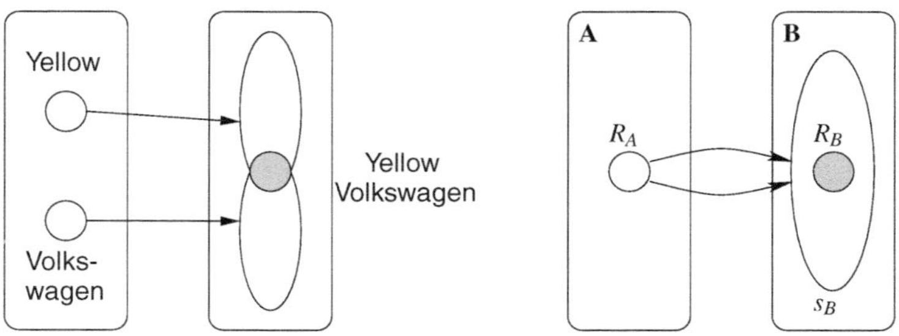

(a) Original case, as a conjunction of two inputs.

(b) General case, resulting from single input.

Fig. 8.3. Type of recruitment learning employed. Original recruitment (a) requires two inputs to create an output concept. (b) We generalize recruitment to require a single set of neuroids to create an output set of neuroids by adjusting the connection probability. The set s_B represents the set of activated synapses in area B.

set (Feldman, 1982, 1990; Valiant, 1994). Synchronous activation from two input sets indicating a temporal binding causes the neuroids receiving inputs from both sets to be recruited, as seen in Figure 8.3a. Recruitment learning requires the connections between the source and the destination to form a random graph with connection probability chosen to satisfy recruitment of a set to be equal in size to that of the input sets. Here we adjust the probability for the random connections such that a set R_A causes the recruitment of a set R_B with equal sizes, $r_A = |R_A| \simeq |R_B| = r_B$.

The approximate probability of having exactly one excitatory connection from a neuroid in A and a neuroid in B is given by

$$p^{+}_{AB} = \sqrt{\lambda / N_B(r_o - 1)}, \tag{8.1}$$

where $0 < \lambda \leq \sqrt{N_B(r_o - 1)}$ is the amplification factor, $N_B = |B|$ is the total number of neuroids in B and $r_o = r_A = r_B$ is the desired size of the neuroid set representing a concept. Equation (8.1) ensures that the expected size of a recruited set has an equilibrium point $r_B = r_o$ when $r_A = r_o$ and $\lambda = 1$. The derivation of this property, and its accuracy, is described in Appendix A.1.3. However, ensuring the expectation does not solve the variance problem in hierarchical learning.

The Boost-and-Limit Mechanism

To address the variance problem, we propose a boost-and-limit mechanism that keeps the recruited set size r_B in a *stable* equilibrium. This mechanism assumes an increased connection density (by manipulating λ) between the source A and destination B to ensure sufficient recruitment at B, and then dynamically limits the recruitment with a negative feedback component (controlled by the current value of r_B) projecting via lateral inhibition within B.

Purpose of Temporally Distributed Delays

It is reasonable to propose that the negative feedback is applied via lateral inhibition when recruited neuroids in B fire. This ensures that neuroids are already recruited before feedback is applied. The negative feedback should inhibit further recruitment candidates after the desired number of neuroids is recruited. Assuming that the initial input is a single synchronous spike volley from the set R_A and that the delays between A and B are uniform and there is no background noise, then the recruitment process in B becomes instantaneous. This does not leave time for the inhibitory feedback mechanism to sense the controlling quantity r_B due to delays (see Figure 8.4a). However, if the recruitment process is temporally dispersed, then the inhibitory feedback strengthens continuously with increasing number of recruited neuroids. This continues until a balance is reached between the input excitation and lateral inhibition to yield a desired recruitment level as in Figure 8.4b, assuming the feedback is fast enough. A realistic dispersion of recruited neuroids can be achieved if the connections between A and B have slightly varying delays. We model these

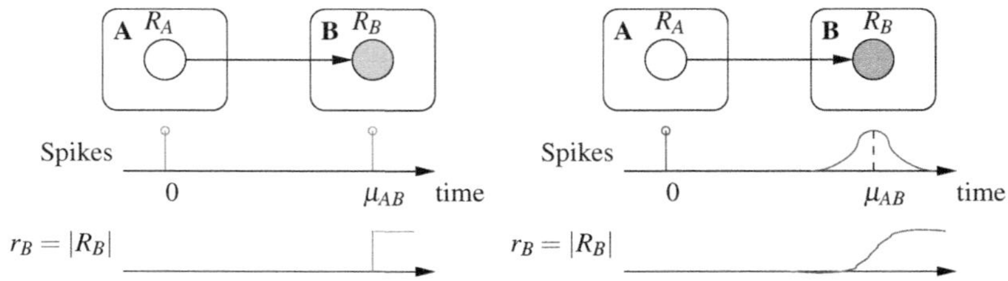

(a) Uniform delays cause an uncontrollable event of instant recruitment.

(b) Temporally distributed delays allow gradual recruitment of neuroids which can be used as a feedback signal.

Fig. 8.4. The cases with (a) and without (b) noisy delays and their effect on the controllability of the recruitment process.

delays with a normal distribution having mean μ_{AB} and standard deviation σ_{AB}.[1] The instantaneous spike rate of activity originating from A and received by excitatory synapses in B, $n_{AB}(t)$, is given by

$$n_{AB}(t) = r_A\, p_{AB}^{+} N_B\, \mathscr{G}(\mu_{AB}, \sigma_{AB}; t)\,, \tag{8.2}$$

where $r_A = |R_A|$, the normal distribution is given by the Gaussian kernel $\mathscr{G}(\mu,\sigma;t) = \frac{1}{\sigma\sqrt{2\pi}} \exp[-\frac{1}{2}(\frac{\mu-t}{\sigma})^2]$, and p_{AB}^{+} is defined in (8.1).

A Non-leaky Spike Integrator

A spiking neuron model is employed, which causes incoming action potentials at the excitatory synapses to have a prolonged excitatory postsynaptic potential (EPSP) on the somatic membrane component of the neuroids. In this model, we assume the decay constant of the membrane is larger than the transmission delays, or no decay is present at all. Therefore, all incoming spikes to a neuroid cause a constant EPSP, and the EPSPs are accumulated over the course of the recruitment process, which is roughly a few tens of milliseconds (namely, the interval $[0,40]$ in the analyses below).

The Recruitment Process

The main variable of interest $r_B(t)$ is the total number of recruited neuroids in area B at time t. According to the definition of recruitment learning (Valiant, 1994), if a neuroid receives two or more spikes at its excitatory synapses, it is recruited and therefore emits an action potential (AP). Therefore the threshold of each neuroid is adjusted to fire at the sum of two input EPSPs. Statistical methods can be used to estimate the number of recruited neuroids under a spatially uniform synapse

[1] Variable delays can also be caused by background noise and higher spike thresholds. To model this, the time to reach threshold would be determined, not only by incoming spike arrival times, but also by a random variable representing background noise.

distribution. It is desired that this number asymptotically approach a maximum level $\bar{r}_B$, and exhibit a stable equilibrium at a fixed point $r_A = \bar{r}_B = r_o$. That is, for perturbations to the input size r_A, the variation of r_B should converge to the fixed point when the process is repeated. This method can be extended to have a larger recruitment threshold to require an arbitrary number of EPSPs before firing (see Section 8.5).

It is assumed in the present model that all effects of the activation caused by excitatory synapses may be disrupted at the soma by an active inhibitory synapse, provided that the inhibitory effect is divisive rather than subtractive (Koch et al, 1983). In other words, an inhibitory synapse positioned on an axon hillock (spike initiation zone) of a neuron can act as a veto mechanism to an arbitrary excitatory input.

The following sections incrementally build models to achieve the stable equilibrium described above. In the analyses, the system parameters are chosen as $N_B = 100$ neuroids, $r_o = 10$ neuroids, $\mu_{AB} = 20\,\text{ms}$, and $\sigma_{AB} = 5\,\text{ms}$, unless otherwise indicated. The time window for the recruitment process is taken as the period $[0, 40]$ ms. The λ values used range from unity, $\lambda = 1$, where $p^+_{AB} = 0.03$, to various degrees of amplified connectivity (e.g., for $\lambda = 20$ causes $p^+_{AB} = 0.15$).

The first model, the *monopolar* synaptic population model, represents the network behavior only with the size s_B of an excitatory synapse population. The block schema of the model is depicted in Figure 8.5.

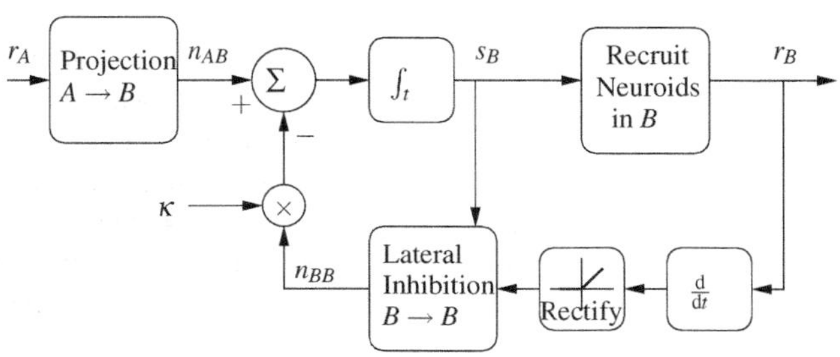

r_A = neuroid count in input R_A [neuroids].
n_{AB}, n_{BB} = instantaneous spike rates of excitatory and inhibitory projections, respectively [spikes/s].
s_B = activated excitatory synapse count in B [synapses].
r_B =recruited/fired neuroid count in B [neuroids].

Fig. 8.5. Block schema of the boost-and-limit control mechanism for the monopolar synaptic population model. The input r_A is a scalar indicating the magnitude of a one-time synchronous input to the system, whereas other quantities are functions of time t. For the instantaneous spike rates, it should be taken into account that source neuroids presumably only fire once during the recruitment process. Therefore the rate indicates the spike throughput of the population.

The total number of activated excitatory synapses $s_B(t)$ at area B increases in proportion to the net sum of spike rates described with the differential equation

$$\frac{\mathrm{d}s_B(t)}{\mathrm{d}t} = n_{AB}(t) - \kappa n_{BB}(t)\,, \tag{8.3}$$

where $n_{AB}(t)$ is defined in (8.2), $n_{BB}(t)$ represents the inhibitory feedback, and κ is the proportionality constant of the effect of an inhibitory synapse. We assume that both the excitatory and inhibitory synapses have equal weights such that they cancel each other on a destination neuroid by taking $\kappa = 1$. Note that, in this model we only observe the cancellation effect of inhibitory synapses on the excitatory synapse population size s_B and not the inhibitory synapse population itself.[2] Another point is that it is not possible to run out of synapses in (8.3) because the feedforward excitatory spike rate distribution n_{AB} is defined in (8.2) according to available synapses from the static connectivity of the network given by (8.1). Thus, each new excitatory spike activates a new excitatory synapse, but then each inhibitory spike will cancel out a fraction of an excitatory synapse. The function n_{BB} will be defined in detail later with the closed-loop system.

Given $s_B(t)$, the number of recruited neuroids in area B, $r_B(t)$, can be obtained by using a statistical expectation operator. Using the synapse-count $s = s_B$ and $N = N_B$, the probability that a neuroid will receive two or more of the excitatory synapses is defined as

$$\begin{aligned} p^* &= \frac{\sum_{k=2}^{s} \binom{s}{2} (N-1)^{s-k}}{N^s} \\ &= 1 - \frac{\sum_{k=0}^{1} \binom{s}{2} (N-1)^{s-k}}{N^s} \\ &= 1 - \left(\frac{N-1}{N}\right)^{s-1} \frac{N+s-1}{N}\,. \end{aligned} \tag{8.4}$$

Thus, $r_B(t)$ can be given as the expected number of neuroids recruited in B,

$$r_B(s) = p^* N_B = N_B - \left(\frac{N_B - 1}{N_B}\right)^{s-1} (N_B + s - 1)\,. \tag{8.5}$$

To test the quality of this population approximation, we compare it to Monte Carlo simulations in Figure 8.6. To assess the behavior of $r_B(s)$ with respect to time and changes in the input r_A, one can observe $s_B(t)$ since these two quantities are directly related by (8.5).

[2] Refer to the dipolar model in Section 8.4 for a more detailed excitatory-inhibitory synapse population model.

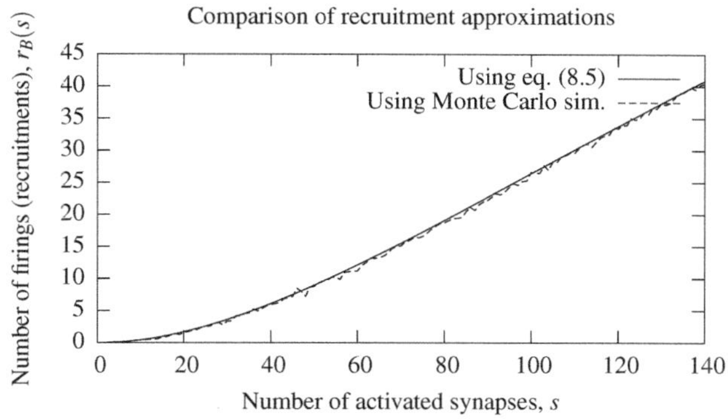

Fig. 8.6. Comparing the population approximation of $r_B(s)$ in (8.5) to a Monte Carlo simulation of a network with the same parameters ($r_o = 10$, $N = 100$), averaged over 50 runs. The number of recruitments in the open-loop system changes almost linearly with the number of activated excitatory synapses s.

8.3.1 The Open-Loop Characteristics

For comparing the performance of the proposed model, we first look at the open-loop system characteristics without the negative feedback. This scenario is similar to the method originally described by Valiant (1994), except that we use the generalized recruitment method defined above.

The open-loop system is obtained by taking $n_{BB}(t) = 0$ in (8.3). Integrating this equation, we get

$$s_B(t) = \int_0^t n_{AB}(\tau)\mathrm{d}\tau \,. \tag{8.6}$$

We can estimate the upper limit asymptote, or steady-state value, of s_B by

$$\bar{s}_B = \lim_{t \to \infty} s_B(t) \simeq r_A\, p_{AB}^{+} N_B \,, \tag{8.7}$$

since $\lim_{t\to\infty} \int_0^t \mathscr{G}(\mu_{AB}, \sigma_{AB}; \tau)\mathrm{d}\tau \simeq 1$ when $\mu_{AB} > 2\sigma_{AB}$, and the term $r_A\, p_{AB}^{+} N_B$ is constant. Using this relation, the final expected number of recruited neuroids $\bar{r}_B$, as a function of the number of activated input neuroids r_A, becomes

$$\bar{r}_B(r_A) = N - \left(\frac{N_B - 1}{N_B}\right)^{r_A\, p_{AB} N - 1} (N_B + r_A\, p_{AB} N_B - 1) \,. \tag{8.8}$$

$\bar{r}_B$ is the maximum of $r_B(t)$ for this recruitment process, since all the spikes initiated at area A have reached area B.

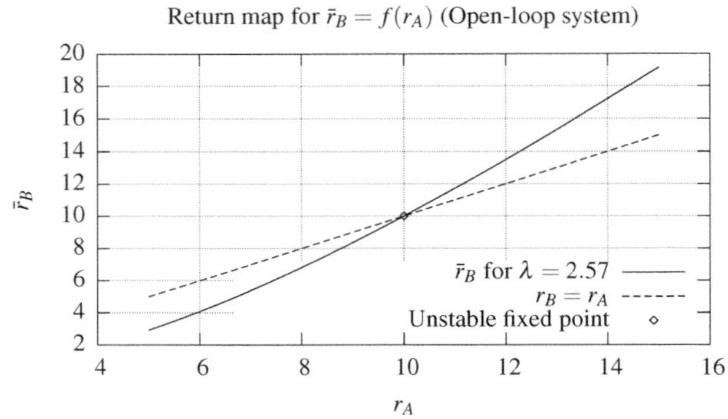

Fig. 8.7. Return map of the change in the r value from r_A to r_B of the open-loop system according to (8.5). Note that the chosen fixed point with the parameters indicated in the plot is unstable.

If we want to find the value of the amplification factor λ, for a desired fixed point $\bar{r}_B = r_o$, there is no simple analytic solution of (8.8). Therefore, we define the function

$$g(\lambda) = N - \left(\frac{N-1}{N}\right)^{r_A p_{AB} N - 1} (N + r_A p_{AB} N - 1) - \bar{r}_B$$

and numerically find λ values that makes $g(\lambda) = 0$. For $r_o = 10$, we get $\lambda = 2.5681$.[3]

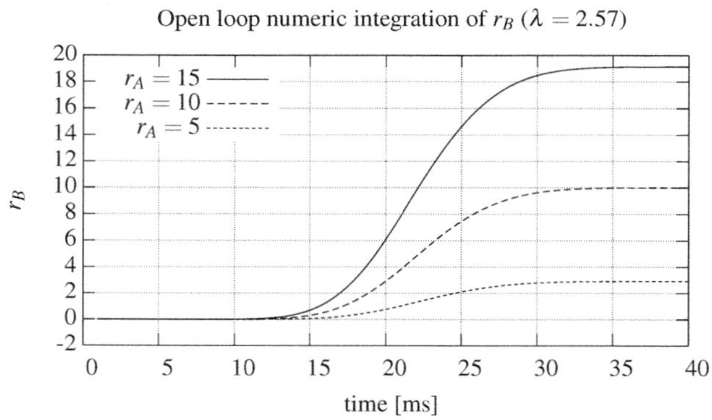

Fig. 8.8. Open-loop simulation for the change in the size of the recruited set r_B with different selections for the size of the input set r_A.

[3] Earlier, it was claimed for the open-loop case that $\lambda = 1$ should give the desired fixed-point. However, here λ compensates for the $O(p^3)$ term ignored in (8.1), discussed in Appendix A.1.3.

Once λ is known, the return map $\bar{r}_B(r_A)$ in (8.8) can be plotted as seen in Figure 8.7. Numerical integration of $r_B(t)$ obtained from (8.5) and (8.6) are plotted in Figure 8.8 for three choices of r_A, which verifies Figure 8.7.[4] As expected, perturbations to the value of r_A cause r_B to diverge from the desired value $r_o = 10$, which is the reason for recruitment instability.

8.3.2 The Closed-Loop System with Negative Feedback

We now proceed to define the negative feedback effect, the instantaneous inhibitory spike rate $n_{BB}(t)$ within B, used earlier in (8.3). This quantity depends on the number of recruited neuroids emitting APs, thus (8.5). However, since $r_B(t)$ is the total number of neuroids recruited, $n_{BB}(t)$ only depends on its derivative representing the instantaneous rate of recruitment at a time t. A recruited neuroid emits a single AP. Therefore, the recurrent inhibitory projection[5] can be represented by

$$\tilde{n}_{BB}(t) = p_{\overline{BB}} N_B \frac{\mathrm{d}r_B(t - \mu_{BB})}{\mathrm{d}t}, \tag{8.9}$$

where μ_{BB} is the magnitude of the uniform recurrent delays,[6] and $p_{\overline{BB}}$ is the density of inhibitory recurrent projections.

$\tilde{n}_{BB}(t)$ gives the expected number of newly activated inhibitory synapses in all of area B. To find the number of activated excitatory synapses that can be vetoed by the activated inhibitory synapses, we use the number of activated excitatory synapses per neuroid in B, $\frac{s_B(t)}{N_B}$. Since, for each activated inhibitory synapse on a distinct neuroid causes all its activated excitatory synapses to be subject to cancellation, we get the total number of activated excitatory synapses in B affected by the new inhibition as

$$n_{BB}(t) = \tilde{n}_{BB}(t) \frac{s_B(t)}{N_B} = s_B(t)\, p_{\overline{BB}} \frac{\mathrm{d}r_B(t - \mu_{BB})}{\mathrm{d}t}. \tag{8.10}$$

The exact number of affected excitatory synapses is obtained according to the proportionality constant κ in (8.3).

The derivative at the right hand side of (8.10) can be obtained by differentiating (8.5) to yield

$$\frac{\mathrm{d}r_B(t)}{\mathrm{d}t} = \frac{1}{N_B} s_B(t) \frac{\mathrm{d}s_B(t)}{\mathrm{d}t},$$

which can be substituted into (8.10); and rewriting (8.3) gives the closed-loop form of the system with fixed-delay feedback

[4] A simple forward-Euler method is used for integration.

[5] Instead of featuring inhibitory interneurons, we assume that projections can be either excitatory or inhibitory. Using actual inhibitory interneurons is a matter of adapting the calculations here.

[6] For simplicity, we assume recurrent delays are of fixed duration.

$$\frac{\mathrm{d}s_B(t)}{\mathrm{d}t} = n_{AB}(t) - \frac{p_{BB}^-}{N_B} s_B(t)\, s_B(t-\mu_{BB}) \frac{\mathrm{d}s_B(t-\mu_{BB})}{\mathrm{d}t}. \tag{8.11}$$

Note that zero initial conditions are assumed for $s_B(t) = 0$ for all $t \leq 0$.

8.3.3 Instantaneous Feedback Condition

From (8.11), taking the instantaneous recurrent feedback condition $\mu_{BB} = 0$ as a simplifying assumption, we get

$$\frac{\mathrm{d}s_B(t)}{\mathrm{d}t} = \frac{n_{AB}(t)}{1 + \frac{p_{BB}^-}{N_B} s_B^2(t)}, \tag{8.12}$$

where $n_{AB}(t)$ is the delay distribution given in (8.2). Note that, s_B increases asymptotically according to (8.12) since $\lim_{t\to\infty} \frac{\mathrm{d}s_B(t)}{\mathrm{d}t} = 0$ and $\frac{\mathrm{d}s_B(t)}{\mathrm{d}t} > 0$, always.

The solution to (8.12) yields two roots. We are only interested in the positive root,

$$s_B(t) = \frac{\left(C(t) + \sqrt{C^2(t)+4}\right)^{\frac{2}{3}} - 2^{\frac{2}{3}}}{\sqrt{\frac{p_{BB}^-}{N_B}} \left(2C(t) + 2\sqrt{C^2(t)+4}\right)^{\frac{1}{3}}} \tag{8.13}$$

where

$$C(t) = 3\, r_A\, p_{AB}^+ \sqrt{p_{BB}^- N_B} \int_0^t \mathscr{G}(\mu_{AB}, \sigma_{AB}; \tau)\mathrm{d}\tau. \tag{8.14}$$

The derivation of this solution can be found in Günay (2003).

The steady-state value of $C(t)$ can be calculated similarly to (8.7),

$$\bar{C} = \lim_{t\to\infty} C(t) = 3\, r_A\, p_{AB}^+ \sqrt{p_{BB}^- N_B}.$$

From $\bar{C}$, steady-state solutions $\bar{s}_B$ and $\bar{r}_B$ can be calculated. The resulting return map of $\bar{r}_B$, given r_A, is plotted in Figure 8.9. Different from the open-loop return map, this time the fixed point at $r_A = \bar{r}_B = 10$ is stable. That is, for variations in the r_A value, the value of $\bar{r}_B$ always approaches the fixed point. Note that the lateral inhibitory connectivity is chosen as $\kappa p_{BB}^- = 0.5$. As seen from the figure, as the $A \to B$ projection density factor λ value is increased, the convergence rate to the fixed point marginally increases. However, this requires increasing the inhibitory feedback density to keep the fixed point at the desired location.

Verifying the closed-loop system with numerical methods is more difficult. This is because (8.12) fails the *Lipschitz condition* (Iserles, 1996) that is required for the convergence proofs for numerical methods based on the integration method (e.g.,

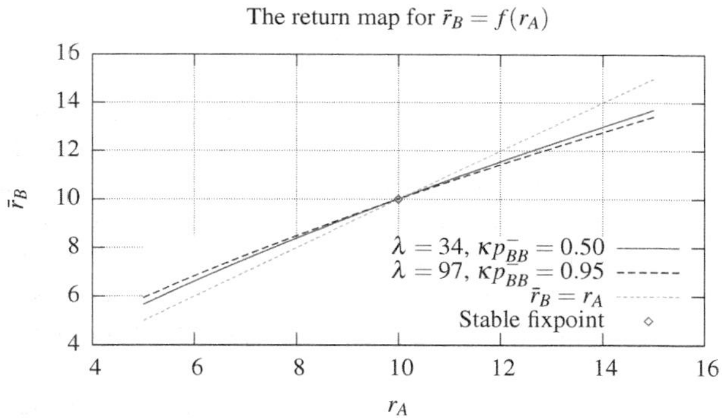

Fig. 8.9. Return map of the change in the r value from r_A to $\bar{r}_B$ of the closed-loop system with the instantaneous feedback condition showing that the chosen fixed point with the parameters indicated in the plot is stable. Convergence speed increases from moderate connectivity case ($\lambda = 34$ and $\kappa p_{\bar{B}B} = 0.5$) to high connectivity case ($\lambda = 97$ and $\kappa p_{\bar{B}B} = 0.95$), but requires higher feedback connectivity.

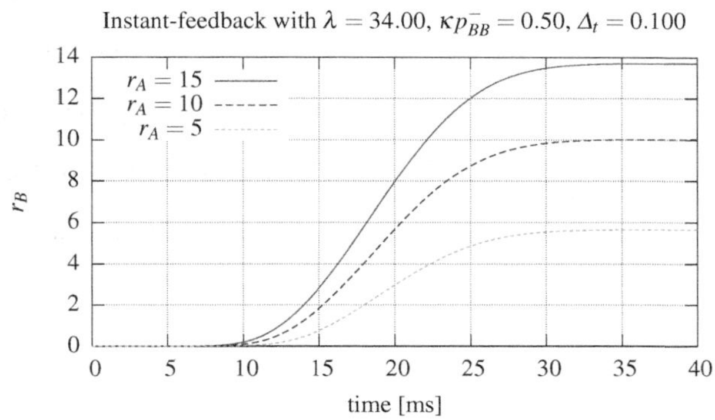

Fig. 8.10. Closed-loop simulation in the instantaneous feedback case for the change in the size of the recruited set r_B with different selections for the size of the input set r_A.

Runge-Kutta). Simulation results obtained with another numerical algorithm[7] to solve the ordinary differential equation (8.12) are given in Figure 8.10.

Note that the instantaneous feedback condition is an ideal case. With delays, more realistic effects of feedback can be observed for stabilizing the system.

[7] Hindmarsh's ODE solver `lsode` (Hindmarsh, 1983) in the GNU Octave package (Eaton, 2002) was used in a step-by-step fashion to solve this system by saving the history in arrays to implement delays. Later equations were integrated with adaptive Simpson quadrature (`quad`) at each time step.

8.3.4 Fixed-Delay Feedback Condition

For a more realistic feedback condition, uniform delays are used by choosing non-zero values for μ_{BB} in (8.11). However, there is no simple way to find a solution of this delayed differential equation. Numerical integration of the equation shows that the model introduces another type of instability to the system which causes r_B to oscillate uncontrollably as seen in Figure 8.11. The oscillations are caused by the feedback delay. Increasing the time resolution or decreasing the feedback delay does not prevent the oscillations.

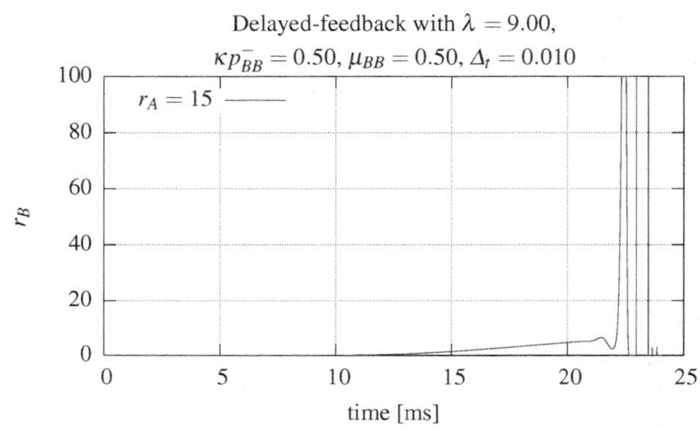

Fig. 8.11. Closed-loop simulation in the uniform-delay feedback case for the change in the size of the recruited set r_B with different selections for the size of the input set r_A. The model is plagued with oscillations.

Since r_B is the total number of recruited neuroids, we assume that it should be a monotonically increasing quantity. Therefore we augment the model by rectifying the feedback as seen in the block schema of Figure 8.5. Even though the oscillations no longer occur in the numerical simulations of Figure 8.12, r_B increases without being affected by the negative feedback as intended. Increasing the time resolution or decreasing the feedback delay does not prevent the oscillations, either. The stabilizing effect from the inhibitory feedback is thus lost.

To analyze this failure, assuming a steady-state is reached eventually, we test if the value of r_B, and therefore s_B, stays at a stable level, i.e. $\lim_{t\to\infty} \frac{\mathrm{d}s_B(t)}{\mathrm{d}t} = 0$. However, applying this to (8.11) yields an inconsistent result of $\lim_{t\to\infty} n_{AB}(t) = 0$. Thus, we proceed to a more detailed model to control the undesired oscillations.

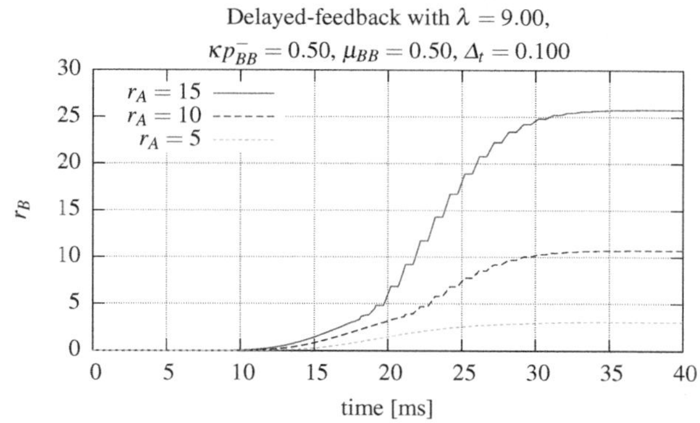

Fig. 8.12. Profiles of r_B with uniform-delay feedback case and rectified $\frac{\mathrm{d}r_B}{\mathrm{d}t}$ with different selections for the size of the input set r_A. r_B no longer decreases but the effects of the oscillatory feedback can still be seen by sudden increases in r_B.

8.4 The Closed-Loop Inhibitory Feedback Population Model

To describe a more realistic model of inhibitory feedback, we change to a dipolar synaptic population model. This model has two populations of synapses, s_P and s_N, for excitatory and inhibitory synapses, respectively. The block schema of this closed-loop model is given in Figure 8.13. Note that the feedback loop via the s_N variable is separated from the excitatory input unlike the monopolar synaptic population model. This simplifies the feedback equations.

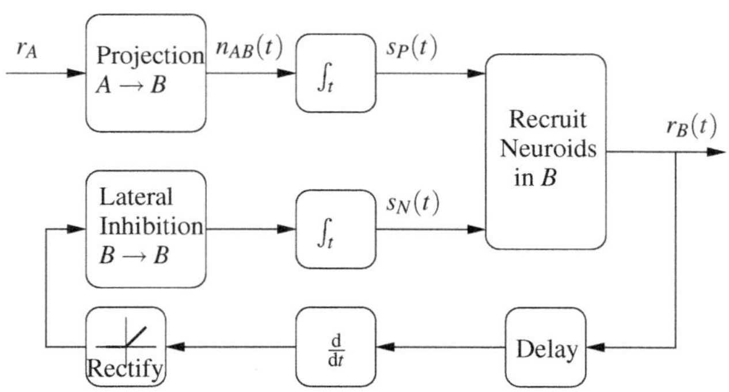

Fig. 8.13. Block schema of the closed-loop synaptic population model. In the diagram, r_A is the neuroid count in the input R_A in [neuroids]; s_P, s_N are the activated excitatory and inhibitory synapse counts in B, respectively, in [synapses]; and r_B is the recruited/fired neuroid count in B in [neuroids]. Note that, the input r_A is a scalar indicating the magnitude of a one-time synchronous input to the system, whereas other quantities are functions of time t.

The model is given by the following equations. The number of activated excitatory synapses is given by

$$s_P(t) = \int_0^t n_{AB}(\tau)\mathrm{d}\tau\,, \tag{8.15}$$

similar to s_B in (8.6). The number of activated inhibitory synapses caused by the feedback is given by

$$s_N(t) = p_{BB}N\int_0^t \Theta\left(\frac{\mathrm{d}r_B(\tau-\mu_{BB})}{\mathrm{d}\tau}\right)\mathrm{d}\tau\,,$$

where

$$\Theta(x) = \begin{cases} x, \ x \geq 0 \\ 0, \ x < 0 \end{cases}$$

is the rectification function, and μ_{BB} is the feedback delay, p_{BB} is the lateral inhibitory connectivity parameter, and $N \equiv N_B$ for simplicity. It is assumed that the inhibitory synapses have a multiplicative effect and thus can veto all excitatory inputs to the neuroid.

The expected number of recruited neuroids r_B can be calculated probabilistically, similar to the method followed with the monopolar model description in Section 8.3. The probability of having one or more inhibitory synapses on a neuron is

$$p_N^* = \frac{N^{s_N} - (N-1)^{s_N}}{N^{s_N}} = 1 - (\frac{N-1}{N})^{s_N}\,.$$

Then, the expected number of recruited neuroids r_B can be calculated as having two or more activated excitatory synapses and no activated inhibitory synapses

$$r_B = p^*(1-p_N^*)N\,, \tag{8.16}$$

where p^* is defined in (8.4). Expanding, it becomes

$$r_B(t) = \left[N - \left(\frac{N-1}{N}\right)^{s_P(t)-1}(N+s_P(t)-1)\right]\left(\frac{N-1}{N}\right)^{s_N(t)}\,. \tag{8.17}$$

As before, the initial conditions are taken as $r_B(t) = 0$ for $t \leq 0$.

Yet a more realistic method of computing the expected number of neuroids is given in Appendix A.1.4, but it is not implemented here. This method counts neuroids with two or more activated excitatory synapses and no inhibitory synapses.

The model already contains a rectifier for managing oscillations, since the purpose of the model was to better control the feedback quantity before it was summed with the excitatory input. However, simulations indicate that the instability imposed by the oscillations cause problems in this model as well. Even with very low delays ($\mu_{BB} = 0.2\,\mathrm{ms}$) and low feedback ($p_{BB}^- = 0.07$) the system does not perform well (see Figure 8.14). If the amplification factor is reduced, the oscillations can be controlled, however the boost-and-limit mechanism loses its advantage in providing stability in r_B with respect to variations in r_A (see Figure 8.15).

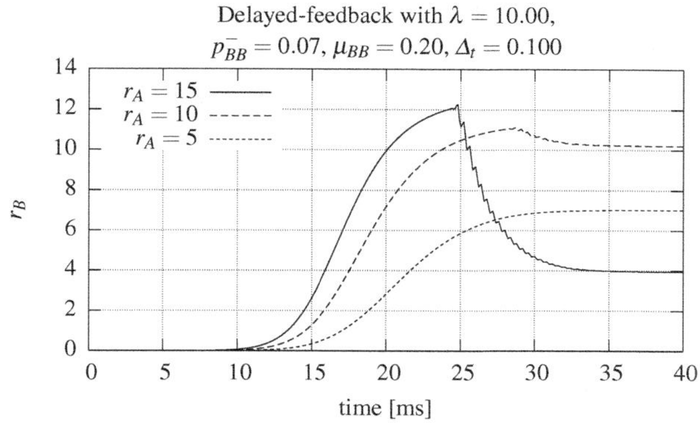

(a) Profiles of r_B with different selections of r_A.

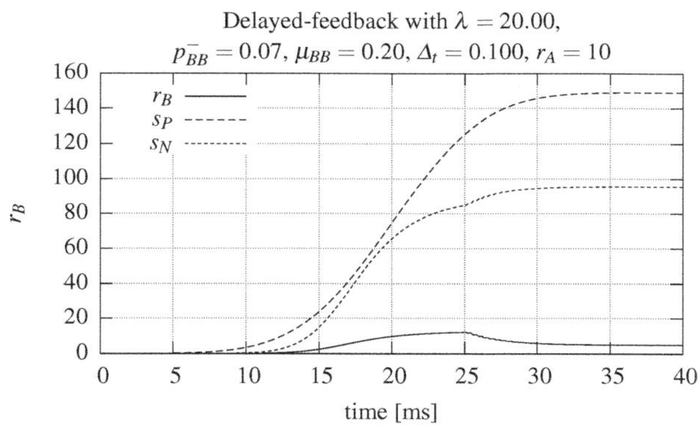

(b) The s_P and s_N components of r_B when $r_A = 10$. The amplification factor is increased for illustration.

Fig. 8.14. The dipolar population model behavior, with uniform-delay feedback and rectified $\frac{\mathrm{d}r_B}{\mathrm{d}t}$. (a) It can be seen that r_B approaches the desired level, but fails to maintain its regime due to the effects of rectified oscillations. (b) The component view show the ladder-stepping effect on s_N and r_B. The effect is not due to the numeric method employed, since lowering the simulation step-size to $\Delta_t = 1\,\mu$s did not prevent the problem.

8.4.1 *Non-rectified Model*

Since simply rectifying is not useful, we fall back to the original model with no rectification, where the inhibitory population equation can be simplified to

$$s_N(t) = p_{BB} N \int_0^t \frac{\mathrm{d}r_B(\tau - \mu_{BB})}{\mathrm{d}\tau} \mathrm{d}\tau = p_{BB} N\, r_B(t - \mu_{BB}) \,. \tag{8.18}$$

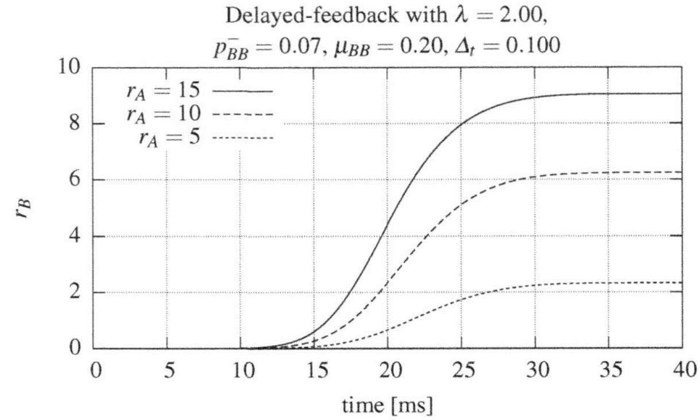

Fig. 8.15. The dipolar population model behavior, with uniform-delay feedback case and rectified $\frac{dr_B}{dt}$. If the feedback delay and amplification is low enough the oscillations disappear, but no convergence can be observed in the profile of r_B with respect to varying r_A.

Thus, (8.17) can be written in the recursive form

$$r_B(t) = \left[N - \left(\frac{N-1}{N}\right)^{s_P(t)-1} (N + s_P(t) - 1)\right] \left(\frac{N-1}{N}\right)^{p_{BB} N r_B(t-\mu_{BB})} . \quad (8.19)$$

If the value of r_B is projected to its steady-state, its behavior can be observed by changing r_A to find the return-map of the iterative hierarchical recruitment process as in Section 8.3.1. Assuming T is sufficiently large, we define the steady-state value of $r_B(t)$ as

$$\bar{r}_B = r_B(t)|_{t>T} ,$$

and the steady-state value of $s_P(t)$ as $\bar{s}_P = \bar{s}_B \simeq r_A\, p_{AB}^{+} N_B$ from (8.7). First, we define

$$\chi(\lambda) = N - \left(\frac{N-1}{N}\right)^{r_A p_{AB} N - 1} (N + r_A p_{AB} N - 1) .$$

Then, we can express the steady-state of (8.19) as

$$\bar{r}_B = \chi \left(\frac{N-1}{N}\right)^{p_{BB} N \bar{r}_B} \quad (8.20)$$

for $t > T + \mu_{BB}$.

A corollary of (8.20) is that the lateral inhibitory connectivity parameter p_{BB} can be calculated in terms of other network parameters for a given fixed point r_o. To calculate p_{BB} and $\bar{r}_B$ numerically, we define the function

$$h(p_{BB}, \bar{r}_B) = \chi \left(\frac{N-1}{N}\right)^{p_{BB} N \bar{r}_B} - \bar{r}_B , \quad (8.21)$$

and look for its zeros. p_{BB} can be found for $\bar{r}_B = r_o$ and given λ. For $\lambda = 20$ and $r_o = 10$, we get $p_{BB} = 0.14739$.

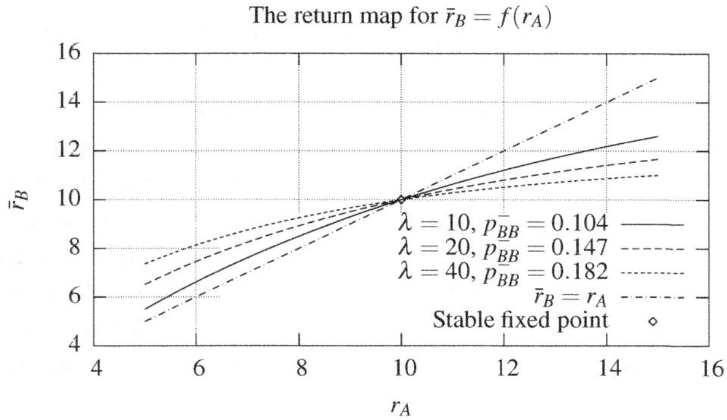

Fig. 8.16. Plots of the return map of $\bar{r}_B(r_A)$ for iterative application of the recruitment described here. Note that the indicated fixed point at $\bar{r}_B(10) = 10$ is stable. Increasing the amplification factor λ that affects the value of p_{AB}^+ from (8.1) results in a more flattened curve, thus a faster convergence to a stable fixed point.

Plots of (8.20) in Figure 8.16 confirm the behavior of $\bar{r}_B$ and show effects of several parameters on the convergence speed. These results show that if oscillations can be avoided, the system behaves as desired. Note that the stability of the fixed point is improved in comparison to the return maps of the monopolar synaptic population model with the instantaneous feedback condition, plotted in Figure 8.9.

8.4.2 *Oscillations in Activity Levels*

The oscillatory behavior in the system is expected from control system analysis as delays in feedback often result in instabilities. However, conventional linear control system tools for analysis of oscillations do not apply because this model is non-linear (Phillips and Harbor, 1991).

For excessive inhibitory feedback, the recruitment level r_B can decrease, causing this instability. It may seem counter-intuitive for r_B to decrease, since it represents a monotonically increasing quantity by definition. However, notice that r_B is simply the expected value of a stochastic variable representing the recruitment level, and no special effort has been made to make this value monotonic in this model. It is important that, during oscillations, the mean value of r_B stays constant, which indicates that it may potentially settle to a stable level as predicted from its steady-state analysis if the oscillations can be prevented.

The oscillations appear because r_B can change infinitely fast; i.e., the change of recruitment rate, dr_B/dt, has no upper bound. To stop the oscillations, it is reasonable to suggest that the model of a physical system must have an upper bound for a parameter like this. A natural candidate for imposing an upper bound is using a low-pass filter which slows down the rate of change.

8.4.3 A Decaying Inhibitory Synapse Population

A filter mechanism can be used to attenuate the unwanted oscillations in the model. Since the oscillatory activity occurs in the negative feedback loop, we can filter either one of the r_B or s_N variables. A low-pass filter, such as the exponential decay, which attenuates higher frequency components in a signal, is appropriate for this task. A biologically justifiable low-pass filter can be modeled as a decaying effect in the size of the inhibitory synapse population, by assuming that the inhibitory effects are lost after a short time span. We choose to apply the decay to s_N rather than r_B, since the latter needs to be a monotonically increasing variable. This decaying effect on s_N is represented by a capacitor and resistor in parallel, shown in the the schema of Figure 8.17.

The filter can be represented by the differential equation

$$\frac{\mathrm{d}s_N(t)}{\mathrm{d}t} = -\frac{s_N(t)}{RC}, \tag{8.22}$$

which models the proportional decay in the population size $s_N(t)$. Under this model, inhibitory synapses in the population lose their potential according to an approximately uniform probability, although the exact duration of activation is not considered. Taking the derivative of the original definition of increase in $s_N(t)$ from (8.18) as

$$\frac{\mathrm{d}s_N(t)}{\mathrm{d}t} = p_{\bar{B}B} N \frac{\mathrm{d}r_B(t-\mu_{BB})}{\mathrm{d}t},$$

and adding it to (8.22) as external input, we get

$$\frac{\mathrm{d}s_N(t)}{\mathrm{d}t} = \frac{p_{\bar{B}B} N}{C} \frac{\mathrm{d}r_B(t-\mu_{BB})}{\mathrm{d}t} - \frac{s_N(t)}{\tau},$$

where $\tau = RC$, the decay time constant. Simulation results in Figure 8.18 indicate that the decay indeed reduces the gain of the oscillation. Nevertheless, it is not sufficient to prevent oscillations altogether, since the best case shown in Figure 8.19 is free of oscillations, but r_B never settles to a stable level. More importantly, the steady-state value of r_B is governed by an unstable balance of the decay parameters with the input.

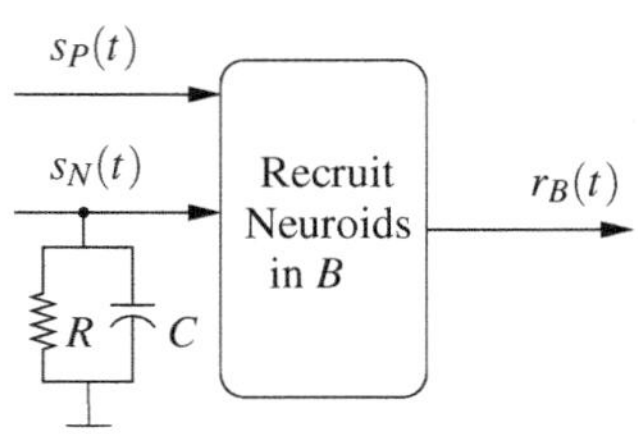

Fig. 8.17. Block schema of the decay effect on s_N for preventing oscillations in the dipolar synaptic population model.

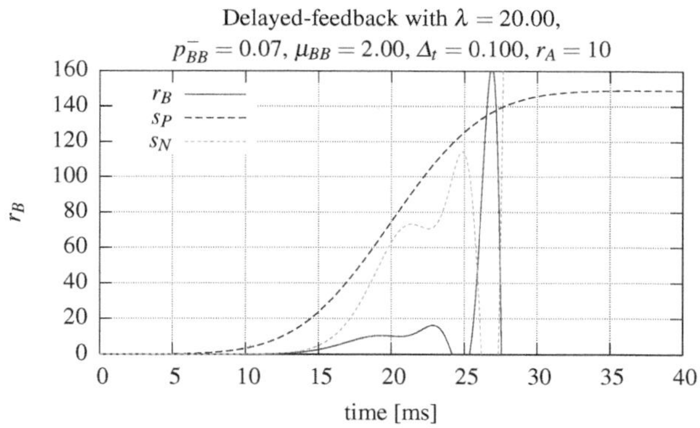

(a) Non-rectified dipolar model, without the decaying s_N.

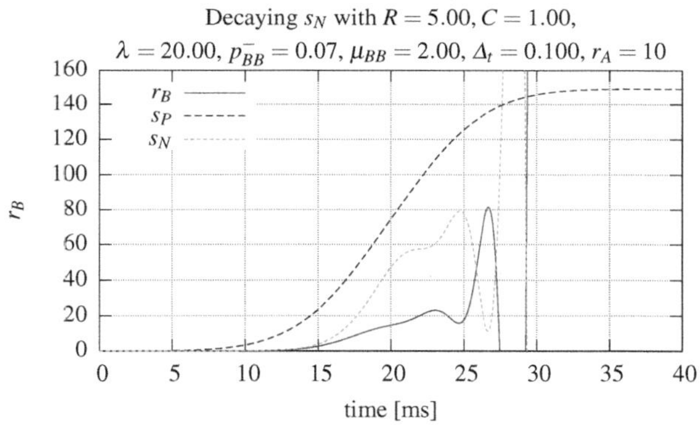

(b) Same conditions with decaying s_N.

Fig. 8.18. Effect of a decaying s_N population on the oscillations. The plot features a dipolar population model, with uniform-delay feedback case and a decaying s_N. It can be seen that the gain of the oscillation in the original model (a) is reduced in the case with the decaying s_N (b).

8.4.4 Lateral Excitatory Feedback

Assuming the only cause of the oscillations is the fragile balance between the s_N and r_B, we test the effects of adding a complementary excitatory feedback. The necessary amount of excitatory feedback p_{BB}^{+} for keeping $\bar{r}_B$ at the desired level is calculated for two different cases. First, we look at the case where there is no external input from A. In this case, the positive feedback input to the system is given as

$$s_P(t) = r_B(t - \mu_{BB}) p_{BB}^{+} N \,.$$

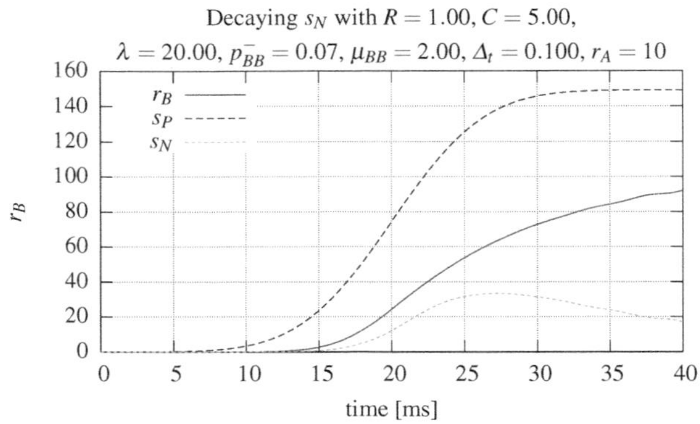

Fig. 8.19. The dipolar population model behavior, with uniform-delay feedback case and a decaying s_N population. This best case to stop oscillations is obtained with $R = 1$ and $C = 5$, that is, a time constant of $\tau = 5\,\text{ms}$.

By looking at the steady-state values, we can calculate the positive feedback amount as

$$p^+_{BB} = 2/r_o N(1 - r_o p^-_{BB}) .$$

Notice that p^+_{BB} is only defined when $r_o p^-_{BB} < 1$, and otherwise becomes negative.

The second case we used to calculate p^+_{BB} includes the external input from r_A, and defines

$$s_P(t) = \int_0^t n_{AB}(\tau)\mathrm{d}\tau + r_B(t - \mu_{BB}) p^+_{BB} N .$$

This time, dependent on r_A, we find

$$p^+_{BB} = \left(-r_A\, p^+_{AB} + \sqrt{\frac{2r_o}{N(1 - r_o p^-_{BB})}} \right) \Big/ \; r_o ,$$

which is also undefined (takes complex values) for $r_o\, p^-_{BB} \geq 1$. For both cases, simulations indicate failure to control the oscillations and no further insight was gained.

8.4.5 Low-Pass Filter

Since none of these more biologically-motivated methods helped prevent the oscillations, we now take the path of applying an effective engineering approach and then discussing its biological implications. Applying a non-decaying low-pass filter on r_B, as seen in the block diagram of Figure 8.20, stops the oscillations. The low-pass filter attenuates high-frequency components of the original r_B signal. The cut-off frequency is inversely proportional to the *RC* constant of the low-pass circuit shown in the figure.

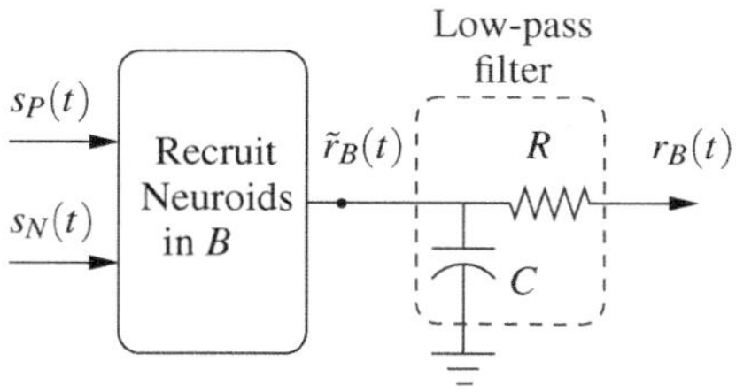

Fig. 8.20. Block schema of the low-pass filter on r_B for preventing oscillations in the dipolar synaptic population model.

The dynamics of the circuit is represented by the differential equation

$$\frac{\mathrm{d}r_B(t)}{\mathrm{d}t} = \frac{\tilde{r}_B(t) - r_B(t)}{\tau},$$

where $\tau = RC$ is the time constant of the circuit, and $\tilde{r}_B$ is the recruitment level as previously defined by (8.17). Simulations plotted in Figure 8.21 indicate that the circuit is attenuating unwanted oscillations.

8.4.6 Monte Carlo Simulations

Finally, we ran Monte Carlo simulations for both the open and closed-loop cases. The results are given in Figure 8.22. The simulations verify both that the size of recruitment is unstable in the open-loop case, and that it is stable in the closed-loop case. However, the closed-loop simulations indicate that the calculated value of p_{BB} from (8.21) for the fixed point $r_o = 10$ gives a higher than expected steady-state value. This is due to the simplification of the population model in modeling the time course of the recruitment process. The population model tends to oscillate and find its optimal steady-state. However, in real neural simulations, such as the Monte Carlo simulations, the number of recruited neuroids can only increase and thus will always provide a slightly higher number than expected. Even though we acknowledge that this indicates a possible point for future improvement in the model, it does not affect the results on stability. In order to achieve the desired fixed point, p_{BB} needs to be chosen higher than originally calculated with (8.21).

8.4.7 The Effect of Feedback Inhibition Speed on Stability

The above stability analysis, based on the return-map, does not address the effect of feedback delays, since it only uses the steady-state values. To test the model's

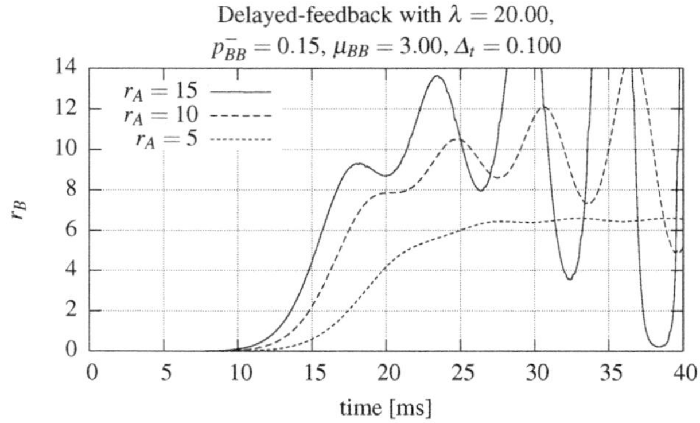

(a) Non-rectified model, without the low-pass filter.

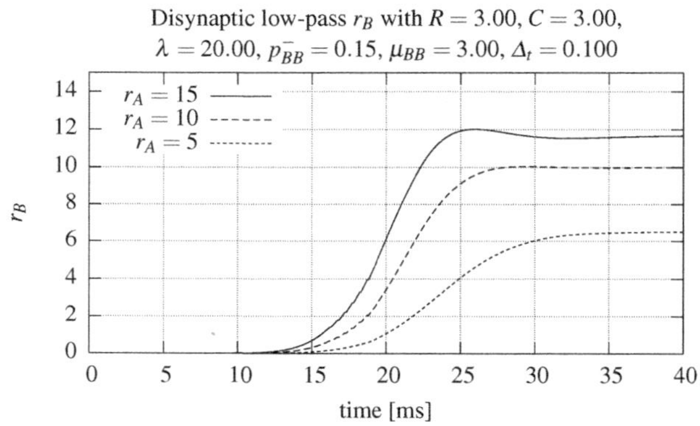

(b) Same conditions with the low-pass filter ($\tau = 9\,\text{ms}$).

Fig. 8.21. Effect of a low-pass filter on the oscillations. The plot features the population model, with uniform-delay feedback case and a low-pass filter on r_B. It can be seen that the gain of the oscillation in the original model (a) is eliminated in the case when the filter is applied (b).

response with varying feedback delays, we ran Monte Carlo simulations shown in Figure 8.23. The figure shows networks with 1, 3 and 5 ms delays for the feedback delay, μ_{BB}. As the feedback delay increases, both the fixed-point value of r_B, when $r_A = r_o = 10$, and the speed of convergence to a stable point changes. For $\mu_{BB} = 5$ ms, the network is only asymptotically stable. For larger delays, the controlling feedback becomes too slow to drive the size of the recruited set towards the fixed-point.

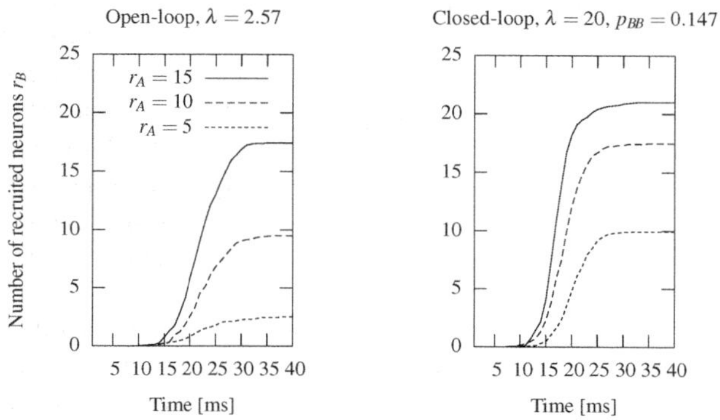

Fig. 8.22. Monte Carlo simulations of networks with parameters emulating the open-loop (left) and closed-loop (right) models, averaged over 50 runs.

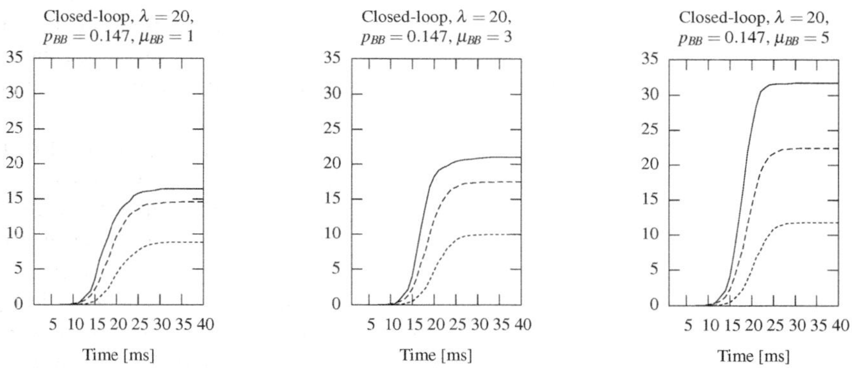

Fig. 8.23. Monte Carlo simulations of networks with different feedback delays, averaged over 50 runs. The legend for r_A values are the same as in Figure 8.22.

8.5 Recruitment Caused by at Least $l > 2$ Synapses

So far we looked at only two or more synapses to be sufficient for recruitment. Considering that real neurons receive on the order of 10,000 synapses and a constant input of background noise and that the effects of individual synaptic inputs are small, two inputs may not be realistic to fire a neuron. To make a more realistic model, we extend the population model definition to require an arbitrary number of synaptic inputs (here denoted as l) to cause a neuroid to fire and be recruited.

8.5.1 *Feedforward Excitation*

In a population of N neurons, the probability of having l or more active synapses on a neuron, out of s active excitatory synapses, is

$$p^* = \frac{\sum_{k=l}^{s} \binom{s}{k} (N-1)^{s-k}}{N^s} = 1 - \frac{\sum_{k=0}^{l-1} \binom{s}{k} (N-1)^{s-k}}{N^s} . \tag{8.23}$$

Then, the expected number of recruited neurons become

$$r_B(s,l) = p^* N = N - \frac{\sum_{k=0}^{l-1} \binom{s}{k} (N-1)^{s-k}}{N^{s-1}} .$$

We rearrange the expression to reduce exponents for simulations

$$r_B(s,l) = N - \sum_{k=0}^{l-1} \binom{s}{k} \left(\frac{N-1}{N}\right)^{s-k} \frac{1}{N^{k-1}} . \tag{8.24}$$

Here, we use a different formula for s,

$$s(t) = r_A f_{AB} N_B \int_0^t \mathcal{G}(\mu_{AB}, \sigma_{AB}; \tau) \mathrm{d}\tau . \tag{8.25}$$

where the normal distribution is given by the Gaussian kernel $\mathcal{G}(\mu,\sigma;t) = \frac{1}{\sigma\sqrt{2\pi}} \exp[-\frac{1}{2}(\frac{\mu-t}{\sigma})^2]$, and f_{AB} is the projection factor from area A to B. The latter is defined as

$$f_{AB} = \sqrt{\varphi}\, p_{AB} , \tag{8.26}$$

where $\varphi = (\rho\nu)^2$ with ρ is the average number of synapses between two neurons and ν is the average number of spikes in a synchronized spike train. In the following figures we use the combined multiplicity factor $\zeta = \varphi\lambda$ to indicate variation from the original projection factor.

Thus, the steady-state value of s becomes

$$\bar{s} = r_A f_{AB} N . \tag{8.27}$$

This can be used to get the steady-state value of r_B as $\bar{r}_B = r_B(\bar{s}, l)$. To find appropriate values of ζ for a fixed-point $\bar{r}_B = r_o$, we define the function

$$g_{ff}(\zeta, \bar{r}_B) = r_B(\bar{s}, l) - \bar{r}_B \tag{8.28}$$

and find its zeros numerically. Given $r_o = 10$, we find the following ζ values for each l:

$$l = 2 \Rightarrow \zeta = 2.56$$
$$l = 3 \Rightarrow \zeta = 11.021$$
$$l = 10 \Rightarrow \zeta = 349.91$$

Time response of r_B in (8.24) is given in Figure 8.24 and return map is given in Figure 8.25.

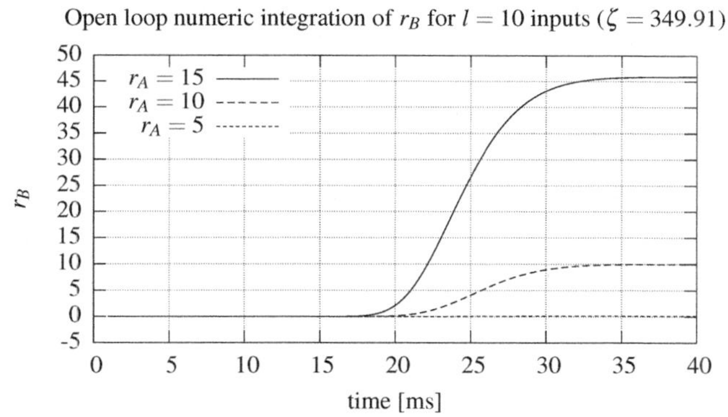

Fig. 8.24. Open-loop simulation for the change in the size of the recruited set r_B with different selections for the size of the input set r_A for $l = 10$ inputs.

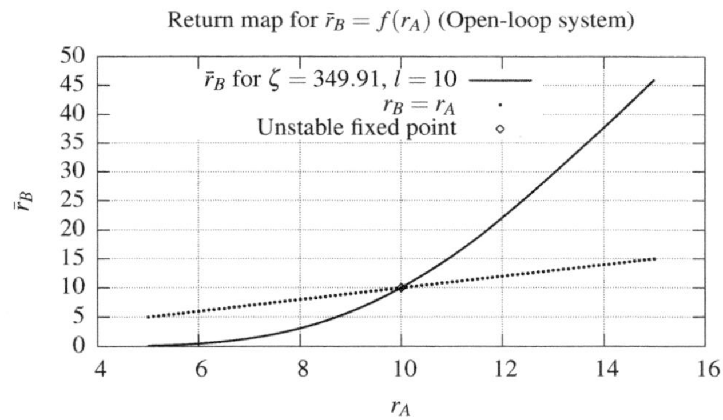

Fig. 8.25. Return map of $\bar{r}_B(r_A)$ in the open-loop model for $l = 10$ inputs.

8.5.2 *Adding Inhibitory Feedback*

The r_B with inhibitory feedback is defined by substituting (8.23) into (8.16),

$$\begin{aligned} r_B(s_P, s_N, l) &= p^*(1 - p_N^*)N \\ &= \left(N - \sum_{k=0}^{l-1} \binom{s_P}{k} \left(\frac{N-1}{N}\right)^{s_P - k} \frac{1}{N^{k-1}}\right) \left(\frac{N-1}{N}\right)^{s_N}. \end{aligned}$$

The steady-state values are defined as

$$\bar{r}_B(l) = r_B(\bar{s}_P, \bar{s}_N, l),$$

where

$$\bar{s}_N = p_{BB} N \bar{r}_B$$

and $\bar{s}_P = \bar{s}$ as in (8.27). Return map simulation results with varying ζ and l values are given in Figure 8.26. p_{BB} values are calculated according to selection of ζ and l parameters. Time response still contains oscillations as seen in Figure 8.27.

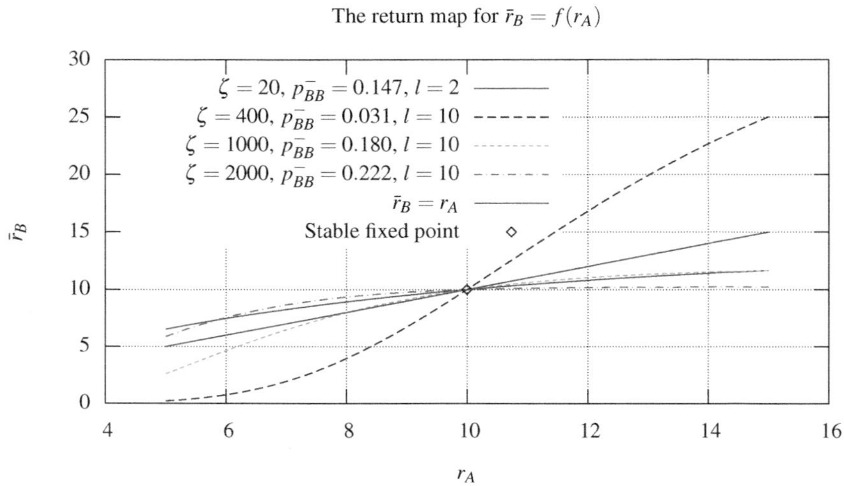

Fig. 8.26. Plots of the return map of $\bar{r}_B(r_A)$ for iterative application of the recruitment described here. The case with $l = 2$ and $\zeta = 20$ replicates results from Figure 8.16. When $l = 10$, r_B is stable only if the combined multiplicity factor ζ, that affects the value of f_{AB} from (8.26), is sufficiently large.

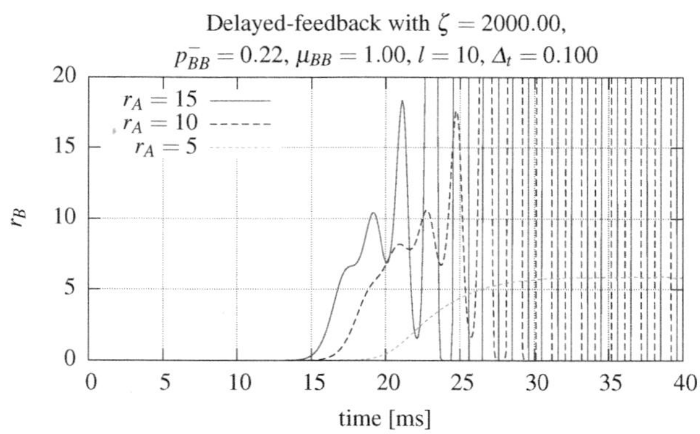

Fig. 8.27. Closed-loop simulation for the change in the size of the recruited set r_B with different selections for the size of the input set r_A for $l = 10$ inputs.

8.5.3 Discussion

With Figures 8.25 and 8.26, we verified that in the special case of requiring l inputs, the population model behaves similarly to the case with two inputs. The time course of activity in the feedback case is not shown because of the oscillations. Also, Monte Carlo simulation results are not shown since the behavior was already explored in the 2-input case.

The l-input model is desirable because it exhibits a higher threshold. This makes the neuroids both more accurate representations of real neurons and also makes them noise resilient. This model can potentially be used when background noise is present, whereas the 2-input model would easily fail.

8.6 Conclusions

The research used a stochastic population approach to model the dynamics of the recruitment process. In particular, this chapter presented the *boost-and-limit* approach (Günay, 2003; Günay and Maida, 2006b) as a mechanism to maintain the stability of the size of recruitment sets as they cross cortical areas iteratively (Günay and Maida, 2006a). This approach primarily proposes that temporally dispersed spike times and use of inhibitory feedback can help stabilize the size of propagating recruitment sets. It is not the first study to use inherent noise in neuronal networks to its advantage (Masuda and Aihara, 2002; Rowland et al, 2006). Other manipulations studied were the inter-area excitatory connectivity, within-area inhibitory connectivity, recruitment threshold, and feedback delay. It was found that if in the closed-loop model, inhibitory feedback was fast enough, then the size of such recruited sets could be maintained at a desired size, r_o, in the presence of perturbations to the input set size, thereby solving the variance problem in recruitment hierarchies. Small networks were studied consisting of 100 neurons per layer with the desired recruitment set size held at 10 neurons. For the open-loop case of only feedforward propagation, we established, using both the stochastic population model and Monte Carlo methods, that recruitment set size is unstable. Then, as the main result of Günay and Maida (2006a), we show that stability can be achieved by use of lateral inhibitory feedback in the output layer by using both the population model and Monte Carlo simulations.

Having achieved stability, the proposed model allows predicting system parameters that enable stable recruitment. The dipolar model allows estimating the lateral inhibitory feedback connectivity parameter p_{BB} in terms of the free parameters of the network. These free parameters are: the replication factor r_o, the number of neuroids N_B at a localized area, and the amplification factor λ for the feedforward excitatory connectivity between areas. The model also shows how the choice of λ affects the speed of convergence to a desired replication factor. The speed of inhibitory feedback is important in achieving stability. For the example network, stability could not be achieved if the feedback delay was longer than 5 ms.

Ultimately, this approach serves to maintain controlled propagation of synchrony across layers. Existing studies of controlled synfire activity use layered feedforward networks in the presence of background noise, but without lateral inhibition (Diesmann et al, 1999; Tetzlaff et al, 2002; Litvak et al, 2003). In our networks, the temporally dispersed transmission delays tend to disrupt the synchrony of inputs to later layers. That is, although the output of layer A is synchronized, the output of layer B is less synchronized because of the temporally dispersed transmission delays. Presumably, the window of synchrony widens at each stage of the hierarchy. Spiking neurons, however, are known to act as band-pass filters because they only fire when incoming spikes are within a short time interval (Diesmann et al, 1999). Thus, the effect of spikes which stray too far from the center of the synchrony window are lost. This stops the synchrony window from widening infinitely. Even though the lost spikes may decrease the size of the spike packet, we showed that, for variations in input, the output size will always converge to the desired fixed-point as the process is repeated. Thus, because of resynchronization effects, temporally dispersed delays may not disrupt recruitment of hierarchies. Additional studies are needed to verify this issue.

It is known from other studies (Diesmann et al, 1999; Tetzlaff et al, 2002) that there are lower and upper bounds on the number of neurons or connection density in layers of feedforward networks to propagate synfire chains in the presence of noise. Our networks were not studied in the presence of background noise. However, we can predict that the network will operate properly under background noise because its convergence property. We extended our model to require an arbitrary number of spikes (such as 10–40) before reaching firing threshold. This demonstrated that the model can exhibit noise resilience that was not possible with the model which required only two spikes to fire. Further simulations are required to verify the conditions of noise resilience.

In summary, the analyses and simulations of this chapter have laid firm foundations for large scale recruitment learning of spiking neuron populations. Some further research is required to refine these technical foundations:

1. To study noise resilience of spike propagation;
2. To explore larger scale simulations to determine the robustness of our results in those contexts; and
3. To improve the quality of population estimations to match results of Monte Carlo simulations more closely.

Nevertheless, the field has advanced sufficiently that large scale simulations of recruitment in neural populations – adressing realistic cognitive tasks – is a natural direction for future research. These issues are considered in more detail in the final chapter of this volume.

Chapter 9
Conclusion

Joachim Diederich, James M. Hogan, and Cengiz Günay

This book provides an overview of recruitment learning, a neural network learning system which exhibits high biological plausibility due to a number of core features - notably sparse connectivity levels and the use of weight updates which both respect synaptic constraints and support rapid concept acquisition. The earlier sections of this volume have given form to these ideas, for both the discrete time systems characteristic of the early wave of connectionism, and the more recent continuous time spiking neural systems - which allow closer alignment with neural observation. In this concluding discussion, it is worthwhile to summarise these ideas and to identify opportunities which remain unexplored, most notably those which arise through the emergence of adult neurogenesis.

The foremost difference between recruitment learning and more traditional connectionist systems is that recruitment operates in sparsely connected neural networks - forsaking the traditional assumptions of dense or full connectivity. Sparse connectivity allows greater fidelity to biological neural processing, and the parameters may be tuned to reflect local variation and function.

However, the plausibility of the approach is not limited to the sparse local structure of the network. The mechanisms that trigger learning are also of deep biological and psychological plausibility. Persistent weight adjustment occurs without backpropagation of error, and is broadly Hebbian in nature. Rapid recruitment commonly takes place through release of inhibition (section 2.3.1), a phenomenon observable in many areas of the brain and realised through interneuron connections. This process is readily modelled computationally through competitive learning mechanisms such as winner-take-all networks. This combination of sparsely connected networks, sparse coding of inputs and rapid concept acquisition through release of inhibition are among the foundations of recruitment learning.

Recruitment learning is further distinguished from more traditional backpropagation networks in its ability to support one-shot learning. While one-shot learning has been found in other mammals, learning based on a single sample becomes uniquely human whenever deep background knowledge and rich experience are involved. It is thus very attractive to combine sparsely connected neural networks with competitive learning mechanisms in order to realise high-level, cognitive function. It is this combination that allows learning outside the purely inductive, supporting rationalist paradigms such as "learning by problem-solving" as proposed in Gestalt Theory.

J. Diederich, C. Günay, and J.M. Hogan: Recruitment Learning, SCI 303, pp. 275 – 281.
springerlink.com

As outlined in the first chapter, recruitment learning approaches have a rich tradition that reaches back to the origins of psychology, but the method has long been informed by refinements in our understanding of neural architecture and processing. It is fascinating, therefore, to look at recruitment learning in light of the most recent developments in the neurosciences. The striking confirmation of the existence of adult neurogenesis presents fundamental challenges and opportunities for the approach. In principle, neurogenesis provides a pool of neurons that are plastic, ready for learning and (possibly) a necessary component for adaptation to the environment.

Unsurprisingly, experimental confirmation of adult neurogenesis has already fuelled a new wave of neural network learning studies specifically aimed at modelling the hippocampus. While these models have exploited the availability of additional processing units and the greater structural plasticity afforded, to date they have given scant attention to the usual characteristics of recruitment. Appleby & Wiskott (2009), for instance, describe a model for hippocampal learning that employs sparse coding and the addition and deletion of units, but provide no consideration of inhibitory mechanisms or sparse connectivity.

In this sense, current models are yet to unify this new paradigm of structural modification with more established neural principles. A core motivation for this book is to combine the rich tradition of recruitment learning with the emerging story of adult neurogenesis in order to explore new opportunities for cognitive modelling. Yet while we embrace these opportunities, some caution is necessary when predicting the impact of adult neurogenesis on recruitment learning and our understanding of cognition. These issues are considered in more detail below.

9.1 The Promise of Neurogenesis

The contribution of adult neurogenesis to learning and memory has been discussed earlier, but significant questions remain unanswered, and the experimental record itself remains inconclusive. For example, one might expect blocking of adult neurogenesis to cause learning deficits, but this effect – while present in some cases - has not been observed consistently. Jaholkowski et al. (2009) used a new approach involving transgenic mice - for which hippocampal adult neurogenesis is *blocked* - to investigate its importance in learning and memory. These mice were tested on tasks for which the importance of adult neurogenesis had been established. The mice showed no impairment in sensorimotor tests, with only sensory impairment in an olfactory task (Jaholkowski et al., 2009). More importantly, the mice showed proper procedural learning and performed well in a range of learning and memory tasks. This work suggests that adult neurogenesis, while potentially important, is not obligatory in learning - including some types of learning for which it has been suggested as a mechanism.

More controversial still is the question of whether neurogenesis extends beyond the areas, for which the evidence is irrefutable, to encompass the neocortex. Gould (2007) reviews adult neurogenesis in the mammalian brain, attempting to distinguish "neurogenic" brain regions that produce or recruit new neurons in the absence of damage from "non-neurogenic" areas that do not produce or recruit new neurons (Gould, 2007, p.481). While a number of authors argue in support of cortical neurogenesis, the neocortex has not yet been accepted as neurogenic and Gould (2007) considers the arguments with an emphasis on methodological reliability. There is the potential for false positive results (i.e., the detection of adult neurogenesis where there is none) and for marking non-neural cells as new if they are in the proximity of neurons. Bromodeoxyuridine labelling, the most frequently used technique, can "falsely identify a new satellite cell lying over the nucleus of a mature neuron as a single double-labelled cell" (Gould, 2007, p.482).[1] Despite these methodological problems, Gould (2007) believes more reliable techniques may confirm a more widely neurogenic adult brain.

From the viewpoint of recruitment learning, there are additional questions about the role of the new neuron in the adult brain. Neurons are not necessarily a scarce resource – indeed frequently neurons may be available in abundance – and hence the question can be asked why new neurons need to be generated at all. Experimental results have shown that new neurons, the youngest first, *are* recruited into functional circuits. For instance, during puberty, brain structures like the centromedial amygdala in the rat increase in size considerably thanks to neurogenesis (Han and De Vries, 1999). Yet it is not quite clear why young and recently generated neurons are required in this way; the plasticity of the new neuron is certainly a factor, but a tighter connection between adult neurogenesis on the one side and learning and memory on the other side is necessary. The core question remains, what is the role of new neurons in learning and memory?

The time course of adult neurogenesis presents an additional problem. It takes about four weeks to generate a new neuron in the hippocampus. It is not possible to generate a new neuron immediately and in the required location if learning demands it. Adult neurogenesis creates a pool of neurons *potentially available* for learning, but only some will be utilized. The role of the hippocampus in learning and memory formation has been known for many decades, yet there are no models to date that clearly demonstrate how a pool of newly generated neurons in the adult hippocampus facilitates the formation of long term memory.

Some neural network models of the hippocampus emphasise the role of this region as a similarity or novelty detector. If, for instance, a new stimulus cannot be linked to learned knowledge, active learning occurs that requires an analysis of the novelty of the stimulus and this process involves the hippocampus. After active learning (quite literally working out what is novel about the stimulus), long term memory formation occurs. The introductory chapters discussed in some detail the notion that learning may be based on the presentation of a single example, for instance, if there is an inconsistency or a discrepancy with what is already known. But how is long term memory established, and what is the role of adult neurogenesis in its formation? There is an urgent need for models that simulate hippocampal learning - including adult neurogenesis - and establish a link to other parts of the brain such as the neocortex. At present there is no clear, coherent picture and the absence of confirmed adult neurogenesis in the neocortex ensures that much of this work will remain speculative for the time being. Models are required

[1] See Table 1 in Gould (2007) for evidence for and against adult neurogenesis in brain regions other than the dentate gyrus and olfactory bulb.

that link adult neurogenesis in the hippocampus with the generation of memory traces in other parts of the brain, over time helping to constrain and thus elucidate the role of neurogenesis in these processes. Models of this nature may be especially important, given the difficulty in studying such processes experimentally – particularly if the laboratory animals do not survive the experiment. Nevertheless, it would be fascinating to learn, for example, that the cortical areas responsible for language processing were neurogenic; or that neurodegenerative disorders such as Alzheimer's and Parkinson's diseases might be treated by replacing dead neurons with new ones. Models can be instrumental in predicting how this could occur.

While the context is potentially revolutionary, the approach described is more measured, building in small increments on the mechanisms developed over many years by a number of authors. We shall return to this notion of *converging constraints* in the following sections.

9.2 Neurally Plausible But Useless vs. Implausible But Useful

The earlier chapters outlined the development of recruitment learning models within cognitive science to address problems related to learning, memorisation and retrieval. As we have seen, the link to the neurosciences has been significantly strengthened in recent times, yet recruitment learning has remained far less prominent than less plausible techniques such as error backpropagation learning. One reason for this lies in the wide availability of the latter method; back propagation has long been a standard component of neural network simulation tools and has frequently been used in data mining applications. Recruitment learning methods, on the other hand, have seldom formed part of publicly available software tools for neural network simulation and data mining. Undoubtedly, this has been an obstacle to a more wide-spread application of the approach, but perhaps the more important issue is the nature of the method itself. Gradient based methods allow simulation – albeit *slow* simulation - directly from the tabula rasa; recruitment has more demanding prerequisites.

Recruitment learning techniques depend critically on prior knowledge and its representation, and are not *general machine learning methods* suitable for inductive data mining applications, as the term has previously been understood. Nevertheless, there has been significant progress over the past two decades, notably in our understanding of capacity control (Vapnik, 1995) and in our ability to collect and manage large scale structured data sets. These advances, and the technology which supports them, allow consideration of problems unimaginable to the Parallel Distributed Processing movement of the mid 1980s.

Moreover, these developments provide significant opportunities for recruitment learning to demonstrate its viability in a broader context, albeit in conjunction with some very stern baselines for comparison. For, if the computational resources are available to support pattern vectors with tens of thousands of components, those resources may also support enormous knowledge bases in a form that allows their rapid combination and exploitation. To a certain extent, the rise of cloud-scale structured data stores may provide the substrate needed for recruitment learning style approaches to become more competitive. Rapid learning based on few examples in conjunction with background knowledge is then an interesting approach to data mining. Yet the earlier point remains valid: recruitment learning is unlikely to become more widely adopted without an appropriate software platform, and there remains an urgent need for a project of this type.

The introduction of a common software platform would facilitate broader experimentation with the approach – both within the traditional cognitive domains and in less traditional environments. Over time, it may be possible to compare and benchmark recruitment learning against more mainstream machine learning techniques. Support vector machines, for instance, can accept high dimensional, sparse input spaces. Generally, text and image classification problems are of this type, especially when low level features such as pixel intensity or bag of words representations are employed. Recruitment learning methods based on sparsely connected, random networks should also be suitable for high dimensional input spaces as long as the fundamental degree of the Boolean relations remains bounded. The models outlined in chapters 3 and 4 are a sensible candidate for such applications since the focus is on the recruitment of the whole networks with limited weight change.

Ultimately, the alternatives are less stark than in the flippant dichotomy of the section title. Neural plausibility may limit the range of computations realizable at small scale, and may be an unforgiving and frustrating to master if pursued blindly. But none of this work can be seen as useless if it offers candidate mechanisms for cognitive computation which respect the limitations of the neural substrate. As Valiant (1994) pointed out so compellingly, it is the role of the computer scientist to bring to these discussions the notion of complexity, the realization that certain classes of algorithm over certain types of data representation will achieve the required cognitive task with psychologically plausible effect in a time consistent with experimental observation. More importantly, inadequate algorithms and data representations will fail and can be shown to fail.

Some will fail on small test problems, some will require unreasonably powerful and informative learning signals, while others will fail utterly to scale in the face of the extraordinary input dimensions characteristic of perceptual processing. Over time, however, this work will have the effect of ruthlessly eliminating implausible and ad hoc computations and representations, with the converging physical and computational constraints bringing us ever closer to the basis of neural processing, and hence of cognition itself.

This volume thus represents an important step in this `argument from converging constraints', bounding the computations realisable in neural networks under constraints of connectivity and data representation, and doing so for both discrete and continuous time local units. Most significantly, this work demonstrates the rapid acquisition of non-trivial Boolean concepts through reliance on existing knowledge while operating under these constraints. Moreover, the work covers a broad range of learning updates, their foundations and their performance – including analyses of generalization for the discrete time case, and of the stability of concept assemblies in the continuous time case.

In the final section of this volume, we consider some of the more promising opportunities which might drive the future of recruitment learning research.

9.3 The Future

While one should not discount the idea of recruitment-like approaches in mining large, structured data sets, it is clear that the most promising avenue for recruitment learning lies in the rapid acquisition of increasingly complex knowledge within sparsely connected networks of spiking neurons. This work is in its infancy, and there remain a number of important steps that must be taken before researchers may entertain the very attractive notion of brain-scale simulations. It is necessary to establish a consistent and effective hierarchy of knowledge learnable in this computational domain – to build upon existing input features to establish relations over these features, which in turn may form the basis of more complex relations.

Essentially these ideas represent a unification of the work presented respectively in parts I and II of this book, extending spiking neural assembly based recruitment to support complex Boolean expressions such as those explored in chapter 5. Some progress has already been made in this area with the successful recruitment of the XOR relation across two assemblies (James and Maida, 2009) and this must be extended to more complex concepts, and the limits of concept formation and generalization over spiking assemblies carefully analysed. In particular, it would be exciting to consider the effect of the concept assembly – and learning rules of various levels of plausibility – on the concept acquired when the system is presented with complex, high dimensional inputs.

This book has considered high level cognitive tasks such as learning by problem solving and early language acquisition, but we have done so in a restricted, manageable domain, abstracting where necessary to allow some progress to be made. We have also, in this chapter in particular, speculated on the role of adult neurogenesis in higher cognition, and on the feasibility of an experimental approach to this question.

We thus conclude with a compelling and increasingly plausible programme for recruitment learning: neural scale simulations of cognitivetasks, relying on the recruitment of existing functional circuits – realized in turn by relations over coherent assemblies of spiking neurons. Recruitment learning has always claimed neural plausibility, but its fidelity has risen spectacularly with the continuous time models of the second half of this volume. A large scale spiking neuron population under sparse connectivity assumptions is a logical progression from this work, and offers the promise of experimentally falsifiable predictions of activity, measurable at a macro-scale and justifying comparison with the results of neuroimaging studies. Supplementing the network with a slowly emerging pool of plastic neurons would similarly allow exploration of the role of neurogenesis in cognition, while integrating such knowledge as can be obtained from experiment.

Overall, the approach continues within the established tradition of converging constraints, bounding the mechanisms of processes such as language acquisition, and shedding light on the complex interactions between the different modalities of perception. The programme outlined is spectacularly presumptuous, but may equally be justified as a natural, incremental extension of earlier success. In some sense those engaged in recruitment learning must become increasingly alert to its principles, carefully constructing each new task from those realized at the previous stage and taking advantage of the opportunities emerging as recruitment learning moves ever closer to the soft-tissue of biological neural systems.

Appendix A
Appendix for Recruitment in Spiking Neural Networks

A.1 Formal Proofs

A.1.1 Derivation of Probability of Connection in Random Multipartite Graphs

We demonstrate here the mathematical deduction to find the probability of connection described in §7.2.2 for random multipartite graphs. This will be similar in essence to that of simple random graphs given by Valiant (1994, pp.70–71). The multipartite graph consisting of two areas, one source area, and a destination area as in Fig. A.1 is considered. The ensembles $\tilde{x}$ and $\tilde{y}$ of size r_s in the source area A_s project to destination area A_d in order to recruit ensemble $\tilde{z}$.

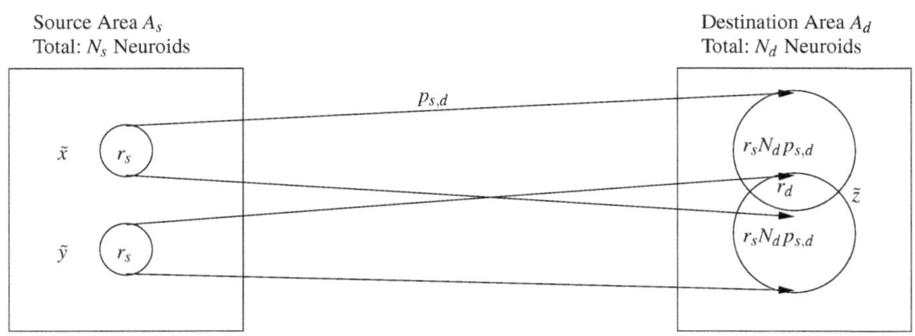

Fig. A.1. Random multipartite graph demonstrating the recruitment process. Ensembles and their projections are shown with circles with their size printed inside.

Definition 1. The probability of a neuroid in A_s being connected to a neuroid in A_d is $p_{s,d}$. That is,

$$p_{s,d} = P((n_i, n_j) \in \tilde{E} \mid \exists n_i \in A_s,\ \exists n_j \in A_d)$$

where n_i, n_j denote neuroids and $\tilde{E}$ is the set of edges, i.e. pairs of neuroids that are connected.

Definition 2. Let the frontier $E_{A_d}(\tilde{x})$ denote the set of neuroids that the ensemble $\tilde{x}$ project to in area A_d. Formally,

$$E_{A_d}(\tilde{x}) = \{n_i | \exists n_i \in A_d,\ \exists n_j \in \tilde{x},\ \exists \tilde{x} \subset A_s,\ (n_j, n_i) \in \tilde{E}\}.$$

The set of neuroids recruited in A_d can then be represented as

$$\tilde{z} = E_{A_d}(\tilde{x}) \cap E_{A_d}(\tilde{y}).$$

We are interested in finding the probability $p_{s,d}$ given in Def. 1 so that the expected size of set $\tilde{z}$, is close or equal to the replication factor r_d of area A_d, that is

$$\varepsilon(|\tilde{z}|) \simeq r_d. \tag{A.1}$$

Definition 3. Let p^* denote the probability that any neuroid in A_d being in set $\tilde{z}$ such that

$$p^* = P(n_i \in \tilde{z} | n_i \in A_d).$$

We construct p^* by defining the following.

Definition 4. The probability of a neuroid in A_d not being connected to a neuroid in A_s is

$$\begin{aligned} P((n_i, n_j) \notin \tilde{E} | \exists n_j \in A_d, \exists n_i \in A_s) &= \bar{p}_{s,d} \\ &= 1 - p_{s,d}. \end{aligned}$$

Definition 5. The probability of a neuroid in A_d not being connected to any of the neuroids of an ensemble of size r_s in A_s is

$$P((n_i, n_j) \notin \tilde{E} | \exists n_j \in A_d, \forall n_i \in \tilde{x}, \exists \tilde{x} \subset A_s) = (1 - p_{s,d})^{r_s}.$$

Definition 6. The inverse of the above quantity is of interest. That is, the probability of a neuroid in A_d being connected to any of the neuroids of an ensemble of size r_s in A_s is

$$P((n_i, n_j) \in \tilde{E} | \exists n_j \in A_d, \forall n_i \in \tilde{x}, \exists \tilde{x} \subset A_s) = 1 - (1 - p_{s,d})^{r_s}.$$

Extending this, the probability of a neuroid in A_d being connected to any of the neuroids of two ensembles of each size r_s in A_s is

$$P((n_i, n_j) \in \tilde{E} | \exists n_j \in A_d, \forall n_i \in \tilde{x} \cup \tilde{y}, \exists \tilde{x}, \tilde{y} \subset A_s) = (1 - (1 - p_{s,d})^{r_s})^2 \tag{A.2}$$

which is p^* given in Def. 3. This binomial can be expanded to yield

$$p^* = r_s^2\, p_{s,d}^2 + \mathrm{O}(p_{s,d}^3) \tag{A.3}$$

in which the second term is a quantity that diminishes proportionately with $p_{s,d}^3$. This is a sufficiently small value with the parameters used here that it can be ignored.

The expectation in (A.1) can now be calculated by

$$\varepsilon(|\tilde{z}|) = N_d\, p^*$$

giving equal probability independently to all N_d neuroids in A_d. Solving for finding a $p_{s,d}$ for the desired expectation yields

$$\begin{aligned} r_d &= N_d\, p^* \qquad &(A.4)\\ &= N_d\, r_s^2\, p_{s,d}^2 \\ p_{s,d} &= \frac{1}{r_s}\sqrt{\frac{r_d}{N_d}} \qquad &(A.5) \end{aligned}$$

resulting in the calculation given in (7.3).

A.1.2 Probability of Merging Connections from Different Areas

We show, in this section, that the result in (A.5) apply in cases such as in Fig. A.2. Notice that we need to calculate two probability parameters, one from each source area. However it can be shown that the earlier result can be used to calculate independently the probability of each connection to give the desired expected size for recruitment at the destination.

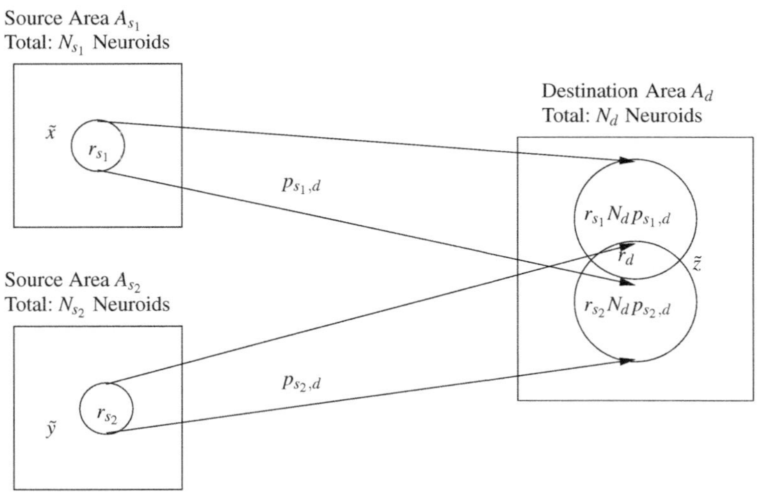

Fig. A.2. Random multipartite graph where connections are merging in from two distinct areas to the destination. The figure legend is the same as in Fig. A.1 on page 283.

Let

$$p_{s_1,d} = \frac{1}{r_{s_1}} \sqrt{\frac{r_d}{N_d}},$$

and

$$p_{s_2,d} = \frac{1}{r_{s_2}} \sqrt{\frac{r_d}{N_d}}.$$

The probability p^* of a unit in A_d to be in the recruited set in Def. 3. We have used in (A.3) the binomial expansion and ignored higher order terms. We can expand the equation as

$$p^* = r_s \, p_{s,d} \, r_s \, p_{s,d}$$

to show the combination of the independent probabilities of choosing the two source ensembles. In the case the connections are merging in from two source areas, A_{s_1} and A_{s_2}, it will be

$$p^* = r_{s_1} \, p_{s_1,d} \, r_{s_2} \, p_{s_2,d}$$

which satisfies the equality in (A.4) yielding the desired output ensemble size.

A.1.3 Connection Probability in the Generalized Recruitment Scenario

Here, we show the derivation of Eq. (8.1), which is the required connection probability p^+_{AB} of a neuroid in area A to a neuroid in area B as shown in Figure 8.3b. This derivation is adapted from the probability calculation for recruitment of conjunctions of two inputs by Valiant (1994). We assume the size of the input set is $|R_A| = r_o$, which is the desired replication factor. We use $p \equiv p^+_{AB}$, $r \equiv r_o$ here for simplicity.

First, we look at the probability of a neuroid in B to be recruited. Recruitment requires two connections from active sources. Since we only have one input set R_A, a recruitment candidate in B needs to have two connections from neuroids in set R_A. The probability of the candidate being in the projection set of R_A can be calculated from its probability of not having connections to any neuroid in set R_A with $(1-p)^r$. Thus, its probability of being in the projection set becomes $1-(1-p)^r$. Since we require the candidate neuroid to receive projections from at least two neuroids in R_A, we can define the probability of a neuroid in B being recruited as

$$p^* = \left(1-(1-p)^r\right)\left(1-(1-p)^{r-1}\right). \tag{A.6}$$

Expanding (A.6), we get

$$p^* = 1-(1-p)^{r-1}(2-p)+(1-p)^{2r-1},$$

where the power terms can be expanded using the binomial theorem. Assuming $p \ll 1$, we can ignore terms with p^k for $k > 3$

$$p^* = 1 - \left(1 - (r-1)p + \frac{(r-1)(r-2)}{2}p^2 + \mathcal{O}(p^3)\right)(2-p)$$
$$+ \left(1 - (2r-1)p + \frac{(2r-1)(2r-2)}{2}p^2 + \mathcal{O}(p^3)\right).$$

Simplifying, we get

$$p^* = -rp^2 + r^2p^2 + \mathcal{O}(p^3). \tag{A.7}$$

The desired property of recruitment is to yield an output set with the same size of the input set. The expected size of the recruited set R_B is

$$|R_B| = p^*N \simeq (-rp^2 + r^2p^2)N.$$

We aim to satisfy the equality $p^*N = r$. Solving for p, we get

$$p \simeq \frac{1}{\sqrt{N(r-1)}}$$

with an error decreasing with $\mathcal{O}(p^3)$ as N increases. Adding an amplification factor $\sqrt{\lambda}$ similar to Valiant's proposals yields the definition in Eq. (8.1).

A.1.4 A More Realistic Probability Calculation for the Dipolar Model

A more general way of computing the expected number of neuroids, given dipolar synaptic population sizes, is counting not just the occurence of necessary number of activated synapses, but also for more. This implies looking for conjunctions of two or more excitatory activated sysnapses from population s_P,and checking for one or more activated inhibitory synapses from s_N, rather than only looking for the existence of one.

As in Section 8.4, we give the expected value calculation with an error $\mathcal{O}(h^3)$, $h = 1/N$, where we use $N \equiv N_B$ for simplicity. The probability of failure to recruit can be explained with having no or only one synapse from s_P, that is

$$\bar{p}_P = (1-h)^{s_P} + h(1-h)^{s_P - 1}.$$

Thus, the probability of finding two or more synapses from s_P is $p_P = 1 - \bar{p}_P$. The probability of finding no synapses from the inhibitory population s_N on a neuroid is

$$\bar{p}_n = (1-h)^{s_N}.$$

The expected number of recruited neuroids is then calculated using the conjunctive probability of p_p and $\bar{p}_n$,

$$r_B = N\bar{p}_n p_p.$$

A.2 Time-to-Spike Shortens If Membrane Time Constant Is Increased by Only Varying the Resistance

We proposed to vary the membrane time constant for adjusting the synchrony tolerance window for delayed spikes at the postsynaptic neuron in Chapter 7 (Günay and Maida, 2001; Günay, 2003; Günay and Maida, 2006b). This can be criticised because changing the time constant may have a significant side effect on the response time of the neuron (Rowland, 2002). This effect can be avoided if the membrane time constant is increased by only varying the cell input resistance and keeping the capacitance constant. Because our earlier work (Günay and Maida, 2001) did not specify how the time constant should be increased, here we review the justification for this solution, where the membrane time constant is increased by manipulating membrane channels to increase the resistance. We also supplement this justification with a formal proof and a computer simulation.

Another possible criticism is the lack of varying time constants in higher cognitive areas of the brain to support our proposal of varying-size tolerance windows. However, studies which show that the membrane time constant dynamically changes in the cortex (Koch et al, 1996) can explain how different time constants may be employed in different parts of the brain. These studies suggest that the change is due to variations in the membrane conductance, which is consistent with our view here.

Membrane Time Constant Dynamics

The membrane equivalent circuit of a leaky Integrate-and-Fire (I/F) neuron is given in Figure A.3. The membrane equation for an I/F neuron is

$$C\frac{\mathrm{d}v}{\mathrm{d}t} = I - \frac{v}{R}, \tag{A.8}$$

where R and C are the resistance and capacitance parameters of the membrane, respectively, and I is the constant external current.

The membrane time constant is then defined as $\tau = RC$. When (A.8) is integrated with initial conditions $v(0) = v_0 = 0$, we get

$$v = RI\left(1 - e^{-t/\tau}\right). \tag{A.9}$$

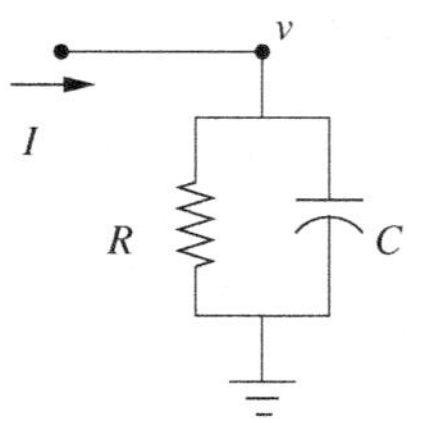

Fig. A.3. Membrane circuit equivalent for I/F model.

In order to find the time-to-spike, from (A.9) we can calculate the time T it takes to reach a fixed threshold V as

$$T = \tau \ln \left(\frac{RI}{RI - V} \right). \tag{A.10}$$

This may look like T depends linearly on the time constant τ. However, we prove here that this is not necessarily true. Especially if only the resistance R is varied and the capacitance C is kept contant in increasing the time constant τ, then the logarithmic term cancels the linearly increasing effect of the first term τ in (A.10). Therefore, in contrast, as R and so τ increases, the time-to-spike T decreases.

We first give an intuitive explanation of this phenomenon with supporting electrophysiology principles. We then give a formal proof and a computer simulation subsequently.

Intuitive Explanation

The membrane time constant, $\tau = RC$, can be increased by only varying the resistance parameter R. It is much more difficult to manipulate the capacitive properties of the membrane, which depends primarily on the insulating properties and the thickness of the membrane, parameters which are static; it is thus unreasonable to suggest they vary over short periods of time.

However, the resistance, or the inversely related conductance parameter ($R = 1/g$), is easily varied by changing the amount of leakage across the membrane. Detailed neuron models, such as the Hodgkin-Huxley model, use variable resistances to model the conductance of membrane channels.

The claim that when the membrane time constant increases, the time it takes to charge the membrane potential is longer can be refuted as follows. When only the resistance is varied, we expect the membrane to only take longer to *leak* and not to *charge*. If the resistance is higher, the current flowing through the resistance will be smaller and it will take a longer time to discharge the capacitor.

However, when filling the capacitor, if the input current is an independent current source (such as a spike received from a remote neuron) then there is a fixed amount of current coming on the membrane (as opposed to a conductance). When the resistance is varied it changes how much of this curent is lost while charging the capacitor. That is, if we increase the resistance, then the leakage will be less and more current will stay on the capacitor.

This behavior becomes clearer by interpreting (A.8). The amount of voltage increase for a capacitor C (left hand side) depends on how much current is supplied externally by I and how much current is lost over the resistance R given the current voltage v.

Proof

In order to mathematically assess the effect of varying R on the expression T in (A.10), we need to investigate the differentiation of T. That is, we calculate the derivative

$$\begin{aligned}
\frac{\mathrm{d}T}{\mathrm{d}R} &= \frac{\mathrm{d}}{\mathrm{d}R}\left[RC\ln\left(\frac{RI}{RI-V}\right)\right] \\
&= C\ln\left(\frac{RI}{RI-V}\right) + R\frac{\mathrm{d}}{\mathrm{d}R}\left[C\ln\left(\frac{RI}{RI-V}\right)\right] \\
&= C\ln\left(\frac{RI}{RI-V}\right) + RC\frac{\mathrm{d}}{\mathrm{d}R}\left(\frac{RI}{RI-V}\right) / \left(\frac{RI}{RI-V}\right) \\
&= C\ln\left(\frac{RI}{RI-V}\right) + RC\left[\frac{I(RI-V)-IRI}{(RI-V)^2}\right]\frac{RI-V}{RI} \\
&= C\ln\left(\frac{RI}{RI-V}\right) + C\frac{(-V)}{RI-V} \\
&= C\left(\ln\left(\frac{RI}{RI-V}\right) - \frac{V}{RI-V}\right). \qquad \text{(A.11)}
\end{aligned}$$

Since RI is the asymptotic value for the membrane potential, it is trivial that the threshold should satisfy

$$V < RI.$$

If we observe the limit case where $V \to RI$ for (A.11), we get

$$\lim_{V \to RI} \frac{\mathrm{d}T}{\mathrm{d}R} \simeq \lim_{\Delta V \to 0} C\left(\ln\left(\frac{V}{\Delta V}\right) - \frac{V}{\Delta V}\right) < 0$$

where $\Delta V = RI - V$, obviously since the logarithmic term is much smaller than the term on the right hand side. Having the derivative less than 0, it shows that increasing the resistance R results in a decrease of the time-to-spike parameter T. □

Computer Simulation

A numerical simulation has been performed to test the above proof using the script in Algorithm A.1. Figure A.4 shows the membrane potential profiles of (A.9) obtained by substituting with two different resistance R parameters.

A.3 Finding the Peak of the Spike Response

This section provides additional proofs supporting the arguments in §7.4. The peak of the EPSP can be found analytically by finding the point where the derivative of (7.5) is zero. The peak time is

Algorithm A.1 MATLAB script used for simulation.

```
function synapse
%### Matlab script for creating a plot of a Pulsed Neural Network
compliant synapse profile
%### changing in time ater a spike has been received from a
presynaptic neuron.
global fontsize

%# Visual parameters
fontsize = 14;

%### Time sweep (ms)
t = 0:1e-1:100;

%### Parameters
R = 1e4;
C = 1e-6;
I = 1e-6;

%### membrane time constant
tm = R * C;

display(sprintf('tm=%f', tm));
e1 = epsilon(1e-3*t, R, C, I);
e2 = epsilon(1e-3*t, 2*R, C, I);
h = figure;
hold on;
decorate('Membrane Rise Profile with Constant Current', R, C, I);
plot(t, 1e3*e1, '-', t, 1e3*e2, '--');
set(gca, 'Fontsize', [ fontsize ]);
legend('R', '2R');

function graf(t, epsilon)
global fontsize
%### normalized plot of epsilon
plot(t, epsilon/max(epsilon));

function decorate(desc, R, C, I)
global fontsize
%### decorations
grid;
title(sprintf('%s\n (R=%.2f k\\Omega, C=%.2f \\mu{}F, I_{ext}=%.2f
\\mu{}A)', \
  desc, R/1e3, C*1e6, I*1e6), 'Fontsize', [fontsize]);
xlabel('time (ms)', 'Fontsize', [fontsize]);
ylabel('potential (mV)', 'Fontsize', [fontsize]);

function e = epsilon(t, R, C, I)
%### epsilon kernel function response
e = R*I - R*I*exp(-t/(R*C));
```

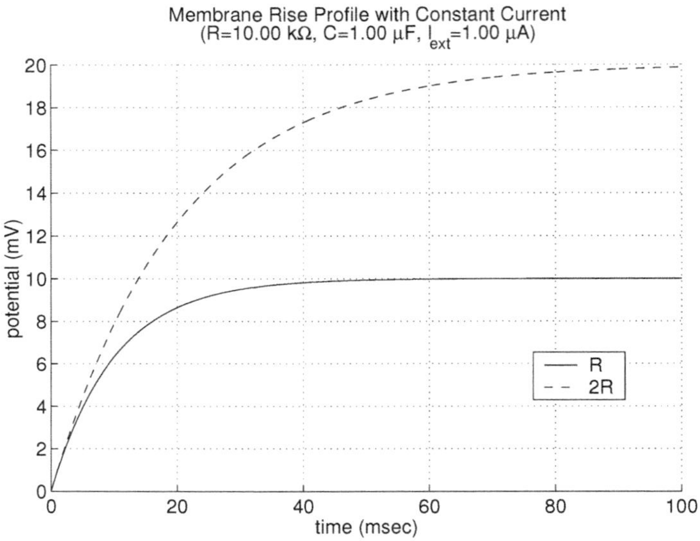

Fig. A.4. Membrane potential profiles given by simulating equation (A.9). They show that it takes less time to charge the membrane to a fixed spike threshold (e.g., 8 mV) if the membrane time constant is increased by doubling the membrane resistance R while keeping the capacitance C constant. Parameter values are displayed on the figure.

$$t_{\max} = \tau_s \tau_m \ln(\tau_m/\tau_s)/(\tau_m - \tau_s) + \Delta_i^{\mathrm{ax}} . \tag{A.12}$$

The rightmost component can be omitted by taking $\Delta_i^{\mathrm{ax}} = 0$, for simplicity.

To show that $t_{\max} \in [\tau_s, \tau_m]$ is always true, we check the conditions $t_{\max} > \tau_s$ and $t_{\max} < \tau_m$. Let

$$x = \tau_s/\tau_m , \tag{A.13}$$

from which it can be deduced that $0 < x < 1$, since both $\tau_s, \tau_m > 0$ and $\tau_s < \tau_m$. Using (A.12), for $t_{\max} > \tau_s$, we get

$$\ln x < x - 1$$

which is true for all values of $x > 0$. Conversely, for $t_{\max} < \tau_m$, we get

$$\ln(1/x) < 1/x - 1$$

which is again true for all values of $x > 0$.

Next, it is shown that the form of $\varepsilon_i(t, \tau_s, \tau_m, \Delta_i^{\mathrm{ax}})$ only depends on the selection of x at the boundaries of the effective window $\omega = [\tau_s, \tau_m]$. Rewriting (7.5) for $t = \tau_s$ and simplifying, we get

$$\varepsilon_i(\tau_s, \tau_s, \tau_m, \Delta_i^{\mathrm{ax}}) = \frac{1}{1-x}\left[\exp(-x) - \exp(-1)\right].$$

For $t = \tau_m$ and simplifying, we get

$$\varepsilon_i(\tau_m, \tau_s, \tau_m, \Delta_i^{\mathrm{ax}}) = \frac{1}{1-x}\left[\exp(-1) - \exp(-1/x)\right].$$

Both boundary expressions are functions of x given in (A.13).

References

Abeles, M.: Corticonics: Neural circuits of the cerebral cortex. Cambridge University Press, Cambridge (1991)

Abeles, M.: Firing rates and well-timed events in the cerebral cortex. In: Domany, et al. (eds.), pp. 121–138 (1994)

Abeles, M., Hayon, G., Lehmann, D.: Modeling compositionality by dynamic binding of synfire chains. Journal of Computational Neuroscience 17, 179–201 (2004)

Aho, A.V., Ullman, J.D.: Foundations of Computer Science. W.H. Freeman and Company, New York (1992)

Aimone, J., Wiskott, L.: Computational modeling of adult neurogenesis. In: Gage, F., Kempermann, G., Song, H. (eds.) Adult Neurogesis, pp. 463–481. Cold Spring Harbor Laboratory Press, Woodbury (2008)

Aimone, J.B., Wiles, J., Gage, F.H.: Computational influence of adult neurogenesis on memory encoding. Neuron 61(2), 187–202 (2009)

Alquézar, R., Sanfeliu, A.: An algebraic framework to represent finite state automata in single-layer recurrent neural networks. Neural Computation 7, 931–949 (1995)

Amari, S.: A method of statistical neurodynamics. Kybernetik 14, 201–215 (1974)

Anderson, J.R.: The architecture of cognition. Harvard University Press, Cambridge (1983)

Appleby, P.A., Wiskott, L.: Additive neurogenesis as a strategy for avoiding interference in a sparsely-coding dentate gyrus. Network: Computation in Neural Systems 20(3), 137–161 (2009)

Arbib, M.A.: Brains, Machines, and Mathematics, 2nd edn. Springer, Berlin (1987)

Arik, S.: A note on the global stability of dynamical neural networks. IEEE Transactions on Circuits and Systems—I: Fundamental Theory and Applications 49(4), 502–504 (2002)

Asuncion, A., Newman, D.: UCI machine learning repository (2007), `http://archive.ics.uci.edu/ml/`

Badel, S., Schmid, A., Leblebici, Y.: A VLSI Hamming artificial neural network with k-winner-take-all and k-loser-take-all capability. In: Hasselmo, Wunsch (eds.), pp. 977–982 (2003)

Baringa, M.: Challenging the "No New Neurons" Dogma. Science 255, 1646 (1992)

Baeg, E., Kim, Y., Kim, J., Ghim, J., Kim, J., Jung, M.: Learning-induced enduring changes in functional connectivity among prefrontal cortical neurons. Journal of Neuroscience 27(4), 909–918 (2007)

Barlow, H.: The neuron doctrine in perception. In: Gazzaniga (ed.), ch. 26, pp. 415–435 (1995)

Barlow, H.B.: Single units and sensation: A neuron doctrine for perceptual psychology. Perception 1, 371–394 (1972)

Bear, M.F.(Rapporteur): Activity-dependent Modification of Functional Circuitry as a Possible Basis for Learning. In: Changeux, J.P., Konishi, M. (eds.) The Neural and Molecular Bases of Learning, pp. 281–300. John Wiley & Sons, Chichester (1987)

Becker, S.: A computational principle for hippocampal learning and neurogenesis. Hippocampus 15(6), 722–738 (2005)

Bi, G., Poo, M.: Synaptic modifications in cultured hippocampal neurons: Dependence on spike timing, synaptic strength, and postsynaptic cell type. J. Neurosci. 18(24), 10,464–10,472 (1998)

Bi, G., Poo, M.: Synaptic modification by correlated activity: Hebb's postulate revisited. Annu. Rev. Neurosci. 24, 139–166 (2001)

Bienenstock, E.: Computing with fast functional links. In: Workshop on principles of information coding and processing in the brain, European Science Foundation, Trieste, Italy (1999), http://www.dam.brown.edu/people/elie/story.html

Boussaoud, D., Ungerleider, L.G., Desimone, R.: Pathways for motion analysis: cortical connections of the medial superior temporal and fundus of the superior temporal visual areas in the macaque. Journal of Comparative Neurology 296(3), 462–495 (1990)

Bower, G.H., Hilgard, E.R.: Theories of Learning, 5th edn. PrenticeHall, Englewood Cliffs (1981)

Bowers, J.S.: On the Biological Plausibility of Grandmother Cells: Implications for Neural Network Theories in Psychology and Neuroscience. Psychological Review 116(1), 220–251 (2009)

Breiman, L., Friedman, J., Olshen, R., Stone, C.: Classification and Regression Trees. Wadsworth International, Belmont (1984)

Browne, A., Sun, R.: Connectionist inference models. Neural Networks 14, 1331–1355 (2001)

Calitoi, D., Nussbaum, D., Oommen, B.J.: Investigating Schizophrenia Using Local Connectivity Considerations within the Piriform Cortex. In: IEEE CCECE/CCGEI, Saskatoon (May 2005)

Calvert, B.D., Marinov, C.A.: Another k-winners-take-all analog neural network. IEEE Transactions on Neural Networks 11(4), 829–838 (2000)

Campbell, S.R., Wang, D.L., Jayaprakash, C.: Synchrony and desynchrony in integrate-and-fire oscillators. Neural Computation 11, 1595–1619 (1999)

Cannon, R., Hasselmo, M.E., Koene, R.A.: From biophysics to behavior: Catacomb2 and the design of biologically plausible models for spatial navigation. Neuroinformatics 1, 3–42 (2002)

Carpenter, G.A., Grossberg, S.: ART3: Hierarchical Search Using Chemical Transmitters in Self-Organizing Pattern Recognition Architectures. Neural Networks 3, 129–152 (1990)

Carrasco, R., Forcada, M., Valdés-Muñoz, M., Ñeco, R.: Stable encoding of nite-state machines in discrete-time recurrent neural nets with sigmoid units. Neural Computation 12(9), 2129–2174 (2000)

Casey, M.: The dynamics of discrete-time computation, with applications to recurrent neural networks and finite-state machine extraction. Neural Computation 8(6), 1135–1178 (1996)

Chandrakasan, A., Bowhill, W.J., Fox, F. (eds.): Design of high-performance microprocessor circuits. IEEE Press, New Jersey (2001)

Chechik, G., Horn, D., Ruppin, E.: Neural Regulation and Hebbian Learning (2010), http://brain.tau.ac.il (Retrieved 18/02/2010)

Chomsky, N.: Aspects of the Theory of Syntax. MIT Press, Cambridge (1965)

Cleeremans, A., Servan-Schreiber, D., McClelland, J.: Finite state automata and simple recurrent networks. Neural Computation 1(3), 372–381 (1989)

Coss, R.G., Perkel, D.H.: The Function of Dendritic Spines: A Review of Theoretical Issues. Behavioral and Neural Biology 44, 151–185 (1985)

Cover, T.: Geometrical and statistical properties of systems of linear inequalities with applications to pattern recognition. IEEE Transactions on Electronic Computers 14, 326–334 (1965)

Crick, F.: Function of the thalamic reticular complex: The searchlight hypothesis. Proceedings of the National Academy of Sciences 81, 4586–4590 (1984)

Dawson, M.R.W., Berkeley, I.S.N.: Making a middling mousetrap. Behavior and Brain Sciences 16(3), 454–455 (1993); commentary on Shastri and Ajjanagadde

Diederich, J.: Instruction and High-Level Learning in Connectionist Networks. Connection Science 1(2), 161–180 (1989)

Diederich, J.: Steps towards knowledge-intensive connectionist learning. In: Barnden, J.A., Pollack, J. (eds.) Advances in Connectionist and Neural Computation Theory, Ablex, Norwood, NJ, vol. 1 (1991)

Diesmann, M., Gewaltig, M.O., Aertsen, A.: Stable propagation of synchronous spiking in cortical neural networks. Nature 402, 529–533 (1999)

Domany, E., van Hemmen, J.L., Schulten, K. (eds.): Models of Neural Networks, Physics of Neural Networks, vol. 2. Springer, New York (1994)

Douglas, R., Koch, C., Mahowald, M., Martin, K., Suarez, H.: Recurrent excitation in neocortical circuits. Science 269, 981–985 (1994)

Duan, X., et al.: Disrupted-In-Schizophrenia 1 Regulates Integration of Newly Generated Neurons in the Adult Brain. Cell 130, 1146–1158 (2007)

Eaton, J.W.: GNU Octave: A Numerical Engineering Software Package (2002), `http://www.che.wisc.edu/octave` (Accessed April 2010)

Ekman, P., Friesen, W.: Unmasking the Face. Prentice Hall, New York (1975)

Elias, S.A., Grossberg, S.: Pattern formation, contrast control, and oscillations in the short term memory of shunting on-center off-surround networks. Biol. Cybern. 20, 69–98 (1975)

Elman, J.: Finding structure in time. Cognitive Science 14, 179–211 (1990)

Engel, A.K., Fries, P., König, P., Brecht, M., Singer, W.: Temporal binding, binocular rivalry, and consciousness. Consciousness and Cognition 8(2), 128–151 (1999)

Fahlman, S.E., Lebierre, C.: The cascade-correlation learning architecture. In: Advances in Neural Information Processing Systems, vol. 2, pp. 524–532. MIT Press, Cambridge (1990)

Fanty, M.A.: Learning in structured connectionist networks. Tech. Rep. 252, Computer Science Department, University of Rochester, Rochester, New York (1988)

Feldman, J.A.: Dynamic connections in neural networks. Biological Cybernetics 46, 27–39 (1982)

Feldman, J.A.: Neural Representation of Conceptual Knowledge. Computer Science Department, University of Rochester, Rochester, N.Y. Technical Report 189 (1986)

Feldman, J.A.: Computational constraints on higher neural representations. In: Schwartz, E.L. (ed.) Computational Neuroscience. System Development Foundation Benchmark Series, ch. 13, pp. 163–178. MIT Press, Cambridge (1990)

Feldman, J., Bailey, D.: Layered hybrid connectionist models for cognitive science. In: Wermter, Sun (eds.), pp. 14–27 (2000)

Feldman, J.A., Ballard, D.H.: Connectionist models and their properties. Cognitive Science 6, 205–254 (1982)

Feldman, J., Fanty, M., Goddard, N.: Computing with structured neural networks. IEEE Computer 21, 91–104 (1988)

Feldman, J., Lakoff, G., Stolcke, A., Hollbach-Weber, S.: Miniature language acquisition: A touchstone for cognitive science. Tech. Rep. TR-90-009, International Computer Science Institute, Berkeley, CA (1990)

Feldman, V., Valiant, L.G.: Experience-induced neural circuits that achieve high capacity. Neural Computation 21(10), 2715–2754 (2009)

Foldiak, P., Endres, D.: Sparse coding. Scholarpedia 3(1), 2984 (2008)

Fransen, E., Tahvildari, B., Egorov, A., Hasselmo, M., Alonso, A.: Mechanism of graded persistent cellular activity of entorhinal cortex layer V neurons. Neuron 49(5), 735–746 (2006), doi:10.1016/j.neuron.2006.01.036

Funahashi, S., Bruce, C., Goldmanrakic, P.: Mnemonic coding of visual space in the monkeys dorsolateral prefrontal cortex. J. Neurophysiol. 61(2), 331–349 (1989)
Ganapathy, S.K., Titus, A.H.: Toward an analog VLSI implementation of adaptive resonance theory (ART2). In: Hasselmo, Wunsch (eds.), pp. 936–941 (2003)
Gao, C., Hammerstrom, D.: Platform performance comparison of PALM network on Pentium 4 and FPGA. In: Hasselmo, Wunsch (eds.), pp. 995–1000 (2003)
Garrigues, P., Olshausen, B.A.: Learning Horizontal Connections in a Sparse Coding Model of Natural Images. In: Platt, J.C., Koller, D., Singer, Y., Roweis, S. (eds.) Advances in Neural Information Processing Systems, vol. 20, pp. 505–551. MIT Press, Cambridge (2008)
Gazzaniga, M.S. (ed.): The cognitive neurosciences. MIT Press, Cambridge (1995)
Ge, S., Yang, C., Hsu, K., Ming, G.L., Song, H.: A critical period for enhanced synaptic plasticity in newly generated neurons of the adult brain. Neuron 54, 559–566 (2007)
Gerbessiotis, A.V.: Topics in parallel and distributed computation. PhD thesis, The Division of Applied Sciences, Harvard University, Cambridge, Massachusetts (1993)
Gerbessiotis, A.V.: A graph-theoretic result for a model of neural computation. Discrete Applied Mathematics 82, 257–262 (1998)
Gerbessiotis, A.V.: Random graphs in a neural computation model. International Journal of Computer Mathematics 80, 689–707 (2003)
Gerstner, W.: Spiking neurons. In: Maass, W., Bishop, C.M. (eds.) Pulsed Neural Networks, ch. 1, pp. 3–54. MIT Press, Cambridge (1999)
Gerstner, W.: A framework for spiking neuron models: The spike response model. In: Moss, F., Gielen, S. (eds.) The Handbook of Biological Physics, ch. 12, vol. 4, pp. 469–516. North-Holland, Amsterdam (2001)
Gilbert, C.D., Wiesel, T.: Receptive field dynamics in adult primary visual cortex. Nature 356, 150–152 (1992)
Goddard, N., Lynne, K., Mintz, T.: The Rochester Connectionist Simulator. Computer Science Department, University of Rochester, Rochester, N.Y. Technical Report 233 (1987)
Gordon, M., Peretto, P.: The statistical distribution of boolean gates in two-inputs, one-output multilayered neural networks. J. Phys. A: Math. Gen. 23, 3061–3072 (1990)
Gould, E.: How widespread is adult neurogenesis in mammals? Nature Reviews Neuroscience 8, 481–488 (2007)
Gould, E., Reeves, A.J., Graziano, M.S., Gross, C.G.: Neurogenesis in the neocortex of adult primates. Science 286, 548–552 (1999)
Gray, C.M., König, P., Engel, A.K., Singer, W.: Oscillatory responses in cat visual cortex exhibit inter-columnar synchronization which reflects global stimulus properties. Nature 338, 334–337 (1989)
Grünbaum, B.: Polyhedra with Hollow Faces. In: Bisztriczky, T., et al. (eds.) Proc of NATO-ASI Conference on Polytopes.. etc., pp. 43–70. Kluwer Academic, Dordrecht (1994)
Günay, C.: Hierarchical learning of conjunctive concepts in spiking neural networks. PhD thesis, Center for Advanced Computer Studies, University of Louisiana at Lafayette, Lafayette, LA 70504-4330, U.S.A (2003)
Günay, C., Maida, A.S.: The required measures of phase segregation in distributed cortical processing. In: Proceedings of the International Joint Conference on Neural Networks, Washington, D.C, vol. 1, pp. 290–295 (2001)

Günay, C., Maida, A.S.: Temporal binding as an inducer for connectionist recruitment learning over delayed lines. Neural Networks 16(5-6), 593–600 (2003b)

Günay, C., Maida, A.S.: Using temporal binding for connectionist recruitment over delayed lines. In: Proceedings of the International Joint Conference on Neural Networks, International Neural Network Society (2003c)

Günay, C., Maida, A.S.: Using temporal binding for robust connectionist recruitment learning over delayed lines. Tech. Rep. TR-2003-2-1, Center for Advanced Computer Studies, University of Louisiana at Lafayette, Lafayette, LA, 70504-4330, U.S.A (2003d)

Günay, C., Maida, A.S.: A stochastic population approach to the problem of stable recruitment hierarchies in spiking neural networks. Biol. Cybern. 94(1), 33–45 (2006a), doi:10.1007/s00422-005-0023-y

Günay, C., Maida, A.S.: Using temporal binding for hierarchical recruitment of conjunctive concepts over delayed lines. Neurocomputing 69(4-6), 317–367 (2006b), doi:10.1016/j.neucom.2005.03.008

Hagan, M.T., Demuth, H.B., Beale, M.: Neural network design. PWS Publishing, Boston, MA (1996)

Hampson, S., Volper, D.: Linear function neurons: structure and training. Biological Cybernetics 53, 203–217 (1986)

Han, T.M., De Vries, G.J.: Neurogenesis of Galanin Cells in the Bed Nucleus of the Stria Terminalis and Centromedial Amygdala in Rats: A Model for Sexual Differentiation of Neuronal Phenotype. Journal of Neurobiology 38(4), 491–498 (1999)

Hasselmo, M., Wunsch, D.C. (eds.): Proceedings of the International Joint Conference on Neural Networks, Portland, Oregon (2003)

Hebb, D.O.: The Organization of behavior. Wiley, New York (1949)

Hertz, J., Krogh, A., Palmer, R.G.: Introduction to the theory of neural computation. Lecture notes of the Santa Fe Institute studies in the sciences of complexity, vol. 1. Addison Wesley, Reading (1991)

Hindmarsh, A.C.: ODEPACK, a systematized collection of ODE solvers. In: Stepleman, R.S., et al. (eds.) Scientific Computing: Applications of Mathematics and Computing to the Physical Sciences, vol. 1. North-Holland, NY (1983)

Hinton, G.E.: Preface to the special issue on connectionist symbol processing. Artificial Intelligence 46(1-2), 1–4 (1990); special issue on connectionist symbol processing

Hinton, G.E., McClelland, J.L., Rumelhart, D.E.: Distributed representations. In: Rumelhart, et al. (eds.), pp. 77–109 (1986)

Hirsch, M.W.: Convergent Activation Dynamics in Continuous Time Networks. Neural Networks 2, 331–349 (1989)

Hogan, J.M.: Recruitment learning in randomly connected neural networks. PhD thesis, Queensland University of Technology (2000)

Hogan, J., Norris, M., Diederich, J.: Classification of facial expressions with domain gaussian rbf networks. In: Howlett, R., Jain, L. (eds.) Radial Basis Function Neural Network Theory and Applications, vol. 2, pp. 143–164. Physica-Verlag, Heidelberg (2000)

Honavar, V.: Perceptual Development and Learning: From Behavioral, Neurophysiological, and Morphological Evidence to Computational Models. Computer Sciences Department, University of Wisconsin-Madison, Madison, WI. Technical Report 818 (1989)

Hooper, S.L., Moulins, M.: Switching of a Neuron from One Network to Another by Sensory-Induced Changes in Membrane Properties. Science 244, 1587–1589 (1989)

Hopcroft, J., Ullman, J.: Introduction to Automata Theory, Languages and Computation. Addison-Wesley, Don Mills (1979)

Hopfield, J.J.: Neural networks and physical systems with emergent collective computational properties. Proceedings of the National Academy of Sciences 79, 2554–2558 (1982)

Hornik, K., Stinchcombe, M., White, H.: Multilayer feedforward networks are universal approximators. Neural Networks 2, 359–366 (1989)

Hummel, J.E., Biederman, I.: Dynamic binding in a neural network for shape recognition. Psychological Review 99(3), 480–517 (1992)

Huyck, C.R.: Variable binding by synaptic strength change. Connect. Sci. 21(4, Sp. Iss. SI), 327–357 (2009), doi:10.1080/09540090902954188

Hyman, S.E., Malenka, R.C., Nestler, E.J.: Neural mechanisms of addiction: The role of reward-related learning and memory. Annual Reviews of Neuroscience 29, 565–598 (2006)

Ikegaya, Y., Aaron, G., Cossart, R., Aronov, D., Lampl, I., Ferster, D., Yuste, R.: Synfire chains and cortical songs: Temporal modules of cortical activity. Science 304(5670), 559–564 (2004)

Indiveri, G.: Modeling selective attention using a neuromorphic analog VLSI device. Neural Computation 12(12), 2857–2880 (2000)

Iserles, A.: A first course in the numerical analysis of differential equations. Cambridge Univ. Press, Cambridge (1996)

Jaholkowski, P., Kiryk, A., Jedynak, P., Ben Abdallah, N.M., Knapska, E., Kowalczyk, A., Piechal, A., Blecharz-Klin, K., Figiel, I., Lioudyno, V., Widy-Tyszkiewicz, E., Wilczynski, G.M., Lipp, H.P., Kaczmarek, L., Filipkowski, R.K.: New hippocampal neurons are not obligatory for memory formation, cyclin D2 knockout mice with no adult brain neurogenesis show learning. Learning and Memory 16, 439–451 (2009)

James, D., Maida, A.S.: Sequential hierarchical recruitment learning in a network of spiking neurons. In: Proceedings of International Joint Conference on Neural Networks, Atlanta, Georgia, USA, pp. 1407–1413 (2009)

Jensen, O., Lisman, J.E.: Novel lists of 7±2 known items can be reliably stored in an oscillatory short-term memory network: Interaction with long-term memory. Learning and Memory 3, 257–263 (1996)

Jeong, W.K., Beyer, J., Hadwiger, M., Vazquez, A., Pfister, H., Whitaker, R.T.: Scalable and interactive segmentation and visualization of neural processes in EM datasets. IEEE Transactions on Visualization and Computer Graphics 15, 1505–1514 (2009)

Kaiser, M., Hilgetag, C.: Nonoptimal component placement, but short processing paths, due to long-distance projections in neural systems. PLoS Computational Biology 2(7), 805–815 (2006)

Kaiser, M.: Brain architecture: a design for natural computation. Phil. Trans. R. Soc. A. 365, 3033–3045 (2007)

Kaszkurewicz, E., Bhaya, A.: On a class of globally stable neural circuits. IEEE Transactions on Circuits and Systems—I: Fundamental Theory and Applications 41(2), 171–174 (1994)

Kempermann, G., Krebs, J., Fabel, K.: The contribution of failing adult hippocampal neurogenesis to psychiatric disorders. Current Opinion in Psychiatry 21, 290–295 (2008)

Keshavan, M.S., Hogarty, G.E.: Brain maturational processes and delayed onset in schizophrenia. Development and Psychopathology 11, 525–543 (1999)

Knoblauch, A., Palm, G.: Pattern separation and synchronization in spiking associative memories and visual areas. Neural Networks 14, 763–780 (2001a)

Knoblauch, A., Palm, G.: Spiking associative memory and scene segmentation by synchronization of cortical activity. In: Wermter, et al. (eds.), pp. 407–427 (2001b)

Koch, C., Poggio, T., Torre, V.: Nonlinear interactions in a dendritic tree: Localization timing and role in information processing. Proc. Natl. Acad. Sci. 80, 2799–2802 (1983)

Koch, C., Rapp, M., Segev, I.: A brief history of time (constants). Cerebral Cortex 6, 93–103 (1996)

Koch, C., Ullman, S.: Shifts in selective visual attention: towards the underlying neural circuitry. Human Neurobiology 4, 219–227 (1985)

Koehler, W.: Intelligenzprüfungen an Menschenaffen. Zweite Auflage der "Intelligenzprüfungen an Anthropoiden I" aus den Abhandlungen der Preussischen Akademie der Wissenschaften, Jahgang 1917. Springer, Berlin (1921)

Koene, R.A., Hasselmo, M.E.: First-in-first-out item replacement in a model of short-term memory based on persistent spiking. Cereb. Cortex 17(8), 1766–1781 (2007), doi:10.1093/cercor/bhl088

König, P., Engel, A.K.: Correlated firing in sensory-motor systems. Current Opinion in Neurobiology 5, 511–519 (1995)

Kremer, S.: On the computational power of Elman-style recurrent networks. IEEE Transactions on Neural Networks 6, 1000–1004 (1995)

Kuipers, B.J.: Qualitative reasoning: Modeling and simulation with incomplete knowledge. MIT Press, Cambridge (1994)

Labov, W.: The boundaries of words and their meanings. In: Bailey, C., Shuy, R. (eds.) New Ways of Analyzing Variation in English, pp. 340–373. Georgetown U. Press, Washington (1973)

Lamme, V.A.F., Roelfsema, P.R.: The distinct modes of vision offered by feedforward and recurrent processing. Trends in Neuroscience 23, 571–579 (2000)

Langacker, R.: Foundations of Cognitive Grammar I: Theoretical Prerequisites. Stanford University Press, Stanford (1987)

Langley, P., Choi, D., Rogers, S.: Acquisition of hierarchical reactive skills in a unified cognitive architecture. Cogn. Syst. Res. 10(4), 316–332 (2009), doi:10.1016/jCogsys.2008.07.003

Lashley, K.: In search of the engram. In: Symposia of the Society for Experimental Biology. Physiological Mechanisms in Animal Behavior, vol. 4, pp. 454–483. Academic, New York (1950)

Lau, P., Bi, G.: Synaptic mechanisms of persistent reverberatory activity in neuronal networks. Proc. Natl. Acad. Sci. USA 102(29), 10,333–10,338 (2005), doi:10.1073/pnas.0500717102

Lazzaro, J., Ryckebusch, S., Mahowald, M.A., Mead, C.: Winner-take-all Networks of O(n) complexity. In: Touretzky, D.S. (ed.) Advances in neural information processing systems, vol. 1. Morgan Kaufmann, San Mateo (1989)

Lehky, S., Sejnowski, T.: Network model of shape from shading: neural function arises from both receptive and projective fields. Nature 333, 452–454 (1988)

Leslie, K., Nelson, S., Turrigiano, G.: Postsynaptic depolarization scales quantal amplitude in cortical pyramidal neurons. J. Neurosci. 21(19), art. no.–RC170 (2001)

Leuner, B., Gould, E., Shors, T.J.: Is There A Link Between Adult Neurogenesis and Learning? Hippocampus 16, 216–224 (2006)

Levelt, W.J.M.: On learnability, empirical foundations and naturalness. Comment on: Hanson SJ, Burr DJ What connectionist models learn: Learning and representation in connectionist networks. Behavioral and Brain Sciences 13, 501 (1990)

Levy, W.B.: A sequence predicting CA3 is a flexible associator that learns and uses context to solve Hippocampal-like tasks. Hippocampus 6, 576–590 (1996)

Levy, W., Colbert, C., Desmond, N.: Elemental adaptive processes of neurons and synapses: a statistical/computational perspective. In: Gluck, M., Rumelhart, D. (eds.) Neuroscience and Connectionist Theory, pp. 187–233. Erlbaum, Hillsdale (1990)

Li, N., DiCarlo, J.J.: Unsupervised natural experience rapidly alters invariant object representation in visual cortex. Science 321(5895), 1502–1507 (2008), doi:10.1126/science.1160028

Lin, L., Osan, R., Shoham, S., Jin, W., Zuo, W., Tsien, J.: Identification of networklevel coding units for real-time representation of episodic experiences in the hippocampus. Proc. Natl. Acad. Sci. USA 102(17), 6125–6130 (2005), doi:10.1073/pnas.0408233102

Lisman, J.E., Idiart, M.A.P.: Storage of 7±2 short-term memories in oscillatory subcycles. Science 267, 1512–1514 (1995)

Littlestone, N.: Learning quickly when irrelevant attributes abound: A new linear threshold algorithm. Machine Learning 2, 285–318 (1988)

Littlestone, N.: Redundant noisy attributes, attribute errors, and linear threshold learning using Winnow. In: Proc. 4th Annu. Workshop on Comput. Learning Theory, pp. 147–156. Morgan Kaufmann, San Meteo (1991)

Litvak, V., Sompolinsky, H., Segev, I., Abeles, M.: On the transmission of rate codes in long feedforward networks with excitatory-inhibitory balance. Journal of Neuroscience 23(7), 3006–3015 (2003)

Livingstone, M., Hubel, D.: Segregation of form, color, movement, and depth anatomy, physiology, and perception. Science 240(4853), 740–749 (1988)

Lo, Y., Poo, M.M.: Activity-Dependent Synaptic Competition in Vitro: Heterosynaptic Suppression of Developing Synapses. Science 254, 1019–1022 (1991)

Lynch, G., Baudry, M.: The Biochemistry of Memory: A New and Specific Hypothesis. Science 224, 1057–1063 (1984)

Lynch, G., Granger, R.: Simulation and analysis of a simple cortical network. In: Hawkins, R.D., Bower, G.H. (eds.) Computational Models of Learning in Simple Neural systems. The Psychology of Learning and Motivation, vol. 23, pp. 205–241. Academic Press, Orlando (1989)

Maass, W.: Networks of spiking neurons: The third generation of neural network models. Neural Networks 10(9), 1659–1673 (1997)

Maass, W.: On the computational power of winner-take-all. Neural Computation 12(11), 2519–2536 (2000)

Majani, E., Erlanson, R., Abu-Mostafa, Y.: On the K-winners-take-all network. In: Touretzky, D.S. (ed.) Advances in Neural Information Processing Systems, vol. 1, pp. 634–642. Morgan Kaufman, San Mateo (1989)

von der Malsburg, C.: The correlation theory of brain function. Tech. Rep., Max-Planck Institute for Biophysical Chemistry, Göttingen (1981)

von der Malsburg, C.: The correlation theory of brain function. In: Domany, et al. (eds.), ch. 2, pp. 95–120 (1994)

von der Malsburg, C.: Binding in models of perception and brain function. Current Opinion in Neurobiology 5, 520–526 (1995)

von der Malsburg, C., Schneider, W.: A neural cocktail-party processor. Biological Cybernetics 54(1), 29–40 (1986)

Mani, D.R., Shastri, L.: A connectionist solution to the multiple instantiation problem using temporal synchrony. In: Proceedings of the Fourteenth Conference of the Cognitive Science Society, Bloomington, Indiana, pp. 974–979 (1992)

Masuda, N., Aihara, K.: Bridging rate coding and temporal spike coding by effect of noise. Phys. Rev. Lett. 88(24) (2002), doi:10.1103/PhysRevLett.88.248101

Mauk, M., Buonomano, D.: The neural basis of temporal processing. Annu. Rev. Neurosci. 27, 307–340 (2004), doi:10.1146/annurev.neuro.27.070203.144247

McClelland, J.L., Rumelhart, D.E.: An interactive activation model of context effects in letter perception. Part I. An account of basic findings. Psychological Review 88, 375–408 (1981)

McCulloch, W.S., Pitts, W.H.: A logical calculus of ideas immanent in nervous activity. Bulletin of Mathematical Biophysics 5, 115 (1943)

Mei, L., Xiong, W.C.: Neuregulin-1 signaling in schizophrenia. Future Neurology 2(5), 477–480 (2007)

Merzenich, M.M., Recanzone, G., Jenkins, W.M., Allard, T.T., Nudo, R.J.: Cortical Representational Plasticity. In: Rakie, P., Singer, W. (eds.) Neurobiology of Neocortex, pp. 41–67. John Wiley & Sons, Chichester (1988)

Milner, P.: A model for visual shape recognition. Psychol. Rev. 816, 521–535 (1974)

Minai, A.A., Levy, W.B.: The dynamics of sparse random networks. Biol. Cybern. 70, 177–187 (1993)

Minai, A.A., Levy, W.B.: Setting the activity level in sparse random networks. Neural Computation 6, 85–99 (1994)

Minsky, M.L.: Neural networks: automata made up of parts. In: Computation: finite and infinite machines, pp. 32–66. Prentice-Hall, Inc, Upper Saddle River (1967)

Minsky, M., Papert, S.: Perceptrons. MIT Press, Cambridge (1969)

Mitchell, T.M., Keller, R.M., Kedar-Cabelli, S.T.: Explanation-Based Generalization: A Unifying View. Machine Learning 1, 47–80 (1986)

Mokeichev, A., Okun, M., Barak, O., Katz, Y., Ben-Shahar, O., Lampl, I.: Stochastic emergence of repeating cortical motifs in spontaneous membrane potential fluctuations in vivo. Neuron 53(3), 413–425 (2007), doi:10.1016/j.neuron.2007.01.017

Muroga, S.: Threshold Logic and its Applications. John Wiley and Sons, New York (1971)

Murray, F., Stanley, R.: Perceptual learning of cartoon faces by young children. Bulletin of the Psychonomic Society 16, 367–370 (1980)

Murre, J., Sturdy, D.: The connectivity of the brain: multi-level quantitative analysis. Biological Cybernetics 73, 529–545 (1995)

Nasrallah, H.A.: Impaired Neuroplasticity in Schizophrenia and the Neuro-regenerative Effects of Atypical Antipsychotics. Highlights in Neuropsychopharmacology (2008)

Natarajan, B.: Machine Learning: A Theoretical Approach. Morgan Kaufmann, San Mateo (1991)

Nelson, S.B., Turrigiano, G.G.: Strength through Diversity. Neuron 60(3), 477–482 (2008), doi:10.1016/j.neuron.2008.10.020

Newell, A.: Unified theories of cognition. Harvard University Press, Cambridge (1990)

Niebur, E., Koch, C.: Control of selective visual attention: modeling the "where" pathway. In: Advances in Neural Information Processing Systems, vol. 8, pp. 802–808. MIT Press, Cambridge (1996)

Niebur, E., Koch, C.: Computational architectures for attention. In: Parasuraman, R. (ed.) The Attentive Brain, pp. 163–186. MIT Press, Cambridge (1998)

Nottebohm, F.: Vom Vogelgesang zur Bildung neuer Nervenzellen. Spektrum der Wissenschaft 4, 112–117 (1989)

Nowak, L.G., Bullier, J.: The timing of information transfer in the visual system. In: Rockland, K.S., Kaas, J.H., Peters, A. (eds.) Cerebral Cortex, vol. 12, pp. 205–240. Kluwer, New York (1997)

Olshausen, B., Koch, C.: Selective visual attention. In: Arbib, M. (ed.) The Handbook of Brain Theory and Neural Networks, pp. 837–840. MIT Press, Cambridge (1995)

Omlin, C., Giles, C.: Constructig deterministic finite-state automata in recurrent neural networks. Journal of the ACM 43, 937–972 (1996a)

Omlin, C., Giles, C.: Stable encoding of large finite-state automata in recurrent neural networks with sigmoid discriminants. Neural Computation 8, 675–696 (1996b)

Pfister, H., Lichtman, J., Reid, C.: The connectome project (2010), `http://cbs.fas.harvard.edu/science/connectome-project`

Page, M.: Connectionist modelling in psychology: A localist manifesto. Behavioral and Brain Sciences 23(4), 443–467 (2000)

Palm, G.: Cell Assemblies as a Guideline for Brain Research. Concepts in Neuroscience 1(1), 133–147 (1990)

Palm, G., Sommer, F.T., et al.: Neural associative memories. In: Krikelis, A., Weems, C.C. (eds.) Associative Processing and Processors, pp. 307–326. IEEE CS Press, Los Alamitos (1997)

Paton, J.A., Nottebohm, F.: Neurons Generated in the Adult Brain are Recruited into Functional Circuits. Science 225, 4666 (1984)

Perera, T.D., Park, S., Nemirovskaya, Y.: Cognitive Role of Neurogenesis in Depression and Antidepressant Treatment. Neuroscientist 14, 326 (2008) (originally published online July 8, 2008)

Phillips, C.L., Harbor, R.D.: Feedback Control Systems, 2nd edn. Prentice Hall, New Jersey (1991)

Plunkett, K., Elman, J.: Exercises in Rethinking Innateness: A Handbook for Connectionist Simulations. MIT Press, Cambridge (1997)

Prickaerts, J., Koopmans, G., Blokland, A., Scheepens, A.: Learning and adult neurogenesis: Survival with or without proliferation? Neurobiology of Learning and Memory 81, 1–11 (2004)

Quiroga, R., Reddy, L., Kreiman, G., Koch, C., Fried, I.: Invariant visual representation by single neurons in the human brain. Nature 435(7045), 1102–1107 (2005), doi:10.1038/nature03687

Rabinowitch, I., Segev, I.: The endurance and selectivity of spatial patterns of long-term potentiation/depression in dendrites under homeostatic synaptic plasticity. J. Neurosci. 26(52), 13–474 (2006), doi:10.1523/jneurosci.4333-06.2006

Ritz, R., Sejnowski, T.J.: Synchronous oscillatory activity in sensory systems: New vistas on mechanisms. Current Opinion in Neurobiology 7, 536–546 (1997)

Rakic, P.: Limits of Neurogenesis in Primates. Science 227, 1054–1055 (1985)

Rakic, P.: Adult Neurogenesis in Mammals: An Identity Crisis. The Journal of Neuroscience 22(3), 614–618 (2002)

Regier, T.: The acquisition of lexical semantics for spatial terms: A connectionist model of perceptual categorization. PhD thesis, University of California, Berkeley (1992)

Reynolds, B.A., Weiss, S.: Generation of Neurons and Astrocytes from Isolated Cells of the Adult Mammalian Central Nervous System. Science 255, 1707–1710 (1992)

Rosenblatt, F.: Principles of neurodynamics: Perceptrons and the theory of brain mechanisms. Spartan Books, Washington (1961)

van Rossum, M., Turrigiano, G., Nelson, S.: Fast propagation of firing rates through layered networks of noisy neurons. Journal of Neuroscience 22(5), 1956–1966 (2002)

Rowland, B.: The direct-indirect problem of phase-segregation networks: Intuitions from the olfactory cortex (2002) (unpublished manuscript)

Rowland, B., Maida, A., Berkeley, I.: Synaptic noise as a means of implementing weight-perturbation learning. Connect Sci. 18(1), 69–79 (2006), doi:10.1080/ 09540090500386551

Rozell, C.J., Johnson, D.H., Baraniuk, R.G., Olshausen, B.A.: Sparse Coding via Thresholding and Local Competition in Neural Circuits. Neural Computation 20, 2526–2563 (2008)

Rumelhart, D.E., Hinton, G.E., Williams, R.J.: Learning Internal Representations by Error Propagation. In: Rumelhart, McClelland (eds.) (1986b)

Rumelhart, D.E., Hinton, G.E., Williams, R.J.: Learning representations by backpropagating errors. Nature 323, 533–536 (1986a)

Rumelhart, D.E., McClelland, J.L.: The PDP Research Group (eds), Parallel distributed processing: Explorations in the microstructure of cognition. Foundations, vol. 1. MIT Press, Cambridge (1986)

Russell, S.J., Norvig, P.: Artificial Intelligence: A Modern Approach. Prentice-Hall Inc., New Jersey (1995)

Schillen, T.B., König, P.: Binding by temporal structure in multiple feature domains of an oscillatory neuronal network. Biological Cybernetics 70, 397–405 (1994)

Schmidt-Hieber, C., Jonas, P., Bischofberger, J.: Enhanced synaptic plasticity in newly generated granule cells of the adult hippocampus. Nature 429(6988), 184–187 (2004)

Schmidt, R.F. (ed.): Grundriss der Sinnesphysiologie, 3rd edn. Springer, Berlin (1977)

Senn, W., Schneider, M., Ruf, B.: Activity-dependent development of axonal and dendritic delays, or, why synaptic transmission should be unreliable. Neural Computation 14, 583–619 (2002)

Servan-Schreiber, D., Cleeremans, A., McClelland, J.: Graded state machines: the representation of temporal contingencies in simple recurrent networks. Machine Learning 7, 161–193 (1991)

Shadlen, M.N., Newsome, W.T.: Noise, neural codes and cortical organization. Current Opinion in Neurobiology 4, 569–579 (1994)

Shadlen, M.N., Newsome, W.T.: The variable discharge of cortical neurons: Implications for connectivity, computation, and information coding. Journal of Neuroscience 18(10), 3870–3896 (1998)

Shastri, L.: Semantic networks: An evidential formalization and its connectionist realization. Research Notes in Artificial Intelligence. Morgan Kaufmann Publishers, Inc., San Meteo (1988)

Shastri, L.: A computational model of tractable reasoning—Taking inspiration from cognition. In: Proceedings of IJCAI 1993, the Thirteenth International Joint Conference on Artificial Intelligence, France, pp. 202–207 (1993)

Shastri, L.: Advances in SHRUTI—A neurally motivated model of relational knowledge representation and rapid inference using temporal synchrony. Applied Intelligence 11, 79–108 (1999a)

Shastri, L.: Recruitment of binding and binding-error detector circuits via long-term potentiation. Neurocomputing 26(7), 865–874 (1999b)

Shastri, L.: Types and quantifiers in SHRUTI: A connectionist model of rapid reasoning and relational processing. In: Wermter, Sun (eds.), pp. 28–45 (2000)

Shastri, L.: Biological grounding of recruitment learning and vicinal algorithms in long-term potentiation. In: Wermter, et al. (eds.), pp. 348–367 (2001a)

Shastri, L.: A computational model of episodic memory formation in the hippocampal system. Neurocomputing, 38–40, 889–897 (2001b)

Shastri, L.: A computationally efficient abstraction of long-term potentiation. Neurocomputing, 44–46, 33–41 (2002a)

Shastri, L.: Episodic memory and cortico-hippocampal interactions. Trends in Cognitive Sciences 6(4), 162–168 (2002b)

Shastri, L., Ajjanagadde, V.: From simple associations to systematic reasoning: A connectionist representation of rules, variables, and dynamic bindings using temporal synchrony. Behavioral and Brain Sciences 16(3), 417–451 (1993)

Shastri, L., Feldman, J.: Semantic Networks and Neural Nets. Computer Science Department, University of Rochester, Rochester, N.Y. Technical Report 131 (June 1984)

Shiffrin, R.M., Schneider, W.: Controlled and automatic human information processing: II. Perceptual learning, automatic attending, and a general theory. Psychological Review 84, 127–190 (1977)

Sholl, D.: The Organization of the Cerebral Cortex. Methuen, London (1956)

Shors, T.J.: From Stem Cells to Grandmother Cells: How Neurogenesis Relates to Learning and Memory. Cell Stem Cell 3 (September 11, 2008)

Shouval, H., Perrone, M.: Post-hebbian learning rules. In: Arbib, M. (ed.) The Handbook of Brain Theory and Neural Networks, pp. 745–748. MIT Press, Cambridge (1995)

Singer, W.: Neuronal Activity as a Shaping Factor in Postnatal Development of Visual Cortex. In: Greenough, W.T., Juruska, J.M. (eds.) Developmental Neuropsychobiology. Behavioral Biology. An International Series, pp. 271–293. Academic Press, Orlando (1986)

Singer, W.: Activity-dependent Self-organization of Synaptic Connections as a Substrate of Learning. In: Changeux, J.P., Konishi, M. (eds.) The Neural and Molecular Bases of Learning, pp. 301–336. John Wiley & Sons, Chichester (1987)

Singer, W.: Time as coding space in neocortical processing: A hypothesis. In: Gazzaniga (ed.), ch. 6, pp. 91–104 (1995)

Singer, W., Gray, C.M.: Visual feature integration and the temporal correlation hypothesis. Annual Review of Neuroscience 18, 555–586 (1995)

Siu, K.Y., Roychowdhury, V., Kailath, T.: Discrete Neural Computation: A Theoretical Foundation. Prentice Hall, Englewood Cliffs (1995)

Sougné, J.: Connectionism and the problem of multiple instantiation. Trends in Cognitive Sciences 2, 183–189 (1998a)

Sougné, J.: Period doubling as a means of representing multiply instantiated entities. In: Proceedings of the Twentieth Annual Conference of the Cognitive Science Society, pp. 1007–1012. Lawrence Erbaum Ass., Mahwah (1998b)

Sougné, J., French, R.M.: A neurobiologically inspired model of working memory based on neuronal synchrony and rythmicity. In: Bullinaria, J.A., Glasspool, D.W., Houghton, G. (eds.) Proceedings of the Fourth Neural Computation and Psychology Workshop: Connectionist Representations, pp. 155–167. Springer, London (1997)

Squire, L.R.: Memory and Brain. Oxford University Press, Oxford (1987)

Squire, L.R., Zola-Morgan, S.: The Medial Temporal Lobe Memory System. Science 253(5026), 1380–1386 (1991)

Stark, E., Globerson, A., Asher, I., Abeles, M.: Correlations between Groups of Premotor Neurons Carry Information about Prehension. J. Neurosci. 28(42), 10–618 (2008), doi:10.1523/jneurosci.3418-08.2008

Steuber, V., Willshaw, D.: A biophysical model of synaptic delay learning and temporal pattern recognition in a cerebellar Purkinje cell. J. Comput. Neurosci. 17(2), 149–164 (2004)

Stevens, C.: How cortical interconnectedness varies with network size. Neural Computation 1, 473–479 (1989)

Terman, D., Wang, D.: Global competition and local cooperation in a network of neural oscillators. Physica D 81, 148–176 (1995)

Tetzlaff, T., Geisel, T., Diesmann, M.: The ground state of cortical feed-forward networks. Neurocomputing, 44-46, 673–678 (2002)

Thorpe, S.T., Imbert, M.: Biological constraints on connectionist modelling. In: Pfeifer, R., Schreter, Z., Fogelman-Soulié, F., Steels, L. (eds.) Connectionism in perspective, pp. 63–92. Elsevier, North-Holland (1989)

Thrun, S.: The monk's problems: A performance comparison of different learning algorithms. Tech. Rep. CMU-CS-91-197, Carnegie-Mellon University (1991)

Toro, C.T., Deakin, J.F.W.: Adult neurogenesis and schizophrenia: A window on abnormal early brain development? Schizophrenia Research 90, 1–14 (2007)

Treisman, A.M.: The binding problem. Current Opinion in Neurobiology 6, 171–178 (1996)

Treisman, A.M., Gelade, G.: A feature integration theory of attention. Cogn. Psychol. 12, 97–106 (1980)

Turrigiano, G., Leslie, K., Desai, N., Rutherford, L., Nelson, S.: Activity-dependent scaling of quantal amplitude in neocortical neurons. Nature 391(6670), 892–896 (1998)

Turrigiano, G., Nelson, S.: Homeostatic plasticity in the developing nervous system. Nat. Rev. Neurosci. 5(2), 97–107 (2004), doi:10.1038/nrn1327

Tymoshchuk, P., Kaszkurewicz, E.: A winner-take-all circuit based on second order Hopfield neural networks as building blocks. In: Hasselmo, Wunsch (eds.), pp. 891–896 (2003)

Urahama, K., Nagao, T.: K-winners-take-all circuit with O(N) complexity. IEEE Transactions on Neural Networks 6(3), 776–778 (1995)

Usher, M., Donnelley, N.: Visual synchrony affects binding and segmentation in perception. Nature 394, 179–182 (1998)

Usher, M., Niebur, E.: A neural model for parallel, expectation-driven attention for objects. J. Cognitive Neuroscience 8, 311–327 (1996)

Valiant, L.G.: A theory of the learnable. Communications of the ACM 27(11), 1134–1142 (1984)

Valiant, L.G.: Functionality in neural nets. In: Proceedings of the 7th National Conference on Artificial Intelligence, AAAI, pp. 629–634. Morgan Kaufmann, San Mateo (1988)

Valiant, L.G.: Circuits of the Mind. Oxford University Press, New York (2000)

Valiant, L.G.: A neuroidal architecture for cognitive computation. Tech. Rep. 11, Division of Computer Science and Engineering, Harvard University (1996)

Valiant, L.G.: A neuroidal architecture for cognitive computation. In: Larsen, K.G., Skyum, S., Winskel, G. (eds.) ICALP 1998. LNCS, vol. 1443, pp. 642–669. Springer, Heidelberg (1998)

Valiant, L.G.: A neuroidal architecture for cognitive computation. Journal of the ACM 47(5), 854–882 (2000a)

Valiant, L.G.: Robust logics. Artificial Intelligence 117, 231–253 (2000b)

Valiant, L.G.: Memorization and association on a realistic neural model. Neural Computation 17(3), 527–555 (2005)

Valiant, L.G.: Knowledge infusion. In: Proc. 21st National Conference on Artificial Intelligence, AAAI 2006, Boston, MA, July 16-20, pp. 1546–1551. AAAI Press, Menlo Park (2006)

Van den Broeck, C., Kawai, R.: Learning in feedforward boolean networks. Phys. Rev. A 42, 6210–6217 (1990)

VanRullen, R., Koch, C.: Is perception discrete or continuous? Trends in Cognitive Sciences 7(5), 207–213 (2003)

Vapnik, V.N.: The Nature of Statistical Learning Theory. Springer, Berlin (1995)

Volman, V., Gerkin, R.C., Lau, P.M., Ben-Jacob, E., Bi, G.Q.: Calcium and synaptic dynamics underlying reverberatory activity in neuronal networks. Phys. Biol. 4(2), 91–103 (2007), doi:10.1088/1478-3975/4/2/003

Watts, D., Strogatz, S.: Collective dynamics of 'small-world' networks. Nature 393(6684), 409–410 (1998)

Weisstein, N.: Beyond the yellow volkswagen detector and the grandmother cell: A general strategy for the exploration of operations in human pattern recognition. In: Solso, R. (ed.) Contemporary Issues in Cognitive Psychology: The Loyola Symposium. W. H. Winston & Sons, Washington (1973)

Wennekers, T.: On the natural hierarchical composition of cliques in cell assemblies. Cogn. Comput. 1(2), 128–138 (2009)

Wennekers, T., Palm, G.: On the relation between neural modelling and experimental neuroscience. Theory in Biosciences 116, 267–283 (1997)

Wennekers, T., Palm, G.: Modelling generic cognitive functions with operational Hebbian cell assemblies. In: Weiss, M.L. (ed.) Neural Network Research Horizons, pp. 225–294. Nova Science Publishers (2007)

Werbos, P.: Beyond regression: New tools for prediction and analysis in the behavioral sciences. PhD thesis, Harvard University (1974)

Wermter, S., Austin, J., Willshaw, D.J. (eds.): Emergent Neural Computational Architectures Based on Neuroscience - Towards Neuroscience-Inspired Computing. LNCS (LNAI), vol. 2036. Springer, Heidelberg (2001)

Wermter, S., Sun, R. (eds.): Hybrid Neural Systems 1998. LNCS, vol. 1778. Springer, Heidelberg (2000)

Wickelgren, W.A.: Chunking and consolidation: A theoretical synthesis of semantic networks. Configuring in conditioning, S-R versus cognitive learning, normal forgetting, the amnestic syndrome, and the hippocampal arousal system. Psychological Review 86, 44–60 (1979)

Williams, R., Zipser, D.: A learning algorithm for continually running fully recurrent neural networks. Neural Computation 1(2), 270–280 (1989)

Wolfe, W.J., Mathis, D., Anderson, C., Rothman, J., Gottler, M., Brady, G., Walker, R., Duane, G., Alaghband, G.: K-Winner Networks. IEEE Transactions on Neural Networks 2(2), 310–315 (1991)

Wyatt Jr., J.L., Standley, D.L.: Criteria for Robust Stability In A Class of Lateral Inhibition Networks Coupled through Resistive Grids. Neural Computation 1, 58–67 (1989)

Yaglom, A.: An Introduction to the Theory of Stationary Random Functions. Dover, New York (1973)

Yerkes, R.M.: The mind of a gorilla: I. Genet. Psychol. Monogr. 2 (1927); Bower, G.H., Hilgard, E.R.: Theories of Learning. Prentice-Hall, Englewood Cliffs (1981)

Author Index

Index

GPSR Compliance

The European Union's (EU) General Product Safety Regulation (GPSR) is a set of rules that requires consumer products to be safe and our obligations to ensure this.

If you have any concerns about our products, you can contact us on ProductSafety@springernature.com

In case Publisher is established outside the EU, the EU authorized representative is:

Springer Nature Customer Service Center GmbH
Europaplatz 3
69115 Heidelberg, Germany

Batch number: 10614940

Printed by Printforce, the Netherlands